Dung Le

Strongly Coupled Parabolic and Elliptic Systems

De Gruyter Series in Nonlinear Analysis and Applications

Volume 28

Dung Le

Strongly Coupled Parabolic and Elliptic Systems

Existence and Regularity of Strong and Weak Solutions

DE GRUYTER

Mathematics Subject Classification 2010
Primary: 35K59, 35J62, 35D35; Secondary: 35B65, 74G25

Author
Prof. Dr. Dung Le
University of Texas at San Antonio
Department of Mathematics
One UTSA Circle
San Antonio, TX 78249
USA

ISBN 978-3-11-060715-4
e-ISBN (PDF) 978-3-11-060876-2
e-ISBN (EPUB) 978-3-11-060717-8
ISSN 0941-813X

Library of Congress Cataloging-in-Publication Data
Names: Le, Dung (Mathematics professor), author.
Title: Strongly coupled parabolic and elliptic systems : existence and regularity of strong and weak solutions / Dung Le.
Description: Berlin ; Boston : De Gruyter, [2018] | Series: De Gruyter series in nonlinear analysis and applications ; volume 28 | Includes bibliographical references and index.
Identifiers: LCCN 2018032555 (print) | LCCN 2018038815 (ebook) | ISBN 9783110608762 (electronic Portable Document Format (pdf)) | ISBN 9783110607154 (print : alk. paper) | ISBN 9783110608762 (e-book pdf) | ISBN 9783110607178 (e-book epub)
Subjects: LCSH: Control theory. | Coupled mode theory. | Differential equations, Parabolic. | Differential equations, Elliptic. | Differential equations, Partial.
Classification: LCC QA402.3 (ebook) | LCC QA402.3 .L3727 2018 (print) | DDC 515/.642--dc23
LC record available at https://lccn.loc.gov/2018032555

Bibliographic information published by the Deutsche Nationalbibliothek
The Deutsche Nationalbibliothek lists this publication in the Deutsche Nationalbibliografie; detailed bibliographic data are available on the Internet at http://dnb.dnb.de.

Typesetting: le-tex publishing services GmbH, Leipzig
Printing and binding: CPI books GmbH, Leck

www.degruyter.com

To my beloveds Thu Nguyen, Uyen Le and Andrew Le.

Preface

This book aims at researchers who are interested in reaction diffusion systems with couplings occurring in both reaction and diffusion parts. The main theme of the book is to demonstrate the crucial role of a priori estimates of the BMO norm of solutions in the local and global existence theory of strong/weak solutions. Several examples are included and show the applicability of the theory to cross diffusion models in mathematical biology/ecology. The content of this book grows out of our published works and extends well beyond the scope of those by new added findings.

I extend my appreciation and thanks to the production staff at De Gruyter who managed to put this book together. They have been so effective and professional.

San Antonio
September 2018

Dung Le

https://doi.org/10.1515/9783110608762-201

Contents

1 Introduction

Since the beginning of the twentieth century, scalar reaction diffusion equations, with constant and then variable coefficients, have been proposed both by applied scientists and pure mathematicians to study physical phenomena. It was soon observed that most of the equations modeling physical phenomena without excessive simplification are nonlinear. They were extensively studied with great success in models involving one unknown function u describing pressure, temperature, density of species... The equation takes the general form

$$u_t = \mathrm{div}(a(u)Du) + f(u, Du) , \quad (x, t) \in \Omega \times (0, T_0) , \tag{1.0.1}$$

where Ω is a bounded domain with smooth boundary $\partial\Omega$ in $\mathbb{R}^N$ for some integer $N \geq 1$. The divergence operator is denoted by div. As usual, $u_t = \frac{\partial u}{\partial t}$ and $Du = \nabla u$ are respectively the temporal derivative and the gradient of u. The term $\mathrm{div}(a(u)Du)$ then reflects the diffusion of u with its diffusitivity $a(u)$, and $f(u, Du)$ is the reaction term with Du included to take into consideration the advection/drifting effects.

The theory of existence of solutions to (1.0.1), a quasilinear parabolic scalar equation of second order, was well developed in the mid twentieth century for instance in the classical book by Ladyzhenskaya *et al.* [25] and a bit more recent one by Liebermann [34]. The first one extends the level set method of DeGeorgi, while the second mainly makes use of the iteration technique of Moser in combination with the first whenever appropriate. These are the two well known standard methods in dealing with scalar equations. Of course, the book of Friedman [13], using semigroup approach, is to be mentioned. One of the crucial facts in these theories is that certain forms of the maximum/comparison principles are true: solutions are bounded under suitable structural conditions of f. From this, (1.0.1) is regular parabolic (or elliptic).

In applications, we have to naturally deal with evolution processes where several species or unknown quantities are involved. They can diffuse (or move) independently but interact with each other via the reaction terms. We then consider the following system

$$(u_i)_t = \mathrm{div}(a_i(u)Du_i) + f_i(u, Du) , \quad (x, t) \in \Omega \times (0, T_0) , \tag{1.0.2}$$

which consists of m equations for m unknowns u_i, $i = 1, \dots, m$. Here, we denoted $u = [u_i]_{i=1}^m$. Of course, Du is now the Jacobian of u. The systems have found their applications in several models in biology, ecology ... since the 70's and are still extensively studied.

We can still apply the classical techniques for scalar equations to each one of (1.0.2). Under suitable assumption on f_i's, the existence theory of this type of systems follows and is considered to be classical nowadays; and much of attention is then devoted to advanced topics such as dynamics or long time behavior of solutions, which have been very active areas in applied mathematics.

https://doi.org/10.1515/9783110608762-001

The interaction among the components of (1.0.2) occurs only in the reaction terms as one assumes that the diffusion (or movement) of each u_i is not sensitive to those of the others. This of course limits the applications of the theory as scientists already raised such questions in the early 50's of the last century (e.g., see [44]). Taking this into consideration, because the movement of a species is reflected by its gradient, one should replace (1.0.2) by

$$(u_i)_t = \mathrm{div}\left(\sum_{j=1}^{m} a_{i,j}(u)Du_j\right) + f_i(u, Du)\,, \quad (x,t) \in \Omega \times (0, T_0)\,. \tag{1.0.3}$$

Introducing the $m \times m$ matrix $A(u) = [a_{ij}(u)]$ and the vector $\hat{f}(u, Du) = [f_i(u, Du)]_{i=1}^m$, we can write (1.0.1)–(1.0.3) in the vectorial form

$$(u)_t = \mathrm{div}(A(u)Du) + \hat{f}(u, Du)\,, \quad (x,t) \in \Omega \times (0, T_0)\,. \tag{1.0.4}$$

It is clear that the scalar (1.0.1) is included in the above when $m = 1$, and (1.0.2) is also a special case of (1.0.4) if $A(u)$ is a *diagonal* matrix $\mathrm{diag}[a_i(u)]$. We refer to (1.0.4) in this case as a *weakly coupled* system and a *strongly coupled* one if $A(u)$ is a *full* matrix.

The strongly coupled system (1.0.4) appears in many physical applications, for instance, Maxwell–Stephan systems describing the diffusive transport of multicomponent mixtures, models in reaction and diffusion in electrolysis, flows in porous media, diffusion of polymers, or population dynamics, among others. A popular model in mathematical biology, that has been widely investigated together with its simplified variances in the last few decades, is the Shigesada–Kawasaki–Teramoto system (see [44]), which consists of two reaction-diffusion equations with variable cross-diffusion and quadratic nonlinearities.

$$\begin{cases} u_t = \Delta(u[d_1 + \alpha_{11}u + \alpha_{12}v]) + u(a_1 + b_1u + c_1v)\,, \\ v_t = \Delta(v[d_2 + \alpha_{21}u + \alpha_{22}v]) + v(a_2 + b_2u + c_2v)\,. \end{cases} \tag{1.0.5}$$

The system (1.0.4) also occurs at the core of mathematical theory as it can be used to describe problems in differential geometry using PDE's approach, e.g., heat flows among Riemann manifolds. In this case, $\hat{f}$ has a quadratic growth in the gradient Du. We refer the readers to the book [22] by Jurgen Jost for an excellent exposition of the theory.

When $A(u)$ in (1.0.4) is a *full* matrix, the existence theory takes a serious turn and reopens many fundamental issues. First of all, many well known techniques for scalar equations are no longer available and cannot be extended to the vectorial case. For example, Moser's iteration technique can not be carried out infinitely many times and, as a consequence, we do not have the Harnack inequalities, which is one of the most important ingredients of the regularity theory of solutions to scalar equations. The maximum and comparison principles, which provide the earliest and simplest apriori estimates, are generally unavailable. Attempts to utilize the classical tools for scalar equations come to a halt immediately.

Another important issue is the global existence problem when the local one is established: will there be a *unique solution* solving (1.0.4) for all $T_0 > 0$? One can choose to work with either weak or strong solutions. In the first case, the local existence of a weak solution can be achieved via Galerkin, time discretization or variational methods but its regularity (e.g., boundedness, Hölder continuity of the solution and its higher derivatives) is then an open and serious issue. Several works have been done along this line to improve the early work [14] of Giaquinta and Struwe and establish *partial regularity* of *bounded* weak solutions to (1.0.4). Uniqueness of weak solutions for the Cauchy problem is still an open question so that the meaning of global existence of a weak solution is vaguely defined in literature.

Otherwise, if strong solutions are considered then their existence can be established via semigroup theories as in the works of Amann [1, 2]. Combining with interpolation theories of Sobolev spaces, Amann established local and global existence of a strong solution u of (1.0.4) under the assumption that one can control $\|u\|_{W^{1,p}(\Omega,\mathbb{R}^m)}$ for some $p > n$. His theory did not seem to apply to the case where $\hat{f}$ has superlinear growth in Du. In addition, the estimate of $W^{1,p}$ norm when $p > n$ is comparable to that of Hölder norms, which is very hard to achieve.

It is always assumed that the matrix $A(u)$ is elliptic in the sense that there exist two scalar positive continuous functions $\lambda_1(u), \lambda_2(u)$ such that

$$\lambda_1(u)|\zeta|^2 \le \langle A(u)\zeta, \zeta\rangle \le \lambda_2(u)|\zeta|^2 \quad \text{for all } u \in \mathbb{R}^m,\ \zeta \in \mathbb{R}^{mN}. \tag{1.0.6}$$

If there exist positive constants c_1, c_2 such that $c_1 \le \lambda_1(u)$ and $\lambda_2(u) \le c_2$ then we say that $A(u)$, or the corresponding system, is *regular elliptic*. If $0 < \lambda_1(u)$ and $\lambda_2(u)/\lambda_1(u) \le c_2$, we say that $A(u)$ is *uniform elliptic*. In addition, if $\lambda_1(u)$ tends to zero (respectively, ∞) when $|u| \to \infty$ then we say that $A(u)$ is *singular* (respectively, *degenerate*) at infinity.

In both forementioned approaches, the assumption on the boundedness of u must be the starting point because the techniques rely heavily on the fact that $A(u)$ is *regular elliptic* in the above terminology. As we mentioned earlier, for strongly coupled systems like (1.0.4) invariant/maximum principles are generally unavailable so the boundedness of the solutions is already a hard problem and perhaps a very restrictive hypothesis. One usually needs to use ad hoc techniques on the case by case basis to show that u is bounded (see [23, 41]). Even so, for bounded weak solutions we know that they are only Hölder continuous almost everywhere (see [14]). In addition (see [21]), there are counter examples for systems ($m > 1$) which exhibit solutions that start smoothly and remain bounded but develop singularities in higher norms in finite times.

In this book, we will work with strong solutions and present a rather simple and unified approach making use of fixed point theories to obtain the local and then global existence of *strong* solutions of *uniform elliptic* (1.0.4), including the degenerate/singular cases. The framework of fixed point methods, the Leray–Schauder fixed point theorem in particular, is fairly standard and consists of two steps. The first step is fairly standard; strong solutions of the system can be viewed as fixed points of a compact map T on a suitable Banach space $\mathbf{X}$. This map is embedded in a continuous family of related compact maps T_σ, $\sigma \in [0, 1]$, on $\mathbf{X}$. The second step is to show that there is a uniform bound for the $\mathbf{X}$ norms of fixed points of T_σ, and this is the hardest and most crucial ingredient of the argument. This step is equivalent to the establishment of some apriori bounds for higher order norm of strong solutions to (1.0.4). We achieve this by combining a new local Gagliardo Nirenberg inequality and a novel finite iteration argument. Our method then provides the existence results under the weakest (and necessary) assumption that some suitable transformation of the strong solution apriori have small BMO norms in small balls. This assumption and the techniques set the sharp difference between this work and others known in literature.

The presentation here is an extensive generalization of our recent works [29, 30] which established the local and global existence results for regularly uniform elliptic (1.0.4) under the assumption that their strong solutions are a priori VMO (Vanishing Mean Oscillation), but not necessarily bounded, and general structural conditions on the data of (1.0.4) which are independent of x. We will allow A depend on x in this work and assume only that $A(x, u)$ is *uniform elliptic*.

Applications were presented in [30] when $\lambda(u)$ has a polynomial growth in $|u|$ and, without the boundedness assumption on the solutions, so (3.0.1) can be degenerate as $\lambda(u) \to \infty$ when $|u| \to \infty$. The singular case, $\lambda(u) \to 0$ as $|u| \to \infty$, was not discussed there. Also, the reaction term $\hat{f}$ was assumed to have linear growth only in gradient described in f.1) of F).

Most importantly, we will present here a very versatile VMO assumption on u. If the system is singular, one may not see immediately or prove that u is VMO. However, it is possible that $K(u)$ is VMO for some suitable and carefully chosen map $K\colon \mathbb{R}^m \to \mathbb{R}^m$.

In Chapter 2 we will establish one of the key ingredients of the proof in this book: the local weighted Gagliardo–Nirenberg inequalities involving BMO norm with general measure. These new inequalities extensively generalize those initially reported in [29] and allow us to replace the VMO assumption on a strong solution u of (1.0.4) there by a more versatile one in order to obtain the key apriori estimates: $K(u)$ is VMO for some general map K. Inequalities of this type, of course, have their own interests in the theory of functional spaces and they beautifully serve our main purpose in this book. The general inequalities allow us to deal with *singular uniform elliptic* systems with suitable and careful choice of the map K. We also show that the inequalities in [29, 30] are just special (and simple) cases of those discussed here.

Chapter 3 is the longest one and the heart of this book. It provides the core framework for the existence theory in the following chapters. The main existence result Theorem 3.2.3 of the chapter relies on the two main conditions M.0) and M.1). We discuss the solvability of (1.0.4) and show that the assumption on the smallness of BMO norm of $K(u)$ for any strong solution u of (1.0.4) for some suitable map K (see the condition M.1) is sufficient (and also necessary) for the existence of strong solutions. We will also consider very mild integrability conditions in M.0) but no boundedness assumption of solutions will be presumed The setting of fixed point argument will be discussed in Section 4.2.1. The required a priori estimates for possible strong solutions will be proved in details in Section 4.2.2. We will establish a general energy estimate for the gradient Du for any strong solution u and then use the Gagliardo Nirenberg inequalities in the previous chapter to obtain a better estimate which allows us to perform an iteration argument in the proof of our main result.

In Section 4.2.2, the reader will see the advantage of our approach in choosing to work with strong solutions throughout this book. Thanks to the differentiability of these solutions we can differentiate the system in order to derive a strong energy estimate for Du. This is not possible if one chooses to deal with weak solutions. Also, we should mention that the Gagliardo–Nirenberg inequalities, which involve with second derivatives of the solutions, are not available for weak solutions.

On the other hand, we will have to assume the spectral gap condition SG) in addition to the structural conditions on the uniform (1.0.4). Roughly speaking, this requires that the ratio between the ellipticity 'constants' of the matrice $A(u)$ is not too big. We only need this assumption in order to obtain the energy estimate. Otherwise, there are counterexamples showing that strong solutions can blow up (or cease to exist) in finite time (see [31]). This also partly explains why Moser's iteration technique for scalar equations cannot be extended to the vectorial case.

The existence result of (1.0.4) in Theorem 3.2.3 is established under the most general assumption that there is a suitable map K such that $K(u)$ is VMO so that some version of the local Gagliardo–Nirenberg inequalities in Chapter 2 is usable for our purpose. However, the construction of such map is quite delicate as we present by examples in Section 3.5. Under slightly stronger structural assumptions, we will show in Section 3.4 another energy estimate for Du and that will allow us to take K to be the identity map, and the condition M.1) will then be considerably simplified.

We conclude Chapter 3 by discussing several examples where the main theorems are applicable in Section 5.1. Motivated by the SKT system (1.0.5), we will introduce the following generalized SKT model

$$u_t = \Delta(P(u)) + \hat{f}(u, Du) , \tag{1.0.7}$$

where $P: \mathbb{R}^m \to \mathbb{R}^m$ is a C^2 map. The ellipticity of $A(u) = P_u(u)$ can have a polynomial growth like $(\lambda_0 + |u|)^k$ for some $k \neq 0$ and $\lambda_0 > 0$. Obviously, the classic (SKT) system (1.0.5) is a special case of (1.0.7) when P is a quadratic map, $m = 2$ and $k = 1$. The solvability of the system for $k > 0$ was reported in our work [29]. Here, we also consider the singular case $k < 0$.

We discuss the existence theory for the elliptic counterpart of (1.0.4) in Chapter 4

$$-\operatorname{div}(A(u)Du) = \hat{f}(u, Du) , \quad x \in \Omega . \tag{1.0.8}$$

This system can be thought of as the stationary (1.0.4) whose solutions are time independent. The proof of the main existence results in this chapter principally follows those of Chapter 3 and is somewhat simpler, of course due to the absence of the temporal derivative. Nevertheless, there are significant differences in the settings of the fixed point argument and a priori estimates in the two cases; and we present the details to keep this chapter as independent as possible.

In Section 4.4, we discuss several examples concerning the solvability of (1.0.8). The absence of the temporal derivative u_t allows us to consider examples with much more general structural conditions and they don't have to be the stationary ones associated to the generalized (SKT) system (1.0.7) considered in Section 5.1. In addition, the spectral gap condition SG) for the parabolic case, which was void if $N \le 2$, is now not needed if $N \le 4$.

We restrain ourself from discussing the elliptic counterpart of generalized (SKT) system

$$-\Delta(P(u)) = \hat{f}(u, Du)$$

in this chapter because this topic will be investigated at great length in the next chapter. In Chapter 5, we will study the parabolic/elliptic generalized (SKT) systems and present several results extending known existence results for the traditional (SKT) under a very weak assumption that some a priori integrability of solutions are given.

Concerning solutions to the generalized *evolution* SKT systems, we will discuss the existence of *weak* solutions and their uniqueness and VMO regularity when the self diffusivities are all zero, that is the system can be singular in Chapter 5. This model was inspired by the combination between the SKT systems and porous media type scalar equations which have been actively investigated in the last few decades, starting with the work by Friedman, Brezis and Crandall (see [6, 12] and the references therein). To the best of our knowledge, this is the first time such a singular and degenerate settings are combined in one model and treated in a unified framework as in this book. The regularity theory of weak solutions for these systems is still a wildly open problem and the best we can do so far is to prove that they are VMO, at least when $N = 2$.

The elliptic counterpart of the generalized *evolution* SKT systems was also studied in the chapter. Thanks to the special form of the diffusion part, we present another simpler way to obtain $L^p(\Omega)$ norm estimates for Du so that the theory and assumptions in Chapter 4 can be greatly simplified and independent from the discussion here. In fact, we can drop the spectral gap condition SG) and ESG) in the preceding Chapter 3 and Chapter 4 so that the theory here applies to all dimension N to establish the existence of a weak solution. Moreover, the regularity of this weak solution is greatly improved: it is in fact Hölder continuous.

Chapter 6 is devoted to the study of solutions to the general elliptic system (1.0.4). In recent years, much effort has been made to the investigation of steady states, besides the obvious zero solution, of the (SKT) system (1.0.5) by several ad hoc methods. The existence of a stronng solution to (1.0.4) is already established in Chapter 4 but, as we see from the example of (SKT) sytem, this solution may be the trivial or other *semi trivial* solutions which can be observed by simple inspections and thus the result in

Chapter 4 is not interesting in this case. We then face with a new question of finding *nontrivial solutions* besides these semi trivial ones. This will be the main topic of the chapter.

We will present a systematic approach to the problem by using fixed point index theory to obtain necessary conditions for such nontrivial solutions to exist. The content of the chapter heavily relies on that of our recent work [30] and we add some new clarifications as well as developments in applications. In particular, we would like to draw the reader attention to the end of the chapter where, in connection with Chapter 5, we will briefly discuss the existence of nontrivial solutions to the singular (SKT) systems.

Finally, we present some necessary facts, which may be elementary to the experts, in the appendix. These facts are in fact fairly important in our proof but briefly mentioned in order to keep the flow of our main ideas in the proof.

Basic notations:

Concerning the basic notations throughout this book, we mention some. Ω is a bounded domain with smooth boundary $\partial\Omega$ in $\mathbb{R}^N$ for some integer $N \geq 1$. In dealing with parabolic problem like (1.0.4) we work with vector valued functions (or matrix-valued) function

$$u(x,t) = (u_1(x,t), \dots, u_m(x,t))^T \quad m > 1\,.$$

The temporal and k order spatial derivatives of u are denoted by u_t and $D^k u$ respectively. The functional spaces, which u and the data of the system belong to, are the standard ones in literature.

For an integer k we shall denote by $C^k(\Omega)$ the space of functions having continuous derivatives up to and including the order k; and with $C^\infty(\Omega)$ the space of infinitely differentiable functions in Ω. The space $C^k(\bar{\Omega})$ consists of functions in $C^k(\Omega)$, whose derivatives up to order k can be continuously extended to the boundary $\partial\Omega$. For $\alpha \in (0,1)$ we will denote by $C^{k,\alpha}(\Omega)$ the subspace of $C^k(\Omega)$ consisting of functions whose k-order derivatives are Hölder continuous with exponent α.

As usual, $W^{1,p}(\Omega, \mathbb{R}^m)$, $p \geq 1$, will denote the standard Sobolev spaces whose elements are vector valued functions $u: , \Omega \to \mathbb{R}^m$ with finite norm

$$\|u\|_{W^{1,p}(\Omega,\mathbb{R}^m)} = \|u\|_{L^p(\Omega)} + \|Du\|_{L^p(\Omega)}\,.$$

Following [25], for some $q, r \geq 1$ and $T_0 > 0$ we denote by $\mathbf{V}_{q,r}(Q, \mathbb{R}^m)$ the Banach space of vector valued functions on $Q = \Omega \times (0, T_0)$ with finite norm

$$\|u\|_{\mathbf{V}_{q,r}(Q)} = \sup_{t \in (0,T_0)} \|u(\cdot,t)\|_{L^2(\Omega)} + \|Du\|_{q,r,Q}\,,$$

where

$$\|v\|_{q,r,Q} := \left(\int_0^{T_0} \left(\int_\Omega |v(x,t)|^q\, dx\right)^{\frac{r}{q}} dt\right)^{\frac{1}{r}}.$$

If the target space $\mathbb{R}^m$ is clear from the context we will omit it from these notations and simply write them by $W^{1,p}(\Omega)$ and $\mathbf{V}_{q,r}(Q)$.

In our statements and proofs, we use $C, C_1, \ldots$ to denote various constants which can change from line to line but depend only on the parameters of the hypotheses in an obvious way. We will write $C(a, b, \ldots)$ when the dependence of a constant C on its parameters is needed to emphasize that C is bounded in terms of its parameters. We also write $a \lesssim b$ if there is a generic constant C, which is clear from the context, such that $a \leq Cb$. In the same way, $a \sim b$ means $a \lesssim b$ and $b \lesssim a$.

2 Interpolation Gagliardo–Nirenberg inequalities

In this chapter we will discuss one of the key ingredients of the proof of our existence theory in this book: the local weighted Gagliardo–Nirenberg inequalities involving BMO norm with general measure.

In [42, 43], for any $p \geq 1$ and C^2 scalar function u on $\mathbb{R}^N$, $N \geq 2$, global and local Gagliardo–Nirenberg inequalities of the form

$$\int_{\mathbb{R}^n} |Du|^{2p+2}\,dx \leq C(n,p)\|u\|^2_{\mathrm{BMO}} \int_{\mathbb{R}^n} |Du|^{2p-2}|D^2u|^2\,dx \tag{2.0.1}$$

were established and applied to the solvability of *scalar* elliptic equations.

More general and vectorial versions of these inequalities were presented in [28, 29] to establish the solvability of strongly coupled parabolic systems of the form *nonregular* but *uniform* parabolic system

$$\begin{cases} u_t = \operatorname{div}(A(u,Du)) + \hat{f}(u,Du) & (x,t) \in Q = \Omega \times (0,T_0)\,, \\ u(x,0) = U_0(x) & x \in \Omega \\ u = 0 \;\text{ on }\; \partial\Omega \times (0,T_0)\,. \end{cases} \tag{2.0.2}$$

We will prove new general global and local versions of (2.0.1) and the results in [29] in many aspects. Roughly speaking, we will establish inequalities of the following type: for any $p \geq 1$ and any C^2 map $U : \Omega \to \mathbb{R}^m$,

$$\int_\Omega \Phi^2(U)|DU|^{2p+2}\,d\mu \leq C\|K(U)\|^2_{\mathrm{BMO}(\mu)} \left\{\int_\Omega \Lambda^2(U)|DU|^{2p-2}|D^2U|^2\,d\mu + \cdots\right\}\,.$$

Here, $K\colon \mathbb{R}^m \to \mathbb{R}^m$ is a map, Φ, Λ are functions on $\mathbb{R}^m$ and $d\mu = \omega dx$ is a doubling measure on Ω. We will only assume that Ω, μ support a Poincaré–Sobolev type inequality.

Of course, inequalities of this type have their own interests and find applications in functional space theories. However, the purpose of such generalizations in this book becomes visible when we apply them to the study of local/global existence of strong solutions to (6.0.1), and its elliptic counterpart in the next two chapters. First of all, by replacing the Lebesgue measure dx with a general measure $d\mu = \omega dx$, we allow the matrices $A, \hat{f}$ in (6.0.1) to depend on x and become degenerate or singular near a subset of Ω.

Secondly, the degeneracy and singularity of (6.0.1) can also come from the behavior of the solution u itself, which is not bounded in general as maximum principles are not available for these systems (i.e., $m > 1$). We replace the factor $\|u\|_{\mathrm{BMO}}$ in (2.0.1) by $\|K(u)\|_{\mathrm{BMO}(\mu)}$ where K is a C^2 map in $\mathbb{R}^m$. This allows us to deal with the case when estimates for $\|u\|_{\mathrm{BMO}}$, but $\|K(u)\|_{\mathrm{BMO}(\mu)}$, are not available. This factor, with different choices of K, will play a crucial role in the a priori estimates in our theory. For example, one of the consequences of our general inequalities in this chapter is the following

https://doi.org/10.1515/9783110608762-002

inequality, which will be useful in dealing with singular systems: if $\|\log(|u|)\|_{BMO(\mu)}$ is sufficiently small then

$$\int_\Omega |u|^{2k-2}|Du|^{2p+2}\,d\mu \le C\|\log(|u|)\|^2_{BMO(\mu)}\int_\Omega (|u|^{2k}|Du|^{2p-2}|D^2u|^2 + |u|^{2k}|Du|^{2p})\,d\mu\,.$$

Various choices of K will be discussed in the applications of our main results in Chapter 3 and Chapter 4.

The hypotheses and preparatory materials will be presented in Section 2.1. The needed theory of RBMO(μ) spaces comes from Tolsa's work [39] and we will discuss it in Section A.1 of the Appendix. The main global and local inequalities are stated in Section 2.2. The proofs of these inequalities are quite technical and we will postpone them to the end of the chapter. In passing, in Section 2.3 we discuss their consequences and show that the results in [29] are just special cases of the main inequalities in this book. This section also contains new inequalities used in the proof of the next two chapters, the heart of this book, on the existence theory of (6.0.1).

2.1 Hypotheses on the measures

Let μ be a measure on Ω. For any μ-measurable subset A of Ω and any locally μ-integrable function $U: \Omega \to \mathbb{R}^m$ we denote by $\mu(A)$ the measure of A and U_A the average of U over A. That is,

$$U_A = \fint_A U(x)\,d\mu = \frac{1}{\mu(A)}\int_A U(x)\,d\mu\,.$$

We now recall some well-known notions from harmonic analysis.

A function $f \in L^1(\mu)$ is said to be in BMO(μ) if

$$[f]_{*,\mu} := \sup_Q \fint_Q |f - f_Q|\,d\mu < \infty\,. \tag{2.1.1}$$

We then define

$$\|f\|_{BMO(\mu)} := [f]_{*,\mu} + \|f\|_{L^1(\mu)}\,.$$

For $\gamma \in (1,\infty)$ we say that a nonnegative locally integrable function w belongs to the class A_γ or w is an A_γ weight if the quantity

$$[w]_\gamma := \sup_{B\subset\Omega}\left(\fint_B w\,d\mu\right)\left(\fint_B w^{1-\gamma'}\,d\mu\right)^{\gamma-1} \quad \text{is finite}\,. \tag{2.1.2}$$

Here, $\gamma' = \gamma/(\gamma-1)$ and the supremum is taken over all cubes B in Ω. For more details on these classes we refer the reader to [11, 40, 45].

We say that Ω and μ support a Poincaré–Sobolev inequality if the following holds:

PS) There are $\sigma \in (0,1)$ and $\tau_* \ge 1$ such that for *some* $q > 2$ and $q_* = \sigma q < 2$ we have

$$\frac{1}{l(B)}\left(\fint_B |u-u_B|^q\,d\mu\right)^{\frac{1}{q}} \le C_{PS}\left(\fint_{\tau_* B} |Du|^{q_*}\,d\mu\right)^{\frac{1}{q_*}} \tag{2.1.3}$$

for some constant C_{PS} and any cube B with side length $l(B)$ and any function $u \in C^1(B)$.

Here and throughout this chapter, we write $B_R(x)$ for a cube centered at x with side length R and sides parallel to the standard axes of $\mathbb{R}^N$. We will omit x in the notation $B_R(x)$ if no ambiguity can arise. We denote by $l(B)$ the side length of B and by τB the cube that is concentric with B and has side length $\tau l(B)$.

We have the following remark on the validity of the assumption PS).

Remark 2.1.1. Suppose that μ is doubling and supports a q_*-Poincaré inequality [19, eqn. (5)]: there are some constants C_P, $q_* \in [1, 2]$ and $\tau_* \geq 1$ the inequality

$$\fint_B |h - h_B|\, d\mu \leq C_P l(B) \left(\fint_{\tau_* B} |Dh|^{q_*}\, d\mu \right)^{\frac{1}{q_*}} \tag{2.1.4}$$

holds true for any cube B with side length $l(B)$ and any function $h \in C^1(B)$.

Assume also that for some $s > 0$ μ satisfies the following inequality:

$$\left(\frac{r}{r_0} \right)^s \lesssim \frac{\mu(B_r(x))}{\mu(B_{r_0}(x_0))}, \tag{2.1.5}$$

where $B_r(x)$, $B_{r_0}(x_0)$ are any cubes with $x \in B_{r_0}(x_0)$. If $q_* = 2$ then [19, Section 3] shows that a q_*-Poincaré inequality also holds for some $q_* < 2$. Thus, we can assume that $q_* \in (1, 2)$.

If $q_* < s$ then [19, 1) of Theorem 5.1] establishes (2.1.3) for $q = sq_*/(s - q_*)$. Thus, $q > 2$ if $s < 2q_*/(2 - q_*)$. This is the case if we choose $q_* < 2$ and close to 2. If $s = q_*$, [19, 2) of Theorem 5.1] shows that (2.1.3) holds true for any $q > 1$. On the other hand, if $q_* > s$ then [19, 3) of Theorem 5.1] gives a stronger version of (2.1.3) for $q = \infty$. In particular, the Hölder norm of u is bounded in terms of $\|Du\|_{L^{q_*}(\mu)}$.

Thus, in order to justify PS), we can assume that Ω, μ support a q_*-Poincaré inequality for some $q_* \in (1, 2)$ and (2.1.5) is valid for some $s > 0$.

We assume the following hypotheses:

M) Let Ω be a bounded domain in $\mathbb{R}^N$ and $d\mu = \omega dx$ for some $\omega \in C^1(\Omega, \mathbb{R}^+)$. Suppose that there are a constant C_μ and a fixed number $n \in (0, N]$ such that for any cube Q_r with side length $r > 0$

$$\mu(Q_r) \leq C_\mu r^n\,. \tag{2.1.6}$$

Furthermore, Ω and μ satisfy PS).

Next, we consider the following condition.

H) $d\mu = \omega_0^2 dx$ supports a Hardy type inequality: there is a constant C_H such that for any function $u \in C_0^1(B)$,

$$\int_\Omega |u|^2 |D\omega_0|^2\, dx \leq C_H \int_\Omega |Du|^2 \omega_0^2\, dx\,. \tag{2.1.7}$$

Remark 2.1.2. A typical choice of ω_0 that satisfies the Hardy type inequality (2.1.7) is $\omega_0(x) = d_\Omega^{\frac{\gamma}{2}}(x)$ for some constant γ. We now recall the following Hardy inequality proved by Necas (see also the paper by Lehrbäck [33] for much more general versions):

$$\int_\Omega |u(x)|^q d_\Omega^{\gamma-q}(x)\,dx \le C\int_\Omega |Du(x)|^q d_\Omega^{\gamma}(x)\,dx\,, \quad \gamma < q-1\,. \tag{2.1.8}$$

Because $|D\omega_0| \sim d_\Omega^{\frac{\gamma-2}{2}}(x)$, (2.1.8), with $q = 2$, shows that (2.1.7) holds true for $\gamma < 1$.

We will also check the condition (2.1.6) of M) for $d\mu = \omega dx \sim d_\Omega^\gamma dx$. If B_r is far away from $\partial\Omega$, we have $\mu(B_r) \lesssim r^N$. Near the boundary, as $\partial\Omega$ is C^1, we easily see that $\mu(B_r) \lesssim r^{N+\gamma}$. If $N+\gamma \ge n$ and r is bounded then $\mu(B_r) \le Cr^n$. This is the case because Ω is bounded. Thus, for any $\gamma > -N$, we define $n = \min\{N, N+\gamma\} \in (0, N]$ to see that $\mu(B_r) \le Cr^n$ for some constant C, which is bounded in terms of diam(Ω).

2.2 The main and technical inequality

In this section we will state our main global and local weighted Gagliardo–Nirenberg interpolation inequalities. The local version and its consequences will be one of our main vehicles for the solvability of strong solutions in this book.

Throughout this section we will always use the following notations and hypotheses.

P.1) Let $K\colon \operatorname{dom}(K) \to \mathbb{R}^m$ be a C^1 map on a domain $\operatorname{dom}(K) \subset \mathbb{R}^m$ such that $K_U(U)^{-1}$ exists for $U \in \operatorname{dom}(K)$. Assume that the map $\mathbb{K}(U) := (K_U(U)^{-1})^T$ is differentiable. Let $\Phi, \Lambda\colon \operatorname{dom}(K) \to \mathbb{R}^+$ be C^1 positive functions. Assume that

$$|\mathbb{K}(U)| \lesssim \Lambda(U)\Phi^{-1}(U) \quad \text{for all} \quad U \in \operatorname{dom}(K)\,. \tag{2.2.1}$$

P.2) Let $U\colon \bar{\Omega} \to \operatorname{dom}(K)$ be a C^2 vector-valued function satisfying

$$\text{either} \quad K(U) = 0 \quad \text{or} \quad \langle \omega\Phi^2(U)\mathbb{K}(U)DU, \vec{\nu}\rangle = 0 \tag{2.2.2}$$

on $\partial\Omega$, where $\vec{\nu}$ is the outward normal vector of $\partial\Omega$.

W) Let

$$\mathbf{W}(x) := \Lambda^{p+1}(U(x))\Phi^{-p}(U(x)) \quad \text{for} \quad x \in \Omega\,. \tag{2.2.3}$$

Assume that $[\mathbf{W}^\alpha]_{\beta+1}$ is finite for some $\alpha > 2/(p+2)$ and $\beta < p/(p+2)$.

We consider the following integrals:

$$I_1 := \int_\Omega \Phi^2(U)|DU|^{2p+2}\,d\mu\,, \quad I_2 := \int_\Omega \Lambda^2(U)|DU|^{2p-2}|D^2U|^2\,d\mu\,, \tag{2.2.4}$$

$$\bar{\mathfrak{I}}_1 := \int_\Omega |\Lambda_U(U)|^2|DU|^{2p+2}\,d\mu\,, \tag{2.2.5}$$

$$\breve{\mathfrak{I}}_0 := \int_\Omega |D\omega|^2\Lambda^2(U)|DU|^{2p}\,dx\,, \tag{2.2.6}$$

and furthermore

$$\hat{\mathfrak{J}}_1 := \int_\Omega (|\Phi_U(U)||\mathbb{K}(U)| + \Phi(U)|\mathbb{K}_U(U)|)^2 |DU|^{2p+2}\, d\mu\,. \tag{2.2.7}$$

The first main result of this section is the following theorem.

Theorem 2.2.1. *Assume M), P.1)–P.2) and W). Suppose that the integrals in* (2.2.4)–(2.2.7) *are finite. Then there are constants* C, $C([\mathbf{W}^\alpha]_{\beta+1})$ *for which*

$$I_1 \le C\|K(U)\|^2_{\mathrm{BMO}(\mu)} \left[I_2 + \bar{\mathfrak{J}}_1 + C([\mathbf{W}^\alpha]_{\beta+1})[I_2 + \hat{\mathfrak{J}}_1 + \breve{\mathfrak{J}}_0]\right]\,. \tag{2.2.8}$$

The constant C depends on C_{PS}, C_μ.

Next, we will establish a local version of Theorem 2.2.1. Let Ω_* be a subset of Ω. We assume that there are two functions ω_*, ω_0 satisfying the following conditions.

L.0) $\omega_* \in C^1_0(\Omega)$ and satisfies

$$\omega_* \equiv 1 \text{ in } \Omega_* \text{ and } \omega_* \le 1 \text{ in } \Omega\,. \tag{2.2.9}$$

L.1) $\omega_0 \in C^1(\Omega)$ and for $d\mu = \omega_0^2 dx$ and some $n \in (0, N]$ we have $\mu(B_r) \le Cr^n$.

L.2) The measure $\omega_0^2 dx$ supports the Poincaré–Sobolev inequality (2.1.3) in PS). In addition, ω_0 also supports a Hardy type inequality: For any function $u \in C^1_0(B)$

$$\int_\Omega |u|^2 |D\omega_0|^2\, dx \le C_H \int_\Omega |Du|^2 \omega_0^2\, dx\,. \tag{2.2.10}$$

Theorem 2.2.2. *Suppose L.0)–L.2) and P.1). Let $U: \bar{\Omega} \to \mathbb{R}^m$ be a C^2 map. For any $\omega_1 \in L^1(\Omega)$ and $\omega_1 \sim \omega_0^2$ we define $d\mu = \omega_1 dx$ and accordingly recall the definitions* (2.2.4)–(2.2.7) *and introduce*

$$I_{1,*} := \int_{\Omega_*} \Phi^2(U)|DU|^{2p+2}\, d\mu\,, \tag{2.2.11}$$

$$\breve{I}_{0,*} := \sup_\Omega |D\omega_*|^2 \int_\Omega \Lambda^2(U)|DU|^{2p}\, d\mu\,. \tag{2.2.12}$$

Then, for any $\varepsilon > 0$ there are constants C, $C([\mathbf{W}^\alpha]_{\beta+1})$ such that

$$I_{1,*} \le \varepsilon I_1 + \varepsilon^{-1} C\mathbf{C}_*^2 [I_2 + \bar{\mathfrak{J}}_1 + C([\mathbf{W}^\alpha]_{\beta+1})[I_2 + \hat{\mathfrak{J}}_1 + \bar{\mathfrak{J}}_1 + \breve{I}_{0,*}]\,. \tag{2.2.13}$$

Here, $\mathbf{C}_* := \|K(U)\|_{\mathrm{BMO}(\mu)}$ and C depends on C_{PS}, C_μ and C_H.

The proof of these two theorems will be postponed to Section 2.4 and Section 2.5 at the end of this chapter. We will move on and discuss various special versions of the above two theorems in the next section.

2.3 Consequences of the technical theorems

Our first consequence of Theorem 2.2.1 is the case where $\Phi = \Lambda$ and K is the identity map. We then have the following main inequality in [29].

Corollary 2.3.1. *Let $U\colon \Omega \to \mathrm{dom}(K)$ be a C^2 vector-valued function. Suppose that either U or $\Phi^2(U)\frac{\partial U}{\partial \nu}$ vanish on the boundary $\partial\Omega$ of Ω.*

We set

$$I_1 := \int_\Omega \Phi^2(U)|DU|^{2p+2}\,dx\,,\quad I_2 := \int_\Omega \Phi^2(U)|DU|^{2p-2}|D^2U|^2\,dx\,, \tag{2.3.1}$$

$$\bar{\mathfrak{J}}_1 := \int_\Omega |\Phi_U(U)|^2|DU|^{2p+2}\,dx\,. \tag{2.3.2}$$

For any $\alpha > 2/(p+2)$ and $\beta < p/(p+2)$ we have

$$I_1 \le C\|U\|^2_{\mathrm{BMO}(\Omega)}\left[I_2 + \bar{\mathfrak{J}}_1 + C([\Phi^\alpha(U)]_{\beta+1})(I_2 + \bar{\mathfrak{J}}_1)\right]\,. \tag{2.3.3}$$

Proof of Corollary 2.3.1. For $\Phi = \Lambda$ and $K(U) = U$ we see that $\mathbb{K}(U) = Id$, the identity matrix, satisfies P.1). The boundary conditions for U here show that P.2) is verified. Also, the weight $\mathbf{W} := \Lambda^{p+1}\Phi_U^{-p} = \Phi$. The integrals in (2.2.4) and (2.2.5) are exactly those defined here. Because $\mathbb{K}(U) = Id$, we have $\mathbb{K}_U(U) = 0$ so that, from the definitions (2.2.5) and (2.2.7), $\hat{\mathfrak{J}}_1 = \bar{\mathfrak{J}}_1$. Finally, with $\omega \equiv 1$, $\breve{\mathfrak{J}}_0 = 0$. We then have (2.3.3) from (2.2.8). □

In particular, if $\Phi = \Lambda \equiv 1$ and K is the identity map then $\hat{\mathfrak{J}}_1 = \bar{\mathfrak{J}}_1 = 0$. Theorem 2.2.1, with general measure $d\mu = \omega dx$, then shows that

$$I_1 \le C\|U\|^2_{\mathrm{BMO}(\mu)}\left[I_2 + \breve{\mathfrak{J}}_0\right]\,, \tag{2.3.4}$$

where

$$I_1 := \int_\Omega |DU|^{2p+2}\,d\mu,\ I_2 := \int_\Omega |DU|^{2p-2}|D^2U|^2\,d\mu\,, \tag{2.3.5}$$

$$\breve{\mathfrak{J}}_0 := \int_\Omega |D\omega|^2|DU|^{2p}\,dx\,. \tag{2.3.6}$$

We see that (2.3.4) is a generalization of the inequality (2.0.1) with general measure.

In the same way, we have the following local version, which will play an important role in the study of existence of strong solutions in the next chapters.

Corollary 2.3.2. *Let $U\colon \Omega \to \mathbb{R}^m$ be a C^2 vector-valued function. Let Ω_* be a subdomain of Ω and $\omega_* \in C^1_0(\Omega)$ such that $\omega_* \equiv 1$ in Ω_* and $\omega_* \le 1$ in Ω.*

Then there is a constant C such that for any $\varepsilon > 0$,

$$I_{1,*} \le \varepsilon I_1 + \varepsilon^{-1}C\|U\|^2_{\mathrm{BMO}(\mu)}[I_2 + \breve{I}_{0,*}]\,. \tag{2.3.7}$$

Here, I_1, I_2 are defined by (2.3.5) and $I_{1,*}$ and $\breve{I}_{0,*}$ are defined by (2.2.11) and (2.2.12) respectively.

In fact, the inequality (2.3.4) can be greatly generalized as follows.

Corollary 2.3.3. *Let $U\colon \Omega \to \mathbb{R}^m$ be a C^2 vector-valued function. Suppose that either U or $\frac{\partial U}{\partial \nu}$ vanish on the boundary $\partial\Omega$ of Ω. For some $k \geq 0$ and $\lambda_0 > 0$ let $\Phi(U) = (\lambda_0 + |U|)^k$ and $d\mu = \omega dx$ satisfying M).*

Then there is a constant C_* depending on λ_0, k, $\|U\|_{\mathrm{BMO}(\mu)}$ and $[\Phi^\alpha(U)]_{\beta+1}$ such that

$$I_1 \leq C_* \|U\|^2_{\mathrm{BMO}(\mu)} \left[I_2 + \breve{\mathfrak{J}}_0\right] . \tag{2.3.8}$$

The integrals in (2.3.8) *are*

$$I_1 := \int_\Omega \Phi^2(U)|DU|^{2p+2}\, d\mu\,, \quad I_2 := \int_\Omega \Phi^2(U)|DU|^{2p-2}|D^2U|^2\, d\mu\,, \tag{2.3.9}$$

$$\breve{\mathfrak{J}}_0 := \int_\Omega |D\omega|^2 \Phi^2(U)|DU|^{2p}\, dx\,. \tag{2.3.10}$$

Proof. If $k = 0$, the corollary is (2.3.4). We need only consider the case $k > 0$. We argue as in the proof of Corollary 2.3.1, without the assumption $\omega \equiv 1$ so that $\breve{\mathfrak{J}}_0 \neq 0$, to get

$$I_1 \leq C\|U\|^2_{\mathrm{BMO}(\mu)} \left[I_2 + \bar{\mathfrak{J}}_1 + C([\Phi^\alpha(U)]_{\beta+1})(I_2 + \bar{\mathfrak{J}}_1 + \breve{\mathfrak{J}}_0)\right] . \tag{2.3.11}$$

We consider the integral $\bar{\mathfrak{J}}_1$ defined by (2.2.5). As $|\Phi_U(U)| \sim (\lambda_0 + |U|)^{k-1}$, we can take

$$I_1^* := \bar{\mathfrak{J}}_1 = \int_\Omega \phi_k^2(U)|DU|^{2p+2}\, d\mu\,, \quad \phi_k(U) := (\lambda_0 + |U|)^{k-1}\,. \tag{2.3.12}$$

We apply (2.3.11) again, with $\Phi(U)$ being $\phi_k(U)$ and the integrals being redefined accordingly, to get

$$I_1^* \leq C\|U\|^2_{\mathrm{BMO}(\mu)} \left[I_2^* + \bar{\mathfrak{J}}_1^* + C([\phi_k^\alpha(U)]_{\beta+1})(I_2^* + \bar{\mathfrak{J}}_1^* + \breve{\mathfrak{J}}_0^*)\right] . \tag{2.3.13}$$

Because $\phi_k(U) \sim \Phi(U)^\delta$ with $\delta = (k-1)/k < 1$, by the definition of weights and Hölder's inequality we have $[\phi_k^\alpha(U)]_{\beta+1} \lesssim [\Phi^\alpha(U)]^\delta_{\beta+1}$. On the other hand, because $\lambda_0 > 0$, $\phi_k(U) \leq C(\lambda_0)\Phi(U)$ for some constant $C(\lambda_0)$. Thus, $I_2^* \leq C(\lambda_0)I_2$ and $\breve{\mathfrak{J}}_0^* \leq C(\lambda_0)\breve{\mathfrak{J}}_0$. Using these facts in (2.3.13) and then applying the result in (2.3.11), we see that (2.3.11) holds again with $\bar{\mathfrak{J}}_1$ being defined as in (2.3.12) with $\phi_k(U)$ being $\phi_{k-1}(U) = (\lambda_0 + |U|)^{k-1}$.

Hence, assume that $k = k_0 + \kappa$ for some integer $k_0 = 0, 1, \ldots$ and $\kappa \in [0, 1)$ then we can repeat the above argument $k_0 + 1$ times to get

$$I_1 \leq C\|U\|^2_{\mathrm{BMO}(\mu)} \left[I_2 + C_*(I_2 + I_1^{**} + \breve{\mathfrak{J}}_0)\right] , \tag{2.3.14}$$

where C_* is a constant depending on λ_0, k, $\|U\|_{\mathrm{BMO}(\mu)}$ and $[\Phi^\alpha(U)]_{\beta+1}$; the integral I_1^{**} is, noting that $\kappa < 1$,

$$I_1^{**} := \int_\Omega |(\lambda_0 + |U|)^{2(\kappa-1)}|DU|^{2p+2}\, d\mu \leq \lambda_0^{2(\kappa-1)} \int_\Omega |DU|^{2p+2}\, d\mu\,. \tag{2.3.15}$$

The last integral can be estimated by (2.3.4), using the definition (2.3.5). Again, because $\lambda_0 > 0$, we can find a constant $C(\lambda_0)$ such that $I_1^{**} \leq C(\lambda_0)\|U\|^2_{\mathrm{BMO}(\mu)}\left[I_2 + \breve{\mathfrak{J}}_0\right]$. Using this in (2.3.14), we obtain (2.3.8). The proof is complete. □

Let us consider general maps K and impose more conditions on Φ, Λ.

Theorem 2.3.4. *Assume as in Theorem 2.2.1. Assume further that there is a constant C_0 such that*

$$|\mathbb{K}_U| \le C_0 , \tag{2.3.16}$$

$$|\Lambda_U| \le C_0\Phi , \quad |\Phi_U|\Phi^{-1} \le C_0|\Lambda_U|\Lambda^{-1} . \tag{2.3.17}$$

Then there are constants $C(C_0)$, $C([\mathbf{W}^\alpha]_{\beta+1})$ for which

$$I_1 \le C\|K(U)\|^2_{\mathrm{BMO}} \left[I_2 + I_1 + C([\mathbf{W}^\alpha]_{\beta+1})[I_2 + I_1 + \breve{\mathfrak{I}}_0]\right] . \tag{2.3.18}$$

Proof. From the assumption $|\Lambda_U| \le C_0\Phi$ in (2.3.17) and the definition (2.2.5), it is clear that $\bar{\mathfrak{I}}_1 \le I_1$.

We now look at $\hat{\mathfrak{I}}_1$, the integral of $(|\Phi_U|\mathbb{K}| + \Phi|\mathbb{K}_U|)^2|DU|^{2p+2}$. First of all, because $\mathbb{K} \lesssim \Lambda\Phi^{-1}$, (2.3.17) implies $|\Phi_U|\mathbb{K}| \lesssim |\Phi_U|\Lambda\Phi^{-1} \lesssim |\Lambda_U| \lesssim \Phi$. Together with (2.3.16), we now see that $|\Phi_U|\mathbb{K}| + \Phi|\mathbb{K}_U| \lesssim \Phi$ and thus have $\hat{\mathfrak{I}}_1 \lesssim I_1$.

The inequality (2.3.18) then follows from (2.2.8) of Theorem 2.2.1. □

The local version Theorem 2.2.2 then implies the following corollary.

Corollary 2.3.5. *Assume* (2.3.16) *and* (2.3.17). *Using the definitions* (2.2.11) *and* (2.2.12) *for $I_{1,*}$ and $\breve{I}_{0,*}$, we have*

$$I_{1,*} \le \varepsilon I_1 + \varepsilon^{-1}C\|K(U)\|^2_{\mathrm{BMO}(\Omega)}[I_2 + I_1 + C([\mathbf{W}^\alpha]_{\beta+1})[I_2 + I_1 + \breve{I}_{0,*}]] . \tag{2.3.19}$$

Remark 2.3.6. In applications, we usually have $\Lambda(U) = \sqrt{\lambda(U)}$ for some C^2 function λ. Furthermore, we also have $\Phi(U) = |\Lambda_U(U)|$. In this case, the first inequality in (2.3.17) is clear. We look at the second one: $|\Phi_U|\Phi^{-1} \le C_0|\Lambda_U|\Lambda^{-1}$. It is clear that

$$\frac{|\Phi_U|}{\Phi} \sim \left(\frac{|\lambda_{UU}|}{\sqrt{\lambda}} + \frac{|\lambda_U|^2}{\lambda^{3/2}}\right)\frac{\sqrt{\lambda}}{|\lambda_U|} = \frac{|\lambda_{UU}|}{|\lambda_U|} + \frac{|\lambda_U|}{\lambda} , \quad \frac{|\Lambda_U|}{\Lambda} \sim \frac{|\lambda_U|}{\lambda} .$$

We thus have $|\Phi_U|\Phi^{-1} \lesssim |\Lambda_U|\Lambda^{-1}$ if $|\lambda_{UU}|\lambda \lesssim |\lambda_U|^2$. This assumption is verified when λ is a polynomial in $|U|$.

We also note that, from the definition $\mathbb{K}(U)$, we have $\mathbb{K}(U)^T = K_U(U)^{-1}$ so that $\mathbb{K}(U)^T_U = K_U^{-1}(U)K_{UU}(U)K_U^{-1}(U)$. Hence

$$|\mathbb{K}_U| \le C_0 \Leftrightarrow |K_U^{-1}(U)K_{UU}(U)K_U^{-1}(U)| \le C_0 .$$

Of course, there are many ways to choose K, Λ, Φ depending on different situations in applications. Let us consider another choice of K where $|K(U)|$ behaves like $\log(|U|)$.

We define for any $k \ne 0$ and $\varepsilon \ge 0$

$$I_1 = \int_\Omega (\varepsilon + |U|)^{2k-2}|DU|^{2p+2}\, d\mu , \quad \breve{\mathfrak{I}}_0 = \int_\Omega (\varepsilon + |U|)^{2k}|DU|^{2p}\, d\mu , \tag{2.3.20}$$

$$I_2 = \int_\Omega (\varepsilon + |U|)^{2k}|DU|^{2p-2}|D^2U|^2\, d\mu . \tag{2.3.21}$$

Corollary 2.3.7. *For $m \geq 1$, any $k \neq 0$ and $\varepsilon \geq 0$ we consider the map*

$$K(U) = [\log(\varepsilon + |U_i|)]_{i=1}^m\,, \quad U = [U_i]_{i=1}^m\,. \tag{2.3.22}$$

With the notations (2.3.20) *and* (2.3.21) *and* $\mathbf{W} = (\varepsilon + |U|)^{k+p}$, *we have*

$$I_1 \leq C\|K(U)\|^2_{\mathrm{BMO}(\mu)}\left[I_2 + I_1 + C([\mathbf{W}^\alpha]_{\beta+1})[I_2 + I_1 + \breve{\mathfrak{I}}_0]\right]\,, \tag{2.3.23}$$

as long as the integrals are finite. Here, C is independent of ε.

Proof of Corollary 2.3.7. We apply (2.3.18) of Theorem 2.3.4 to this case. We consider $\Lambda(U) = (\varepsilon + |U|)^k$ and $\Phi(U) \sim |\Lambda_U(U)|$ and validate the assumptions of Theorem 2.3.4.

Concerning the condition (2.3.16), as $K(U) = [\log(\varepsilon + |U_i|)]_{i=1}^m$ so that $K_U(U) = \mathrm{diag}[(\varepsilon + |U_i|)^{-1}]$ and $K_{UU} = \mathrm{diag}[-(\varepsilon + |U_i|)^{-2}\frac{U_i}{|U_i|}]$. It is clear that (2.3.16) is justified.

Next, for $\Lambda(U) = (\varepsilon + |U|)^k$ and $\Phi(U) = |\Lambda_U(U)|$ we let $\lambda(U) = (\varepsilon + |U|)^{2k}$ in Remark 2.3.6 and see that $|\lambda_{UU}(U)|\lambda(U) \sim |U|^{4k-2} \sim |\lambda_U(U)|^2$. We see that the assumption (2.3.17) is verified.

Hence, (2.3.18) of Theorem 2.3.4 applies here. As $\Phi(U) \sim (\varepsilon + |U|)^{k-1}$, we have $\mathbf{W} = \Lambda^{p+1}\Phi^{-p} \sim (\varepsilon + |U|)^{k+p}$. We then obtain (2.3.23) from (2.3.18) and the proof is complete. □

Let us discuss the connection between the two quantities $\|K(U)\|_{\mathrm{BMO}(\mu)}$, $[\mathbf{W}^\alpha]_{\beta+1}$. We consider the case $m = 1$. As $\mathbf{W} \sim |k|^{-p}(\varepsilon+|U|)^{k+p}$, we have $[\mathbf{W}^\alpha]_{\beta+1} = [(\varepsilon+|U|)^{\alpha(k+p)}]_{\beta+1}$ and

$$[\log(\mathbf{W}^\alpha)]_{*,\mu} = \alpha|k+p|[\log(\varepsilon + |U|)]_{*,\mu}\,.$$

Via a simple use of Jensen's inequality, it is well known (e.g., [17, Chapter 9]) that $\|\log w\|_{\mathrm{BMO}} \leq [w]_{A_q}$ for $1 < q \leq 2$. In our case, $q = \beta + 1 < 2$ so that $\|\log \mathbf{W}^\alpha\|_{\mathrm{BMO}} \leq [\mathbf{W}^\alpha]_{A_q}$. Thus, the term $[\log(\varepsilon+|U|)]_{\mathrm{BMO}(\mu)}$ can be controlled by $[\mathbf{W}^\alpha]_{\beta+1}$. However, this type of result is not helpful for our purpose in this book.

On the other hand, if $\log \mathbf{W}$ is BMO then we also know that $\mathbf{W}$ is a weight. We recall the John–Nirenberg inequality (e.g., [40] or [17, Corollary 7.1.7]): if μ is doubling then for any function v there are constants $\mathbf{c}_1, \mathbf{c}_2$, which depend only on the doubling constant of μ, such that

$$⨍_B e^{\frac{\mathbf{c}_1}{[v]_{*,\mu}}|v - v_B|}\,d\mu \leq \mathbf{c}_2\,. \tag{2.3.24}$$

We then have the following result.

Corollary 2.3.8. *In addition to the assumptions of Corollary 2.3.7 we suppose that*

$$|k+p|[\log(\varepsilon + |U|)]_{*,\mu} \leq \mathbf{c}_1\beta\alpha^{-1}\,. \tag{2.3.25}$$

Then there is a constant C, which depends also on $\mathbf{c}_2$, for which

$$I_1 \leq C\|\log(\varepsilon + |U|)\|^2_{\mathrm{BMO}(\mu)}\left[I_2 + I_1 + \breve{\mathfrak{I}}_0\right]\,. \tag{2.3.26}$$

It is clear that if $\|\log(\varepsilon+|U|)\|_{\mathrm{BMO}(\mu)}$ is sufficiently small then (2.3.25) and (2.3.26) imply

$$I_1 \le C\|\log(\varepsilon + |U|)\|^2_{\mathrm{BMO}(\mu)} \left[I_2 + \breve{\mathfrak{J}}_0\right] .$$

To prove Corollary 2.3.8 we have the following estimate for $[\mathbf{W}^\alpha]_{\beta+1}$.

Lemma 2.3.9. *For any scalar function w on Ω and α, β > 0 and* $\mathbf{c}_1, \mathbf{c}_2$ *as in* (2.3.24), *we have*

$$[\log(w)]_{*,\mu} \le \mathbf{c}_1\beta\alpha^{-1} \Rightarrow [w^\alpha]_{\beta+1} \le \mathbf{c}_2^{1+\beta} . \tag{2.3.27}$$

Proof. For any $\beta > 0$ and any scalar function v we know that e^v is an $A_{\beta+1}$ weight with $[e^v]_{\beta+1} \le \mathbf{c}_2^{1+\beta}$ [17, Chapter 9] if

$$\sup_B \fint_B e^{(v-v_B)}\, d\mu \le \mathbf{c}_2 , \quad \sup_B \fint_B e^{-\frac{1}{\beta}(v-v_B)}\, d\mu \le \mathbf{c}_2 . \tag{2.3.28}$$

It is clear that (2.3.28) follows from the John–Nirenberg inequality (2.3.24) if $\mathbf{c}_1[v]^{-1}_{*,\mu} \ge \max\{1, \beta^{-1}\}$. Using these facts with $v = \alpha\log(w)$, we see that (2.3.27) holds. □

Proof of Corollary 2.3.8. From the definition of $\mathbf{W} = |k|^{-p}(\varepsilon + |U|)^{k+p}$, we then have $[\log(\mathbf{W})]_{*,\mu} = |k+p|[\log(\varepsilon + |U|)]_{*,\mu}$. Therefore the assumption (2.3.25), that $|k+p|[\log(\varepsilon + |U|)]_{*,\mu} \le \mathbf{c}_1\beta\alpha^{-1}$, and (2.3.27) (with $w = \mathbf{W}$) imply $[\mathbf{W}^\alpha]_{\beta+1} \le \mathbf{c}_2^{1+\beta}$. The corollary then follows from the inequality (2.3.23) of Corollary 2.3.7. □

2.4 Proof of Theorem 2.2.1

One of the key ingredients of our proof is the duality between the Hardy space $H^1(\mu)$ and BMO(μ) space. If μ is the Lebesgue measure then this is the famous Fefferman–Stein theorem [8]. It is well known that the norm of the Hardy space can be defined by

$$\|F\|_{H^1(\Omega)} = \|F\|_{L^1(\Omega)} + \|M_*F\|_{L^1(\Omega)} .$$

Here,

$$M_*F(y) = \sup_\phi \int_\Omega \phi(x)f(x)dx ,$$

where the supremum is taken over all $\phi \in C^1(\mathbb{R}^N)$, which has support in some cube $Q \subset \mathbb{R}^N$ centered at y with side length $l(Q)$ and satisfies

$$0 \le \phi(x) \le l(Q)^{-N} , \quad |D\phi(x)| \le l(Q)^{-N-1} \text{ for all } x \in Q .$$

For a general measure μ, this theory was generalized by Tolsa and we will discuss parts of his works in Appendix A. Roughly speaking, Tolsa showed that there is a constant $\mathbf{c}_H$ depending on n, N and C_μ such that we can similarly define the norm of

$H^1(\mu)$ by considering the class $\check{\Phi} = \cup_{y\in\mathbb{R}^N}\check{\Phi}_y$ of functions $\phi \in C^1(\mathbb{R}^N)$ satisfying the following properties. For any $y \in \mathbb{R}^N$ and *some* cube $Q \subset \mathbb{R}^N$ centered at y with side length $l(Q)$:
f.1) $0 \le \phi(x) \le \mathbf{c}_H l(Q)^{-n}$ for all $x \in Q$;
f.2) $\phi \in C_0^1(Q)$ and $|D\phi(x)| \le \mathbf{c}_H l(Q)^{-n-1}$ for all $x \in Q$.

For any $F \in L^1(\mu)$ we can define a $H^1(\mu)$ norm of F by

$$\|F\|_{\check{\Phi}} = \|F\|_{L^1(\mu)} + \|M_{\check{\Phi}}F\|_{L^1(\mu)}\,, \tag{2.4.1}$$

where

$$M_{\check{\Phi}}F(y) = \sup_{\phi\in\check{\Phi}_y}\int_\Omega \phi(x)f(x)d\mu(x) \in L^1(\mu)\,. \tag{2.4.2}$$

We now have from Lemma A.1.3 in Appendix A the following result.

Lemma 2.4.1. *Let $f \in \mathrm{BMO}(\mu)$ and $F \in L^1(\mu)$ such that*

$$\int_\Omega F\,d\mu = 0\,. \tag{2.4.3}$$

Then there is a constant $C(\mathbf{c}_H)$ such that

$$|\langle F, f\rangle| \le C(\mathbf{c}_H)\|F\|_{\check{\Phi}}[f]_{*,\mu}\,. \tag{2.4.4}$$

We will also use the definition of the *centered* Hardy–Littlewood maximal operator acting on functions $F \in L^1_{\mathrm{loc}}(\mu)$,

$$M(F)(y) = \sup_{\varepsilon}\left\{\int_{B_\varepsilon(y)} F(x)\,d\mu : \varepsilon > 0 \text{ and } B_\varepsilon(y) \subset \Omega\right\}\,. \tag{2.4.5}$$

We also note here the Muckenhoupt theorem for nondoubling measures. By [40, Theorem 3.1], we have that if w is an $A_q(\mu)$ weight then for any $F \in L^q(\mu)$ with $q > 1$,

$$\int_\Omega M(F)^q w\,d\mu \le C(C_\mu, [w]_q)\int_\Omega F^q w\,d\mu\,. \tag{2.4.6}$$

In particular,

$$\int_\Omega M(F)^q\,d\mu \le C_\mu\int_\Omega F^q\,d\mu\,. \tag{2.4.7}$$

We start the proof of Theorem 2.2.1. First of all, let $W = K(U)$. We then have $DU = K_U(U)^{-1}DW$ so that, from the definition of $\mathbb{K}(U) = (K_U^{-1})^T$, $|DU|^2 = \langle\mathbb{K}\rangle(U)DU, DW$. Hence, we can write

$$\begin{aligned} I_1 &= \int_\Omega \langle |DU|^{2p}\Lambda(U)\Phi^2(U)\Lambda^{-1}(U)\mathbb{K}(U)DU, DW\rangle\omega\,dx \\ &= \int_\Omega \langle |DU|^{2p}\Lambda(U)\omega\mathbb{P}(U)DU, DW\rangle\,dx\,, \end{aligned} \tag{2.4.8}$$

where

$$\mathbb{P}(U) := \Phi^2(U)\Lambda^{-1}(U)\mathbb{K}(U)\,. \tag{2.4.9}$$

Applying integration by parts to the last integral in (2.4.8), we see that the boundary integral is zero because of the boundary assumption (2.2.2) in P.2). We then have

$$I_1 = -\int_\Omega \langle \mathrm{div}(|DU|^{2p}\Lambda(U)\omega\mathbb{P}(U)DU), W\rangle\, dx\,.$$

Therefore, for $G := \mathrm{div}(|DU|^{2p}\Lambda(U)\omega\mathbb{P}(U)DU)\omega^{-1}$,

$$I_1 = -\int_\Omega \langle G, W\rangle\, d\mu\,. \tag{2.4.10}$$

From (2.2.2) and integrations by parts again, we see that

$$\int_\Omega G\, d\mu = \int_\Omega \mathrm{div}(|DU|^{2p}\Lambda(U)\omega\mathbb{P}(U)DU)\, dx = 0\,.$$

We will establish bounds for $\|G\|_{L^1(\mu)}$, $\|M_{\breve{\Phi}}G\|_{L^1(\mu)}$ and show that

$$\|G\|_{\breve{\Phi}} \le C\left[I_2^{\frac12} + \bar{\mathfrak{I}}_1^{\frac12} + C([\mathbf{W}^\alpha]_{\beta+1})[\hat{\mathfrak{I}}_1^{\frac12} + I_2^{\frac12} + \breve{\mathfrak{I}}_0^{\frac12}]\right] I_1^{\frac12}\,. \tag{2.4.11}$$

Once this is proved, we obtain from (2.4.10) and (2.4.4) of Lemma 2.4.1 that $I_1 \le C\|K(U)\|_{\mathrm{RBMO}(\mu)}\|G\|_{\breve{\Phi}}$. As we are assuming that μ is doubling, $\|K(U)\|_{\mathrm{RBMO}(\mu)} \sim \|K(U)\|_{\mathrm{BMO}(\mu)}$. We then deduce

$$I_1 \le C\|K(U)\|_{\mathrm{BMO}(\mu)}\left[I_2^{\frac12} + \bar{\mathfrak{I}}_1^{\frac12} + C([\mathbf{W}^\alpha]_{\beta+1})[I_2^{\frac12} + \hat{\mathfrak{I}}_1^{\frac12} + \breve{\mathfrak{I}}_0^{\frac12}]\right] I_1^{\frac12}\,,$$

which yields (2.2.8) of Theorem 2.2.1 via a simple use of Young's inequality. The proof will then be complete.

To prove (2.4.11), we first estimate $\|M_{\breve{\Phi}}G\|_{L^1(\mu)}$ and note that

$$M_{\breve{\Phi}}G(y) \le \sup_{\phi\in\breve{\Phi}_y}\left|\int_\Omega \phi G\, d\mu\right| = \sup_{\phi\in\breve{\Phi}_y}\left|\int_\Omega \phi g\, dx\right|\,,$$

where

$$g := G\omega = \mathrm{div}(|DU|^{2p}\Lambda(U)\omega\mathbb{P}(U)DU)\,. \tag{2.4.12}$$

Therefore, we need to establish that there are constants C, $C([\mathbf{W}^\alpha]_{\beta+1})$ for which

$$\int_\Omega \sup_{\phi\in\breve{\Phi}_y}\left|\int_\Omega \phi g\, dx\right| d\mu \le C\left[I_2^{\frac12} + \bar{\mathfrak{I}}_1^{\frac12} + C([\mathbf{W}^\alpha]_{\beta+1})[\hat{\mathfrak{I}}_1^{\frac12} + I_2^{\frac12} + \breve{\mathfrak{I}}_0^{\frac12}]\right] I_1^{\frac12}\,. \tag{2.4.13}$$

From (2.4.12) we can write $g = g_1 + g_2$ with $g_i = \mathrm{div}\, V_i$, setting

$$h := \Lambda(U)|DU|^{p-1}DU\,,\quad J_{0,\varepsilon} := h_{B_\varepsilon} = \fint_{B_\varepsilon} \Lambda(U)|DU|^{p-1}DU\, d\mu\,, \tag{2.4.14}$$

$$V_1 = \omega|DU|^{p+1}\mathbb{P}(U)\left(h - J_{0,\varepsilon}\right)\,, \tag{2.4.15}$$

$$V_2 = \omega|DU|^{p+1}\mathbb{P}(U)J_{0,\varepsilon}\,. \tag{2.4.16}$$

We will establish (2.4.13) for g being g_1, g_2 in the following lemmas.

In the sequel, for any $y \in \mathbb{R}^N$ and $\phi_\varepsilon \in \check{\Phi}_y$ we denote by $B_\varepsilon = B_\varepsilon(y)$ the corresponding cube centered at y with side length ε.

Let us consider g_1 first.

Lemma 2.4.2. *There is a constant C depending on C_{PS}, C_μ such that*

$$\int_\Omega \sup_{\phi_\varepsilon \in \check{\Phi}_y} \left| \int_\Omega \phi_\varepsilon g_1 \, dx \right| d\mu \le C \left[I_2^{\frac{1}{2}} + \bar{\jmath}_1^{\frac{1}{2}} \right] I_1^{\frac{1}{2}} . \tag{2.4.17}$$

Proof. We use integration by parts (the boundary integral is zero because $\phi_\varepsilon \in C_0^1(B_\varepsilon)$) to get

$$\begin{aligned} \left| \int_{B_\varepsilon(y)} \phi_\varepsilon(x) g_1 \, dx \right| &= \left| \int_{B_\varepsilon(y)} D\phi_\varepsilon(x) \mathbb{P}(U)(h - J_{0,\varepsilon}) |DU|^{p+1} \, d\mu \right| \\ &\le \frac{C}{\varepsilon} \fint_{B_\varepsilon(y)} |h - h_{B_\varepsilon(y)}| |\mathbb{P}(U)| |DU|^{p+1} \, d\mu . \end{aligned} \tag{2.4.18}$$

Here, we used the property of $D\phi_\varepsilon$, as $\phi_\varepsilon \in \check{\Phi}_y$, that $|D\phi_\varepsilon| \lesssim \varepsilon^{-n-1}$, and the assumption M) that $\mu(B_\varepsilon) \lesssim \varepsilon^n$.

Note that (2.2.1) implies $|\mathbb{P}(U)| \le \Phi^2(U) \Lambda^{-1}(U) |\mathbb{K}(U)| \lesssim \Phi(U)$. This and a simple use of Hölder's inequality for $q > 2$ yield that the last integral in (2.4.18) is bounded by

$$\frac{C}{\varepsilon} \left(\fint_{B_\varepsilon(y)} |h - h_{B_\varepsilon(y)}|^q \, d\mu \right)^{\frac{1}{q}} \left(\fint_{B_\varepsilon(y)} [\Phi(U) |DU|^{p+1}]^{q'} \, d\mu \right)^{\frac{1}{q'}} .$$

Applying the Poincaré–Sobolev inequality (2.1.3) to each component of h and noting that there is a constant C such that

$$|Dh| \lesssim |\Lambda_U(U)| |DU|^{p+1} + \Lambda(U) |DU|^{p-1} |D^2 U| ,$$

we find a constant C depending on C_{PS} such that

$$\begin{aligned} \frac{1}{\varepsilon} \left(\fint_{B_\varepsilon} |h - h_{B_\varepsilon}|^q \, d\mu \right)^{\frac{1}{q}} &\le C \left(\fint_{\tau_* B_\varepsilon} |Dh|^{q_*} \, d\mu \right)^{\frac{1}{q_*}} \\ &\le C \left[\fint_{\tau_* B_\varepsilon} (|\Lambda_U(U)|^{q_*} |DU|^{(p+1)q_*} + \Lambda^{q_*}(U) |DU|^{(p-1)q_*} |D^2 U|^{q_*}) \, d\mu \right]^{\frac{1}{q_*}} . \end{aligned} \tag{2.4.19}$$

Using the definition of maximal functions (2.4.5) and combining the above estimates, we get from (2.4.18)

$$\sup_{\phi_\varepsilon \in \check{\Phi}} \left| \int_\Omega \phi_\varepsilon g_1 \, dx \right| \le C [\Psi_1(y) + \Psi_2(y)] \Psi_3(y) , \tag{2.4.20}$$

where $\Psi_i(y) = (M(F_i^{q_i}(y)))^{\frac{1}{q_i}}$ with $q_1 = q_2 = q_*$ and $q_3 = q'$ and

$$F_1 = \Lambda_U(U) |DU|^{p+1} , \quad F_2 = \Lambda(U) |DU|^{p-1} |D^2 U| , \quad F_3 = \Phi(U) |DU|^{p+1} .$$

Because $q_i < 2$ (as $q > 2$ and $q_* = q\sigma < 2$), Muckenhoupt's inequality (2.4.7) implies

$$\left(\int_\Omega \Psi_i^2\,d\mu\right)^{\frac{1}{2}} = \left(\int_\Omega M(F_i^{q_i})^{\frac{2}{q_i}}\,d\mu\right)^{\frac{1}{2}} \le C_\mu\left(\int_\Omega F_i^2\,d\mu\right)^{\frac{1}{2}}.$$

Therefore, applying Hölder's inequality to (2.4.20) and using the above estimates and the notations (2.2.4) and (2.2.5), we obtain (2.4.17). □

Remark 2.4.3. We remark that (2.4.19) is the only place where we need the assumption PS) that Ω, μ support a Poincaré–Sobolev inequality.

We now turn to g_2.

Lemma 2.4.4. *For any $p \ge 1$ and $r \in (\frac{1}{p+1}, 1)$ we denote*

$$\alpha(r) = \frac{r+1}{rp+r+1}, \quad \beta(r) = \frac{r(p+1)-1}{r(p+1)+1}.$$

Then there is $C([\mathbf{W}^{\alpha(r)}]_{\beta(r)+1}) \sim [\mathbf{W}^{\alpha(r)}]_{\beta(r)+1}^{\frac{1}{\alpha(r)(p+1)}}$ such that

$$\int_\Omega \sup_{\phi_\varepsilon \in \breve{\Phi}_y}\left|\int_\Omega \phi_\varepsilon g_2\,dx\right| d\mu \le C\left(\left[\mathbf{W}^{\alpha(r)}\right]_{\beta(r)+1}\right)\left[\hat{\mathfrak{J}}_1^{\frac{1}{2}} + \bar{\mathfrak{J}}_1^{\frac{1}{2}} + I_2^{\frac{1}{2}} + \breve{\mathfrak{J}}_0^{\frac{1}{2}}\right] I_1^{\frac{1}{2}}. \tag{2.4.21}$$

Proof. From (2.4.16), we have $\operatorname{div} V_2 \le C(J_1 + J_2 + J_3)$ for some constant C and

$$J_1 := \omega|\mathbb{P}_U||DU|^{p+2}J_{0,\varepsilon}, \quad J_2 := \omega|\mathbb{P}(U)||DU|^p|D^2U|J_{0,\varepsilon},$$
$$J_3 := D\omega|\mathbb{P}(U)||DU|^{p+1}J_{0,\varepsilon},$$

with $J_{0,\varepsilon}$ being defined in (2.4.14).

In the sequel, for any $r > 1/(p+1)$ we denote $r^* = 1 - \frac{1}{r(p+1)}$. We also write $f = \Phi|DU|^{p+1}$.

We consider $J_{0,\varepsilon}$. From the notation $\mathbf{W} := \Lambda^{p+1}\Phi^{-p}$ ((2.2.3)),

$$J_{0,\varepsilon}(y) \le \left|\fint_{B_\varepsilon} \Lambda\Phi^{\frac{-p}{(p+1)}}\Phi^{\frac{p}{p+1}}|DU|^p\,d\mu\right| = \left|\fint_{B_\varepsilon} \mathbf{W}^{\frac{1}{(p+1)}} f^{\frac{p}{p+1}}\,d\mu\right|.$$

If $r_1 > 1/(p+1)$ we apply Hölder's inequality to the last integral to get the following estimate for $J_{0,\varepsilon}$:

$$J_{0,\varepsilon} \le \left(\fint_{B_\varepsilon} \mathbf{W}^{\frac{1}{r_1^*(p+1)}}\,d\mu\right)^{r_1^*}\left(\fint_{B_\varepsilon} f^{pr_1}\,d\mu\right)^{\frac{1}{r_1(p+1)}}. \tag{2.4.22}$$

For J_1, we write $J_1 = \omega L_* L J_{0,\varepsilon}$ with

$$L_* = |\mathbb{P}_U|\Lambda\Phi^{-1}|DU|^{p+1}, \quad L = \Lambda^{-1}\Phi|DU|.$$

Because $\phi_\varepsilon(x) \lesssim \varepsilon^{-n} \sim \mu(B_\varepsilon)^{-1}$ so that we can use Hölder's inequality to get for any $s > 1$,

$$\left|\int_\Omega \phi_\varepsilon J_1\,dx\right| \le \left(\fint_{B_\varepsilon} L_*^{s'}\,d\mu\right)^{\frac{1}{s'}}\left(\fint_{B_\varepsilon} L^s\,d\mu\right)^{\frac{1}{s}} J_{0,\varepsilon}.$$

We write $L^s = \Lambda^{-s}\Phi^{\frac{-sp}{(p+1)}}\Phi^{\frac{s}{p+1}}|DU|^s$ and use Hölder's inequality to get for any $r > 1/(p+1)$ the following estimate:

$$\left(\fint_{B_\varepsilon} L^s\,d\mu\right)^{\frac{1}{s}} \le \left(\fint_{B_\varepsilon} |\Lambda|^{\frac{-s}{r_*}}\Phi^{\frac{sp}{r^*(p+1)}}\,d\mu\right)^{\frac{r^*}{s}}\left(\fint_{B_\varepsilon} f^{sr}\,d\mu\right)^{\frac{1}{rs(p+1)}}.$$

Combining these estimates with (2.4.22) we then have

$$\sup_{\phi_\varepsilon\in\check{\Phi}_y}\left|\int_\Omega \phi_\varepsilon J_1\,dx\right| \le C_1 M(L_*^{s'})^{\frac{1}{s'}} M(f^{sr})^{\frac{1}{rs(p+1)}} M(f^{pr_1})^{\frac{1}{r_1(p+1)}}, \tag{2.4.23}$$

where the constant C_1 can be estimated by, as $\mathbf{W} := \Lambda^{p+1}\Phi^{-p}$,

$$\begin{aligned} C_1 &\lesssim \left(\fint_{B_\varepsilon} \mathbf{W}^{\frac{1}{r_1^*(p+1)}}\,d\mu\right)^{r_1^*}\left(\fint_{B_\varepsilon} |\Lambda|^{\frac{-s}{r_*}}\Phi^{\frac{sp}{r^*(p+1)}}\,d\mu\right)^{\frac{r^*}{s}} \\ &= \left[\left(\fint_{B_\varepsilon} \mathbf{W}^{\frac{1}{r_1^*(p+1)}}\,d\mu\right)\left(\fint_{B_\varepsilon} \mathbf{W}^{\frac{-s}{r^*(p+1)}}\,d\mu\right)^{\frac{r^*}{r_1^* s}}\right]^{r_1^*}. \end{aligned}$$

We now choose s, r, r_1 such that $s' = sr = pr_1$ and $sr < 2$. This is the case if $r \in (\frac{1}{p+1}, 1)$, $s = (r+1)/r$ then $s' = r+1$ and $r_1 = (r+1)/p > 1/(p+1)$. Let $\alpha(r) = \frac{1}{r_1^*(p+1)}$ and $\beta(r) = \frac{r^*}{r_1^* s}$. With such a choice of s, r, r_1 we have

$$\alpha(r) = \frac{r+1}{rp+r+1}, \quad \beta(r) = \frac{r(p+1)-1}{r(p+1)+1}, \quad \frac{\alpha(r)}{\beta(r)} = \frac{s}{r^*(p+1)}. \tag{2.4.24}$$

It is clear that $C_1 \lesssim C_{1,r}^{r_1^*}$ with $r_1^* = \frac{1}{\alpha(r)(p+1)}$ and

$$C_{1,r} = \sup_{B\subset\Omega}\left(\fint_B \mathbf{W}^{\alpha(r)}\,d\mu\right)\left(\fint_B \mathbf{W}^{\frac{-\alpha(r)}{\beta(r)}}\,d\mu\right)^{\beta(r)}, \tag{2.4.25}$$

where the supremum is taken over all cubes B in Ω. Clearly, the definition of weight (3.2.2) implies

$$[w]_{\nu+1} = \sup_{B\subset\Omega}\left(\fint_B w\,d\mu\right)\left(\fint_B w^{-\frac{1}{\nu}}\,d\mu\right)^{\nu} \quad \text{for all } \nu > 0. \tag{2.4.26}$$

This and (2.4.25) imply $C_{1,r} = [\mathbf{W}^{\alpha(r)}]_{\beta(r)+1}$. We then have

$$C_1 \lesssim C_{1,r}^{r_1^*} \lesssim [\mathbf{W}^{\alpha(r)}]_{\beta(r)+1}^{\frac{1}{\alpha(r)(p+1)}}. \tag{2.4.27}$$

As $sr = pr_1$, we then obtain from (2.4.23) the following:

$$\sup_{\phi_\varepsilon\in\check{\Phi}_y}\left|\int_\Omega \phi_\varepsilon J_1\,dx\right| \lesssim C_{1,r}^{r_1^*} M(L_*^{sr})^{\frac{1}{rs}} M(f^{sr})^{\frac{1}{rs}}.$$

Applying Hölder's inequality to the right-hand side, we get

$$\int_\Omega \sup_{\phi_\varepsilon\in\check{\Phi}_y}\left|\int_\Omega \phi_\varepsilon J_1\,dx\right|\,d\mu \lesssim C_{1,r}^{r_1^*}\left(\int_\Omega M(L_*^{sr})^{\frac{2}{rs}}\,d\mu\right)^{\frac{1}{2}}\left(\int_\Omega M(f^{sr})^{\frac{2}{rs}}\,d\mu\right)^{\frac{1}{2}}.$$

Because $q = 2/(rs) > 1$, we can apply (2.4.7) to the integrals on the right and then use the definitions of $L_*, f, \hat{\mathrm{J}}_1$ to see that

$$\int_\Omega \sup_{\phi_\varepsilon \in \check{\Phi}_y} \left| \int_\Omega \phi_\varepsilon J_1 \, dx \right| d\mu \lesssim C_{1,r}^{r_1^*} \|L_*\|_2 \|f\|_2 . \tag{2.4.28}$$

Concerning the term $\|L_*\|_2$, we recall the definitions $L_* = \mathbb{P}_U \Lambda \Phi^{-1} |DU|^{p+1}$ and $\mathbb{P}(U) = \Phi^2(U)\Lambda^{-1}(U)\mathbb{K}(U)$. It is clear that $|\mathbb{P}_U|\Lambda\Phi^{-1}$ is bounded by

$$|\Phi_U||\mathbb{K}| + \Phi|\Lambda_U|\Lambda^{-1}|\mathbb{K}| + \Phi|\mathbb{K}_U(U)| \lesssim |\Phi_U||\mathbb{K}| + \Phi|\mathbb{K}_U(U)| + |\Lambda_U| .$$

Here, we used (2.2.1) to get $\Phi|\Lambda_U|\Lambda^{-1}|\mathbb{K}| \lesssim |\Lambda_U|$. Therefore, from the definition of $\hat{\mathrm{J}}_1$, $\bar{\mathrm{J}}_1$, $\|L_*\|_2 \lesssim \hat{\mathrm{J}}_1^{\frac{1}{2}} + \bar{\mathrm{J}}_1^{\frac{1}{2}}$. We then obtain from (2.4.28) that

$$\int_\Omega \sup_{\phi_\varepsilon \in \check{\Phi}_y} \left| \int_\Omega \phi_\varepsilon J_1 \, dx \right| d\mu \lesssim C(C_{1,r}) \left(\hat{\mathrm{J}}_1^{\frac{1}{2}} + \bar{\mathrm{J}}_1^{\frac{1}{2}} \right) I_1^{\frac{1}{2}} . \tag{2.4.29}$$

Next, we write $J_2 = \omega|\mathbb{P}(U)||DU|^{p-1}|D^2U||DU|J_{0,\varepsilon} = \omega L_* L J_{0,\varepsilon}$ with

$$L_* = \Lambda|DU|^{p-1}|D^2U| , \quad L = \Lambda^{-1}(U)|\mathbb{P}(U)||DU| .$$

We repeat the argument for J_1. Note that $|\mathbb{P}(U)| \le \Phi(U)$, by (2.2.1), and therefore $L^s \le \Lambda^{-s}\Phi^{\frac{-sp}{(p+1)}}\Phi^{\frac{s}{p+1}}|DU|^s$. We have the following inequality:

$$\left(\fint_{B_\varepsilon} L^s \, d\mu \right)^{\frac{1}{s}} \le \left(\fint_{B_\varepsilon} |\Lambda|^{\frac{-s}{r_*}} \Phi^{\frac{sp}{r^*(p+1)}} \, d\mu \right)^{\frac{r^*}{s}} \left(\fint_{B_\varepsilon} f^{sr} \, d\mu \right)^{\frac{1}{rs(p+1)}} .$$

The estimate (2.4.23) for J_1 now applies to J_2 and yields

$$\sup_{\phi_\varepsilon \in \check{\Phi}_y} \left| \int_\Omega \phi_\varepsilon J_2 \, dx \right| \le C_1 M(L_*^{s'})^{\frac{1}{s'}} M(f^{sr})^{\frac{1}{rs(p+1)}} M(f^{pr_1})^{\frac{1}{r_1(p+1)}} . \tag{2.4.30}$$

As $sr = pr_1$, we have as before

$$\sup_{\phi_\varepsilon \in \check{\Phi}_y} \left| \int_\Omega \phi_\varepsilon J_2 \, dx \right| \lesssim C_{1,r}^{r_1^*} M(L_*^{sr})^{\frac{1}{rs}} M(f^{sr})^{\frac{1}{rs}} .$$

The same argument in (2.4.29) for J_1 with the new definition of L_* yields

$$\int_\Omega \sup_{\phi_\varepsilon \in \check{\Phi}_y} \left| \int_\Omega \phi_\varepsilon J_2 \, dx \right| d\mu \lesssim C_{1,r}^{r_1^*} \|L_*\|_2 \|f\|_2 = C(C_{1,r}) I_2^{\frac{1}{2}} I_1^{\frac{1}{2}} . \tag{2.4.31}$$

Concerning J_3, we write $J_3 = D\omega|\mathbb{P}(U)||DU|^p|DU|J_{0,\varepsilon} = \omega L_* L J_{0,\varepsilon}$ with

$$L_* = D\omega\omega^{-1}\Lambda|DU|^p , \quad L = \Lambda^{-1}|\mathbb{P}(U)||DU| .$$

A similar argument for J_2 applied to this case then implies

$$\int_\Omega \sup_{\phi_\varepsilon \in \check{\Phi}_y} \left| \int_\Omega \phi_\varepsilon J_3 \, dx \right| d\mu \lesssim C_{1,r}^{r_1^*} \|L_*\|_2 \|f\|_2 = C(C_{1,r}) \bar{\mathrm{J}}_0^{\frac{1}{2}} I_1^{\frac{1}{2}} . \tag{2.4.32}$$

Combining the estimates (2.4.29), (2.4.31), (2.4.32) and (2.4.27) (for the constant $C_{1,r}$), we derive (2.4.21). □

Finally, we easily estimate $\|G\|_{L^1(\mu)}$.

Lemma 2.4.5. *We have*

$$\int_\Omega |G|\,d\mu \le C\left[\hat{\mathfrak{J}}_1^{\frac12} + I_2^{\frac12} + \breve{\mathfrak{J}}_0^{\frac12}\right] I_1^{\frac12}\,.$$

Proof. Recall that $G := g\omega^{-1}$ ((2.4.12)), thus $\|G\|_{L^1(\mu)} = \|g\|_{L^1(dx)}$. We write $g = \operatorname{div}(B|DU|^{2p}DU)$ with $B = \omega\Lambda(U)\mathbb{P}(U)$. First of all,

$$|\operatorname{div}(B|DU|^{2p}DU)| \lesssim |B_U||DU|^{2p+2} + |B||DU|^{2p}|D^2U|\,.$$

Because $|B_U|$ is bounded by a multiple of

$$\{|D\omega|\omega^{-1}\Lambda(U)|\mathbb{P}(U)| + |\Lambda_U||\mathbb{P}(U)| + \Lambda(U)|\mathbb{P}_U(U)|\}|DU|^{p+1}\Lambda|Du|^p\omega\,,$$

we see that a simple use of Hölder's inequality as in the proof of Lemma 2.4.4, treating the last factor $\omega\Lambda|Du|^p$ as $J_{0,\varepsilon}$, implies

$$\int_\Omega |B_U||DU|^{2p+2}\,dx \le C[\hat{\mathfrak{J}}_1^{\frac12} + I_2^{\frac12} + \breve{\mathfrak{J}}_0^{\frac12}]I_1^{\frac12}\,.$$

As $|\mathbb{P}(U)| \lesssim \Phi$, we have $|B||DU|^{2p}|D^2U| \lesssim \Phi|DU|^{p+1}\Lambda|DU|^{p-1}|D^2U|\omega$. By Hölder's inequality we then obtain

$$\int_\Omega B|DU|^{2p+2}\,dx \le CI_2^{\frac12}I_1^{\frac12}\,.$$

Combining the above estimates, we prove the lemma. □

Proof of Theorem 2.2.1. It is now clear that the above lemmas yield

$$\|G\|_{\breve{\Phi}} \le C\left[I_2^{\frac12} + \bar{\mathfrak{J}}_1^{\frac12} + C([\mathbf{W}^{\alpha(r)}]_{\beta(r)+1})[\hat{\mathfrak{J}}_1^{\frac12} + I_2^{\frac12} + \breve{\mathfrak{J}}_0^{\frac12}]\right] I_1^{\frac12}\,. \tag{2.4.33}$$

Recall that $\alpha(r) = \frac{r+1}{rp+r+1}$ and $\beta(r) = \frac{r(p+1)-1}{r(p+1)+1}$. We see that $\alpha(r)$ decreases to $2/(p+2)$ and $\beta(r)$ increases to $p/(p+2)$ as $r \to 1^-$.

From the definition of weights, a simple use of Hölder's inequality gives

$$[w^\delta]_\gamma \le [w]_\gamma^\delta \quad \forall\delta \in (0,1)\,. \tag{2.4.34}$$

Thus, if $\alpha > 2/(p+2)$ and $\beta < p/(p+2)$ then for r close to 1 we have $\alpha(r) < \alpha$ and $\beta(r) > \beta$. Hence, by choosing r close to 1 and using (2.4.34) and the open-end property of weights, we see that

$$[\mathbf{W}^{\alpha(r)}]_{\beta(r)+1} \lesssim C[\mathbf{W}^\alpha]_{\beta+1}^{\frac{\alpha(r)}{\alpha}}\,. \tag{2.4.35}$$

As $C([\mathbf{W}^{\alpha(r)}]_{\beta(r)+1}) \sim [\mathbf{W}^{\alpha(r)}]_{\beta(r)+1}^{\frac{1}{\alpha(r)(p+1)}} \lesssim [\mathbf{W}^\alpha]_{\beta+1}^{\frac{1}{\alpha(p+1)}}$, we can replace the constant $C([\mathbf{W}^{\alpha(r)}]_{\beta(r)+1})$ in (2.4.33) by a multiple of $[\mathbf{W}^\alpha]_{\beta+1}^{\frac{1}{\alpha(p+1)}}$ and obtain (2.4.11). Namely,

$$\|G\|_{\breve{\Phi}} \le C\left[I_2^{\frac12} + \bar{\mathfrak{J}}_1^{\frac12} + C([\mathbf{W}^\alpha]_{\beta+1})[\hat{\mathfrak{J}}_1^{\frac12} + I_2^{\frac12} + \breve{\mathfrak{J}}_0^{\frac12}]\right] I_1^{\frac12}\,.$$

As we explained earlier, the above and (2.4.4) of Lemma 2.4.1 yield

$$I_1 \le C\|K(U)\|_{\mathrm{BMO}(\mu)}\left[I_2^{\frac{1}{2}} + \bar{\mathfrak{J}}_1^{\frac{1}{2}} + C([\mathbf{W}^\alpha]_{\beta+1})[\hat{\mathfrak{J}}_1^{\frac{1}{2}} + I_2^{\frac{1}{2}} + \breve{\mathfrak{J}}_0^{\frac{1}{2}}]\right] I_1^{\frac{1}{2}} .$$

This gives

$$I_1 \le C\|K(U)\|^2_{\mathrm{BMO}(\mu)}\left[I_2 + \bar{\mathfrak{J}}_1 + C([\mathbf{W}^\alpha]_{\beta+1})[I_2 + \hat{\mathfrak{J}}_1 + \breve{\mathfrak{J}}_0]\right] . \tag{2.4.36}$$

The proof of the theorem is complete. □

Remark 2.4.6. If $\Phi = \Lambda \equiv 1$ and K is the identity map then it is clear that $\mathbb{K} = 0$ so that $J_1 = 0$ and $\mathbf{W} \equiv 1$. The proof then gives ((2.4.36))

$$I_1 \le C\|U\|^2_{\mathrm{BMO}(\mu)}\left[I_2 + \breve{\mathfrak{J}}_0\right] .$$

Remark 2.4.7. The only place we use the assumption PS) is (2.4.19). We just need to assume that PS) holds true for $h = \Lambda(U)|DU|^{p-1}DU$ and some measure μ satisfying M). Combining this with [19, Theorem 5.1] (Remark 2.1.1), which deals only with a pair u, Du, we need only that some Poincaré inequality (2.1.4) holds for the pair h, Dh. That is, we do not need (2.1.4) to hold for any h but the function $h = \Lambda(U)|DU|^{p-1}DU$ in the consideration.

2.5 Proof of Theorem 2.2.2

We provide the proof of Theorem 2.2.2.

We consider first the case $\omega_1 = \omega_0^2$. Clearly, from the definition of $I_{1,*}$ and (2.4.8), we have for $W = K(U)$

$$I_{1,*} \le \int_\Omega \Phi^2(U)|DU|^{2p+2}\omega_* \, d\mu = \int_\Omega \langle |DU|^{2p}\Lambda(U)\omega_*\omega_0^2\mathbb{P}(U)DU, DW\rangle \, dx .$$

Applying integration by parts to the last integral, where the boundary integral is zero because $\omega_* \in C_0^1(\Omega)$, we obtain

$$I_{1,*} \le -\int_\Omega \langle G, W\rangle \, d\mu ,$$

where $G := \operatorname{div}(|DU|^{2p}\Lambda(U)\omega_*\omega_0^2\mathbb{P}(U)DU)\omega_0^{-2}$ and $W = K(U)$.

We now follow the proof of Theorem 2.2.1 to establish a similar version of (2.4.11), with $d\mu = \omega_0^2 dx$ and G being defined above, to complete the proof. We see that (2.4.11) holds true if (2.4.13) does. We then need only establish a similar version of (2.4.13). Again, we can write $g = g_1 + g_2$ with $g_i = \operatorname{div} V_i$, setting

$$V_1 = \omega_*\omega_0^2|DU|^{p+1}\mathbb{P}(U)\left(h - J_{0,\varepsilon}\right), \quad V_2 = \omega_*\omega_0^2|DU|^{p+1}\mathbb{P}(U)J_{0,\varepsilon} ,$$

where $h := \Lambda(U)|DU|^{p-1}DU$ and

$$J_{0,\varepsilon} := h_{B_\varepsilon} = \fint_{B_\varepsilon} \Lambda(U)|DU|^{p-1}DU \, d\mu . \tag{2.5.1}$$

We revisit the lemmas giving the proof of (2.4.13) and estimate

$$\int_\Omega \sup_{\phi_\varepsilon\in\breve{\Phi}} \left|\int_\Omega \phi_\varepsilon g_i \, dx\right| \, d\mu\,, \quad i = 1, 2\,. \tag{2.5.2}$$

Since $\omega_* \le 1$ we can discard it in the estimates for g_1 after the use of integration by parts (2.4.18) in the proof of Lemma 2.4.2. Because the measure μ supports a Poincaré–Sobolev inequality (2.1.3), we can repeat the argument in the proof of Lemma 2.4.2 to obtain the same estimate for the integral in (2.5.2) with $i = 1$. Similarly, we drop ω_* in J_i's, with the exception of J_3, in the proof of Lemma 2.4.4 to estimate the integral in (2.5.2) with $i = 2$. Therefore,

$$\int_\Omega \sup_{\phi_\varepsilon\in\breve{\Phi}} \left|\int_\Omega \phi_\varepsilon g \, dx\right| \, d\mu \le \left[I_2^{\frac{1}{2}} + \bar{\mathfrak{J}}_1^{\frac{1}{2}} + C([\mathbf{W}^\alpha]_{\beta+1})[\hat{\mathfrak{J}}_1^{\frac{1}{2}} + I_2^{\frac{1}{2}} + \breve{I}_*^{\frac{1}{2}}]\right] I_1^{\frac{1}{2}}\,.$$

Here, the term $\breve{I}_*$, replacing $\breve{\mathfrak{J}}_0$ in (2.4.21), comes from the estimate for

$$J_3 = D(\omega_*\omega_0^2)|\mathbb{P}(U)||DU|^p|DU|J_{0,\varepsilon}\,.$$

In fact, we write $J_3 = \omega_0^2 L_* L J_{0,\varepsilon}$ for $L_* = D(\omega_*\omega_0^2)\omega_0^{-2}\Lambda|DU|^p$ and $L = \Lambda^{-1}|\mathbb{P}(U)||DU|$ to obtain the following version of (2.4.32) (with $C_{1,r}$ being replaced by $[\mathbf{W}^\alpha]_{\beta+1}$):

$$\int_\Omega \sup_{\phi_\varepsilon} \left|\int_\Omega \phi_\varepsilon J_3 \, dx\right| \, d\mu \le C([\mathbf{W}^\alpha]_{\beta+1})\breve{I}_*^{\frac{1}{2}} I_1^{\frac{1}{2}}\,,$$

with $\breve{I}_* = \|L_*\|_2^2$. That is,

$$\breve{I}_* = \int_\Omega |D(\omega_*\omega_0^2)|^2\omega_0^{-4}\Lambda^2|DU|^{2p} \, d\mu = \int_\Omega |D(\omega_*\omega_0^2)|^2\omega_0^{-2}\Lambda^2|DU|^{2p} \, dx\,.$$

Because $|D(\omega_*\omega_0^2)|^2 \lesssim |D\omega_*|^2\omega_0^4 + \omega_*^2|D\omega_0|^2\omega_0^2$, we have

$$\breve{I}_* \lesssim \int_\Omega |D\omega_*|^2\omega_0^2\Lambda^2|DU|^{2p} \, dx + \int_\Omega \omega_*^2|D\omega_0|^2\Lambda^2|DU|^{2p} \, dx\,.$$

The first integral on the right-hand side is less than $\breve{I}_{0,*}$, defined by (2.2.12). Meanwhile, we apply the Hardy inequality (2.2.10) in L.2) to the second integral for $u = \omega_*\Lambda|DU|^p$, which belongs to $C_0^1(\Omega)$, and note that (as $\omega_* \le 1$)

$$|Du|^2 \lesssim \Lambda^2|DU|^{2p-2}|D^2U|^2 + |\Lambda_U|^2|DU|^{2p} + |D\omega_*|^2\Lambda^2|DU|^{2p}\,.$$

We then have

$$\int_\Omega \omega_*^2|D\omega_0|^2\Lambda^2|DU|^{2p} \, dx \le C\int_\Omega |Du|^2\omega_0^2 \, dx \lesssim I_2 + \bar{\mathfrak{J}}_1 + \breve{I}_{0,*}\,. \tag{2.5.3}$$

Thus, we get the following version of (2.4.13):

$$\int_\Omega \sup_{\phi_\varepsilon\in\breve{\Phi}} \left|\int_\Omega \phi_\varepsilon g \, dx\right| \, d\mu \le C\left[I_2^{\frac{1}{2}} + \bar{\mathfrak{J}}_1^{\frac{1}{2}} + C([\mathbf{W}^\alpha]_{\beta+1})[\hat{\mathfrak{J}}_1^{\frac{1}{2}} + I_2^{\frac{1}{2}} + \breve{I}_{0,*}^{\frac{1}{2}}]\right] I_1^{\frac{1}{2}}\,.$$

The constant C depends on C_{PS}, C_μ and C_H. In the same way, Lemma 2.4.5 gives a similar estimate for $\|G\|_{L^1(\mu)}$. Hence,

$$\|G\|_{\breve{\Phi}} \le C\left[I_2^{\frac{1}{2}} + \bar{\mathfrak{J}}_1^{\frac{1}{2}} + C([\mathbf{W}^\alpha]_{\beta+1})[\hat{\mathfrak{J}}_1^{\frac{1}{2}} + I_2^{\frac{1}{2}} + \breve{I}_{0,*}^{\frac{1}{2}}]\right] I_1^{\frac{1}{2}} .$$

We then apply Lemma 2.4.1 as before and use Young's inequality to prove (2.3.7) for the case $d\mu = \omega_0^2 dx$.

Finally, if $\omega_1 \sim \omega_0^2$ then the integrals in (2.3.7) with respect to the two measures are comparable, because $D\omega_1$, $D\omega_0$ are not involved, so that (2.3.7) holds true as well. The proof is complete. □

Remark 2.5.1. For simplicity we assumed in L.0) that $\omega_* \in C_0^1(\Omega)$. More generally, we need only that $u = \omega_* \Lambda |DU|^p \in C_0^1(\Omega)$ so that the Hardy inequality can apply in (2.5.3).

3 The parabolic systems

In this chapter, for any $T_0 > 0$ and bounded domain Ω with smooth boundary in $\mathbb{R}^N$, $N \geq 2$, we consider the following parabolic system of m equations ($m \geq 2$) for the unknown $u\colon Q \to \mathbb{R}^m$, where $Q = \Omega \times (0, T_0)$:

$$\begin{cases} u_t - \operatorname{div}(A(x,u)Du) = \hat{f}(x,u,Du)\,, & (x,t) \in Q\,, \\ u = 0 \text{ or } \frac{\partial u}{\partial \nu} = 0 \text{ on } \partial\Omega\,, & t \in (0, T_0)\,, \\ u(x,0) = U_0(x) & x \in \Omega\,. \end{cases} \tag{3.0.1}$$

Here, $A(x,u)$ is a $m \times m$ matrix in $x \in \Omega$ and $u \in \mathbb{R}^m$, $\hat{f}\colon \Omega \times \mathbb{R}^m \times \mathbb{R}^{mN} \to \mathbb{R}^m$ is a vector-valued function. The initial data U_0 is given in $W^{1,p_0}(\Omega, \mathbb{R}^m)$ for some $p_0 > N$, in the dimension of Ω.

We shall be concerned with the solvability of this system and the existence of its *strong* solutions. We say that u is a strong solution if u solves (3.0.1) a.e. on Q with $Du \in L^\infty_{\mathrm{loc}}(Q)$ and $D^2u \in L^2_{\mathrm{loc}}(Q)$.

The first fundamental problem in the study of (3.0.1) is the local and global existence of its solutions. One can decide to work with either weak or strong solutions. In the first case, the existence of a weak solution can be achieved via Galerkin, time discretization or variational methods but its regularity (e.g., boundedness, Hölder continuity of the solution and its higher derivatives) is then an open and serious issue. Several works have been done along this line to improve on the early work [14] of Giaquinta and Struwe and establish *partial regularity* of *bounded* weak solutions to (3.0.1).

Otherwise, if strong solutions are considered then their existence can be established via semigroup theories as in the works of Amann [1, 2]. Combining this with interpolation theories of Sobolev's spaces, Amann established local and global existence of a strong solution u of (3.0.1) under the assumption that one can control $\|u\|_{W^{1,p}(\Omega,\mathbb{R}^m)}$ for some $p > n$. His theory did not apply to the case where $\hat{f}$ has a superlinear growth in Du.

In both aforementioned approaches, the assumption on the boundedness of u must be the starting point because the techniques rely heavily on the fact that $A(x,u)$ is *regular elliptic*. For strongly coupled systems like (3.0.1), as invariant/maximum principles are generally unavailable, the boundedness of the solutions is already a hard problem. One usually needs to use ad hoc techniques on a case by case basis to show that u is bounded [23, 41]). Even so, for bounded weak solutions we know that they are only Hölder continuous almost everywhere [14]. In addition, there are counter examples by John and Stará [21] for systems ($m = 2$) that exhibit solutions that start smoothly and remain bounded for all time but develop singularities in higher norms in finite times.

In our recent work [29, 30], we chose a different approach making use of fixed-point theory and discussed the existence of *strong* solutions of (3.0.1) under the weak-

https://doi.org/10.1515/9783110608762-003

est assumption that they are a priori VMO (vanishing mean oscillation), but not necessarily bounded, and general structural conditions on the data of (3.0.1) that are independent of x. We assumed only that $A(u)$ is *uniformly elliptic*. Applications were presented in [30] when $\lambda(u)$ has a polynomial growth in $|u|$ and, without the boundedness assumption on the solutions, so (3.0.1) can be degenerate as $\lambda(u) \to \infty$ when $|u| \to \infty$. The singular case, $\lambda(u) \to 0$ as $|u| \to \infty$, was not discussed there. Also, the reaction term $\hat{f}$ was assumed to only have linear growth in gradient.

In this book, we will establish much stronger results than those in [29] under more general assumptions on the structure of (3.0.1) as described in A) and F) of Section 3.1. Besides the minor fact that the data can depend on x, we allow further that:

- $A(x, u)$ can be either degenerate or singular as $|u|$ tends to infinity.
- $\hat{f}(x, u, Du)$ can have a superlinear growth in Du as in f.2). Even quadratic growth in gradient is considered in the low-dimensional case.
- No a priori boundedness of solutions is assumed but a very weak a priori integrability of strong solutions of (3.0.1) is considered.

Most remarkably, the key assumption in [29, 30] that the BMO norm of u is small in small balls will be replaced by a more versatile one in this paper: $K(u)$ is has small BMO norm in small balls for some suitable map $K\colon \mathbb{R}^m \to \mathbb{R}^m$. This allows us to consider the singular case where one may not be able to estimate the BMO norm of u but can estimate that of $K(u)$. Examples in applications when the systems are singular will be provided later in Section 3.5 where $|K(u)| \sim \log(|u|)$.

One of the key ingredients in the proof in [29, 30] is the *local* weighted Gagliardo–Nirenberg inequality involving BMO norm [29, Lemma 2.4]. In this book, we make use of a new version of this inequality developed in Chapter 2 replacing the BMO norm of u by that of $K(u)$. Combining this with a new and nontrivial iteration argument, starting with a very weak integrability assumption on the solutions to (3.0.1), we establish uniform estimates for their derivatives so that a homotopy argument in fixed-point theories can be used here to provide the existence of strong solutions.

We organize this chapter as follows. We describe the general structural conditions of (3.0.1) in Section 3.1. These conditions will be referred to throughout the book, with the exception of Chapter 5. In Section 3.2 we state the main result of this chapter, Theorem 3.2.3, and present its proof in Section 3.3. A simpler case of this theorem, which however finds many applications, will be discussed in Section 3.4. Applications of these theorems will be given in Section 3.5 for the general SKT models (see also Chapter 5 for more general results).

3.1 The main structural conditions

It is always assumed, with the exception of Chapter 5, that the matrix $A(x, u)$ is elliptic in the sense that there exist two scalar *positive* continuous functions $\lambda_1(x, u)$, $\lambda_2(x, u)$

such that

$$\lambda_1(x,u)|\zeta|^2 \le \langle A(x,u)\zeta,\zeta\rangle \le \lambda_2(x,u)|\zeta|^2 \tag{3.1.1}$$

for all $x \in \Omega$, $u \in \mathbb{R}^m$, $\zeta \in \mathbb{R}^{mN}$.

If there exist positive constants c_1, c_2 such that $c_1 \le \lambda_1(x,u)$ and $\lambda_2(x,u) \le c_2$ then we say that $A(x,u)$ is *regular elliptic*. If $0 < \lambda_1(x,u)$ and $\lambda_2(x,u)/\lambda_1(x,u) \le c_2$, we say that $A(x,u)$ is *uniform elliptic*. In addition, if $\lambda_1(x,u)$ tends to zero (respectively ∞) when $|u| \to \infty$ then we say that $A(x,u)$ is *singular* (respectively *degenerate*) at infinity.

Without the boundedness assumption on the solutions of (3.0.1), we will consider the cases when $A(x,u)$ is *uniformly elliptic* and *degenerate/singular* at infinity. The following structural conditions on the data of (3.0.1) will be assumed throughout the book.

A) $A(x,u)$ is C^1 in $x \in \Omega$, $u \in \mathbb{R}^m$ and there exist a constant $C_* > 0$ and scalar C^1 positive functions $\lambda(u)$, $\omega(x)$ such that for all $u \in \mathbb{R}^m$, $\zeta \in \mathbb{R}^{mN}$ and $x \in \Omega$,

$$\lambda(u)\omega(x)|\zeta|^2 \le \langle A(x,u)\zeta,\zeta\rangle \quad \text{and} \quad |A(x,u)| \le C_*\lambda(u)\omega(x)\,. \tag{3.1.2}$$

In addition, there is a constant C such that

$$|A_u(x,u)| \le C|\lambda_u(u)|\omega(x)\,, \quad |A_x(x,u)| \le C|\lambda(u)||D\omega|\,. \tag{3.1.3}$$

Here and throughout this book, if B is a C^1 (vector-valued) function in $u \in \mathbb{R}^m$ then we abbreviate its derivative (or Jacobian matrix) $\frac{\partial B}{\partial u}$ by B_u and its second derivative (or Hessian matrix) by B_{uu}. Also, with a slight abuse of notations, $A(x,u)\zeta$, $\langle A(x,u)\zeta,\zeta\rangle$ in (3.1.1), (5.0.3) should be understood in the following way: For $A(x,u) = [a_{ij}(x,u)]$, $\zeta \in \mathbb{R}^{mN}$ we write $\zeta = [\zeta_i]_{i=1}^m$ with $\zeta_i = (\zeta_{i,1}, \dots \zeta_{i,N})$ and

$$A(x,u)\zeta = [\Sigma_{j=1}^m a_{ij}\zeta_j]_{i=1}^m\,, \quad \langle A(x,u)\zeta,\zeta\rangle = \Sigma_{i,j=1}^m a_{ij}\langle\zeta_i,\zeta_j\rangle\,.$$

The positivity of λ then implies that for any bounded set $K \subset \mathbb{R}^m$ there are positive constants $\lambda_*(K)$, $\lambda_{**}(K)$ such that

$$\lambda_*(K) \le \lambda(u) \le \lambda_{**}(K) \quad \forall u \in K\,. \tag{3.1.4}$$

We also assume that $A(x,u)$ is regular elliptic with respect to $x \in \Omega$.

AR) $\omega \in C^1(\Omega)$ and there are positive numbers μ_*, μ_{**} such that

$$\mu_* \le \omega(x) \le \mu_{**}, \quad |D\omega(x)| \le \mu_{**} \quad \forall x \in \Omega\,. \tag{3.1.5}$$

Combining (3.1.4) and AR), we see that $A(x,u)$ is regular elliptic for *bounded* u. When $|u| \to \infty$, the matrix $A(x,u)$ can be *degenerate/singular*.

Concerning the reaction vector $\hat{f}(x,u,Du)$, which may have linear or *subquadratic* growth in Du, we assume the following condition.

F) There exist a constant C and a nonnegative differentiable function $f\colon \mathbb{R}^m \to \mathbb{R}$ such that either:

f.1) $\hat{f}$ has a linear growth in ζ

$$|\hat{f}(x,u,\zeta)| \le C\lambda(u)|\zeta|\omega(x) + f(u)\omega(x)\,, \tag{3.1.6}$$

and

$$\begin{cases} |\partial_x \hat{f}(x,u,\zeta)| \le C[\lambda(u)|\zeta| + f(u)]|D\omega(x)|\,, \\ |\partial_u \hat{f}(x,u,\zeta)| \le C[|\lambda_u(u)||\zeta| + |f_u(u)|]\omega(x)\,, \\ |\partial_\zeta \hat{f}(x,u,\zeta)| \le C\lambda(u)|\omega(x)\,; \end{cases}$$

or that

f.2) $\hat{f}$ has a superlinear growth in ζ. There is $\delta \in (0, 2/N)$ and $\lambda_{uu}(u)$ exists such that

$$|\hat{f}(x,u,\zeta)| \le C|\lambda_u(u)||\zeta|^{1+\delta}\omega(x) + f(u)\omega(x)\,, \tag{3.1.7}$$

$$\begin{cases} |\partial_x \hat{f}(x,u,\zeta)| \le C[|\lambda_u(u)||\zeta|^{1+\delta} + f(u)]|D\omega(x)|\,, \\ |\partial_u \hat{f}(x,u,\zeta)| \le C[|\lambda_{uu}(u)||\zeta|^{1+\delta} + |f_u(u)|]\omega(x)\,, \\ |\partial_\zeta \hat{f}(x,u,\zeta)| \le C|\lambda_u(u)||\zeta|^{\delta}|\omega(x)\,. \end{cases}$$

Furthermore, we assume that

$$|\lambda_{uu}(u)|\lambda(u) \le C|\lambda_u(u)|^2\,. \tag{3.1.8}$$

If $\hat{f}$ explicitly depends on x then we also assume that there is a constant C such that

$$|\lambda_u(u)||u| \le C\lambda(u)\,. \tag{3.1.9}$$

By a formal differentiation of (3.1.6) and (3.1.7), one can see that the growth conditions for $\hat{f}$ naturally imply those of its partial derivatives in the above assumptions. The condition (3.1.8) is obviously verified if $\lambda(u)$ has a polynomial growth in $|u|$.

3.2 Preliminaries and main results

We state the main results of this paper in this section. The key assumption of these results is some uniform a priori estimate for the BMO norm of $K(u)$ where K is some suitable map on $\mathbb{R}^m$ and u is any strong solution to (3.0.1). To begin, we recall some basic definitions in harmonic analysis.

Let $\omega \in L^1(\Omega)$ be a nonnegative function and define the measure $d\mu = \omega(x)dx$. For any μ-measurable subset A of Ω and any locally μ-integrable function $U\colon \Omega \to \mathbb{R}^m$ we denote by $\mu(A)$ the measure of A and U_A the average of U over A. That is,

$$U_A = \fint_A U(x)\,d\mu = \frac{1}{\mu(A)}\int_A U(x)\,d\mu\,.$$

We define the measure $d\mu = \omega(x)dx$ and recall that a vector-valued function $f \in L^1(\Omega, \mu)$ is said to be in $BMO(\Omega, \mu)$ if

$$[f]_{*,\mu} := \sup_{B_R \subset \Omega} \fint_{B_R} |f - f_{B_R}|\, d\mu < \infty\,, \quad f_{B_R} := \frac{1}{\mu(B_R)} \int_{B_R} f\, d\mu\,. \tag{3.2.1}$$

We then define

$$\|f\|_{BMO(\Omega,\mu)} := [f]_{*,\mu} + \|f\|_{L^1(\Omega,\mu)}\,.$$

For $\gamma \in (1, \infty)$ we say that a nonnegative locally integrable function w belongs to the class A_γ or w is an A_γ weight on Ω if the quantity

$$[w]_{\gamma,\Omega} := \sup_{B \subset \Omega} \left(\fint_B w\, d\mu \right) \left(\fint_B w^{1-\gamma'}\, d\mu \right)^{\gamma - 1} \quad \text{is finite}\,. \tag{3.2.2}$$

Here, $\gamma' = \gamma/(\gamma - 1)$. For more details on these classes we refer the reader to [40, 45]. If the domain Ω is specified we simply denote $[w]_{\gamma,\Omega}$ by $[w]_\gamma$.

To begin, as in [29] with A independent of x, we assume that the eigenvalues of the matrix $A(x, u)$ are not too far apart. Namely, for C_* defined in (5.0.3) of A) we assume
SG) $(N - 2)/N < C_*^{-1}$.

Here C_* is, in a certain sense, the ratio of the largest and smallest eigenvalues of $A(x, u)$. This condition seems to be necessary as we deal with systems [31].

First of all, we will assume that the system (3.3.1) satisfies the structural conditions A) and F). Additional assumptions then follow so that the local weighted Gagliardo–Nirenberg inequality of Chapter 2 can apply here.

K) There is a C^1 map $K\colon \mathbb{R}^m \to \mathbb{R}^m$ such that $\mathbb{K}(u) = (K_u(u)^{-1})^T$ exists and $\mathbb{K}_u \in L^\infty(\mathbb{R}^m)$. Furthermore, for all $u \in \mathbb{R}^m$,

$$|\mathbb{K}(u)| \lesssim \lambda(u)|\lambda_u(u)|^{-1}\,. \tag{3.2.3}$$

We consider the following system:

$$\begin{cases} u_t - \operatorname{div}(A(x, u)Du) = \hat{f}(x, u, Du)\,, & (x, t) \in \Omega \times (0, T_0)\,, \\ u = 0 \text{ or } \frac{\partial u}{\partial \nu} = 0 \text{ on } \partial\Omega \times (0, T_0) \\ u(x, 0) = U_0(x)\,. \end{cases} \tag{3.2.4}$$

We embed this system in the following family of systems:

$$\begin{cases} u_t - \operatorname{div}(A(x, \sigma u)Du) = \hat{f}(x, \sigma u, \sigma Du)\,, & (x, t) \in \Omega \times (0, T_0), \sigma \in [0, 1]\,, \\ u = 0 \text{ or } \frac{\partial u}{\partial \nu} = 0 \text{ on } \partial\Omega \times (0, T_0) \\ u(x, 0) = U_0(x)\,. \end{cases} \tag{3.2.5}$$

For any strong solution u of (3.2.5) we will consider the following assumptions.

M.0) There exists a constant C_0 such that for some $r_0 > 1$ and $\beta_0 \in (0,1)$

$$\sup_{\tau\in(0,T_0)} \||f_u(\sigma u)|\lambda^{-1}(\sigma u)\|_{L^{r_0}(\Omega,\mu)} \le C_0\,, \tag{3.2.6}$$

$$\sup_{\tau\in(0,T_0)} \|u^{\beta_0}\|_{L^1(\Omega,\mu)}\,, \qquad \sup_{\tau\in(0,T_0)} \|\lambda^{\beta_0}(u)\|_{L^1(\Omega,\mu)} \le C_0\,, \tag{3.2.7}$$

$$\iint_Q |f_u(\sigma u)|(1+|u|^2)\,d\mu d\tau \le C_0\,, \tag{3.2.8}$$

$$\iint_Q (|f_u(\sigma u)| + \lambda(\sigma u))|Du|^2\,d\mu d\tau \le C_0\,. \tag{3.2.9}$$

M.1) For any given $\mu_0 > 0$ there is positive R_{μ_0} sufficiently small in terms of the constants in A) and F) such that

$$\sup_{x_0\in\bar\Omega,\tau\in(0,T_0)} \|K(u)\|^2_{BMO(B_R(x_0)\cap\Omega,\mu)} \le \mu_0\,. \tag{3.2.10}$$

Furthermore, for $\mathbf{W}_p(\sigma,x,\tau) := \lambda^{p+\frac{1}{2}}(\sigma u)|\Lambda_u(\sigma u)|^{-p}$ and any p such that $1-1/p < C_*^{-1}$ we have $\sup_{\tau\in(0,T_0)}[\mathbf{W}_p^{\frac{2}{3}}]_{\frac{4}{3}} \le C_0$.

Remark 3.2.1. The conditions (3.2.7) and (3.2.8) are not needed if $\hat f$ is independent of x. Also, in many physical models, the ellipticity 'constant' $\lambda(u)$ in A) and the 'reaction' $f(u)$ have polynomial growths like $\lambda(u) \sim (\lambda_0 + |u|)^k$ and $f(u) \sim |u|^{k_0}\lambda(u)$ for some positive constants λ_0, k and k_0. In these cases, the condition (3.2.6) is clearly true if $\|u\|_{L^{l_0}(\Omega)} \le C_0$ for some $l_0 > k_0$. The conditions in (3.2.7) then come from this if we choose β_0 to be sufficiently small. Furthermore, in connection with M.1), especially when $K(u) = u$, we see that (3.2.10) implies $|u|$ is BMO so that $|u| \in L^q_{\mathrm{loc}}(\Omega)$ for any $q > 1$ and hence M.1) implies M.0), with the exception of (3.2.9).

Remark 3.2.2. The condition on $\mathbf{W}_p$ in M.1) allows us to make use of the Gagliardo–Nirenberg inequalities in Chapter 2 for any $p \ge 1$. In fact, from the basic properties of A_γ weights, we have

$$[w^\delta]_\gamma \le [w]_\gamma^\delta\,, \quad [w]_{\gamma_1} \le [w]_{\gamma_2}\,, \quad \text{if } \delta\in(0,1) \text{ and } 1<\gamma_2<\gamma_1\,.$$

Furthermore, from the open-end property of weights [17, Theorem 9.2.5], if w is an A_γ weight then $w^{1+\delta}$ is also an A_γ weight for some $\delta > 0$.

Therefore, if $[\mathbf{W}_p^{\frac{2}{3}}]_{\frac{4}{3}} \le C_0$ then we can find $\alpha_0 > 2/3$ and $\beta_0 < 1/3$ such that $[\mathbf{W}_p^{\alpha_0}]_{\beta_0+1} \le C(C_0)$. For any $p \ge 1$ we can find α, β such that $\alpha_0 > \alpha > 2/(p+2)$ and $\beta_0 < \beta < p/(p+2)$ and thus

$$[\mathbf{W}_p^\alpha]_{\beta+1,B_{R_{\mu_0}}(x_0)\cap\Omega} \le [\mathbf{W}_p^{\alpha_0}]^{\frac{\alpha}{\alpha_0}}_{\beta_0+1,B_{R_{\mu_0}}(x_0)\cap\Omega} \le C(C_0)\,.$$

It follows that the condition W) on the weight $\mathbf{W}$ in Chapter 2 holds and the Gagliardo–Nirenberg inequalities are available.

The main theorem of this chapter is the following.

Theorem 3.2.3. *Assume A), AR) and K). In addition to F) we assume*

$$f(u) \leq C|f_u(u)|(1 + |u|) . \tag{3.2.11}$$

Suppose also that any strong solution u to (3.2.5) *satisfies M.0), M.1) uniformly in* $\sigma \in [0, 1]$.

Then the system (3.2.4) *has a unique strong solution on* $\Omega \times (0, T_0)$.

The condition (3.2.10) on the smallness of the BMO norm of $K(u)$ in small balls is the most crucial one in applications. In [29, 30], we consider the case $\lambda(u) \sim (\lambda_0 + |u|)^k$ with $k > 0$ and assume that $K = Id$, the identity matrix. We assumed that $K(u) = u$ has small BMO norm in small balls, which can be verified by establishing that $\|Du\|_{L^N(\Omega)}$ is bounded. These results already improve on those of Amann in [1, 2] where boundedness of solutions was assumed and uniform estimates for $\|Du\|_{L^p(\Omega)}$ for some $p > N$ was needed. Both such conditions seem to be very difficult to verify in applications.

We should remark that all the assumptions on strong solutions of the family (3.2.5) can be checked by considering the case $\sigma = 1$ (i.e., (3.0.1)) because these systems satisfy the same structural conditions uniformly with respect to the parameter $\sigma \in [0, 1]$.

3.3 Proof of the main theorem

In this section, we prove Theorem 3.2.3. We recall the definitions of the systems (3.2.4)

$$\begin{cases} u_t - \operatorname{div}(A(x, u)Du) = \hat{f}(x, u, Du) , & (x, t) \in \Omega \times (0, T_0) , \\ u = 0 \text{ or } \frac{\partial u}{\partial \nu} = 0 \text{ on } \partial\Omega \times (0, T_0) \\ u(x, 0) = U_0(x) . \end{cases} \tag{3.3.1}$$

and the family (3.2.5), $\sigma \in [0, 1]$,

$$\begin{cases} u_t - \operatorname{div}(A(x, \sigma u)Du) = \hat{f}(x, \sigma u, \sigma Du) , & (x, t) \in \Omega \times (0, T_0), \sigma \in [0, 1] , \\ u = 0 \text{ or } \frac{\partial u}{\partial \nu} = 0 \text{ on } \partial\Omega \times (0, T_0) \\ u(x, 0) = U_0(x) . \end{cases} \tag{3.3.2}$$

The proof of Theorem 3.2.3, which asserts the existence of strong solutions u to (3.3.1), relies on the Leray–Schauder fixed-point index theorem. Such a strong solution u of (3.3.1) is a fixed point of a nonlinear map defined on an appropriate Banach space **X**. The proof will be based on several lemmas and we will sketch the main steps below.

We will show in Lemma 3.3.9 that there exist $p > N/2$ and a constant M_* depending only on the constants in A) and F) such that any strong solution u of (3.3.2) will satisfy

$$\sup_{\tau \in (0, T_0)} \|Du\|_{L^{2p}(\Omega)} \leq M_* \ \|u_t\|_{L^{q_0}(Q)} \leq M_* . \tag{3.3.3}$$

We will show that there are positive constants α, M_0 such that

$$\|u\|_{C^{\alpha,\alpha/2}(Q)} \le M_0 \,. \tag{3.3.4}$$

For $\sigma \in [0,1]$ and any $u \in \mathbb{R}^m$ and $\zeta \in \mathbb{R}^{mN}$ we define the vector-valued functions $F^{(\sigma)}$ and $f^{(\sigma)}$ by

$$F^{(\sigma)}(x,u,\zeta) := \int_0^1 \partial_\zeta F(\sigma,u,t\zeta)\,dt\,, \quad f^{(\sigma)}(x,u) := \int_0^1 \partial_u F(\sigma,x,tu,0)\,dt\,. \tag{3.3.5}$$

For any given $u, w \in \mathbf{V}_{q,r}(Q)$ we write

$$\hat{\mathbf{f}}(\sigma,x,u,w) = F^{(\sigma)}(x,u,Du)Dw + f^{(\sigma)}(x,u)w + \hat{f}(x,0,0)\,. \tag{3.3.6}$$

We will define a suitable Banach space $\mathbf{X}$ and for each $u \in \mathbf{X}$ we consider the following linear systems, noting that $\hat{\mathbf{f}}(\sigma,x,u,w)$ is linear in w, Dw:

$$\begin{cases} w_t - \operatorname{div}(A(x,\sigma u)Dw) = \hat{\mathbf{f}}(\sigma,x,u,w) & (x,t) \in \Omega \times (0,T_0)\,,\\ w = 0 \text{ or } \frac{\partial w}{\partial \nu} = 0 \text{ on } \partial\Omega \times (0,T_0)\,, \\ w(x,0) = U_0(x) \text{ on } \Omega\,. \end{cases} \tag{3.3.7}$$

We will show that the above system has a unique weak solution w if u satisfies (3.3.4). We then define $T_\sigma(u) = w$ and apply the Leray–Schauder fixed-point theorem to establish the existence of a fixed point of T_1. It is clear from (3.3.6) that $\hat{f}(x,\sigma u,\sigma Du) = \hat{\mathbf{f}}(\sigma,x,u,u)$. Therefore, from the definition of T_σ we see that a fixed point of T_σ is a weak solution of (3.3.2). By choosing an appropriate choice of $\mathbf{X}$, we will see that these fixed points are strong solutions of (3.3.2), and so a fixed point of T_1 is a strong solution of (3.3.1).

From the proof of the Leray–Schauder fixed-point theorem in [15, Theorem 11.3], we need to find some ball B_M of radius M centered at 0 of $\mathbf{X}$ such that $T_\sigma : \bar{B}_M \to \mathbf{X}$ is compact and that T_σ has no fixed point on the boundary of B_M. The topological degree $\operatorname{ind}(T_\sigma, B_M)$ is then well defined and invariant by homotopy so that $\operatorname{ind}(T_1, B_M) = \operatorname{ind}(T_0, B_M)$. It is easy to see that the latter is nonzero because the linear system

$$\begin{cases} u_t - \operatorname{div}(A(x,0)Du) = \hat{\mathbf{f}}(x,0,0) & x \in \Omega \times (0,T_0)\,,\\ u = 0 \text{ or } \frac{\partial u}{\partial \nu} = 0 \text{ on } \partial\Omega \times (0,T_0)\,, & u(x,0) = U_0(x) \text{ on } \Omega \end{cases}$$

has a unique solution in B_M. Hence, T_1 has a fixed point in B_M.

Therefore, the theorem is proved as we will establish the following claims.

Claim 1 There exist a Banach space $\mathbf{X}$ and $M > 0$ such that the map $T_\sigma : \bar{B}_M \to \mathbf{X}$ is well defined and compact.

Claim 2 T_σ has no fixed point on the boundary of $\bar{B}_M$. That is, $\|u\|_{\mathbf{X}} < M$ for any fixed points of $u = T_\sigma(u)$.

3.3.1 The setting of fixed-point theory

The following lemma defines the space $\mathbf{X}$, the map T_σ and establishes the aforementioned claims. The map T_σ is defined in a quite standard way but the number M will be intricately chosen so that Claim 2 is also satisfied. The whole setting is based on the crucial assumption that a priori estimates for strong solutions like (3.3.8) below are available. This will be established in the next section.

Lemma 3.3.1. *Suppose that there exist $p > N/2$, $q_0 > 1$ and a constant M_* such that any strong solution u of* (3.3.2) *satisfies*

$$\sup_{\tau\in(0,T_0)} \|Du\|_{W^{1,2p}(\Omega)} \le M_* \,, \qquad \|u_t\|_{L^{q_0}(Q)} \le M_* \,. \tag{3.3.8}$$

Then, there exist $M, \beta > 0$ and $q, r \ge 1$ such that for

$$\mathbf{X} = C^{\beta,\beta/2}(Q, \mathbb{R}^m) \cap \mathbf{V}_{q,r}(Q)$$

the map $T_\sigma \colon \bar{B}_M \to \mathbf{X}$ is well defined and compact for all $\sigma \in [0,1]$. Moreover, T_σ has no fixed points on ∂B_M.

Proof. For some constant $M_0 > 0$, to be determined in terms of M_* later, we consider $u\colon Q \to \mathbb{R}^m$ satisfying

$$\sup_{\tau\in(0,T_0)} \|u\|_{C(\Omega)} \le M_0 \,, \qquad \iint_Q |Du|^2 \, d\mu d\tau \le M_0 \,, \tag{3.3.9}$$

and write the system (3.3.7) as a linear parabolic system for w

$$w_t = \operatorname{div}(\mathbf{a}(u)Dw) + \mathbf{b}(u)Dw + \mathbf{g}(u)w + \mathbf{f} \,, \tag{3.3.10}$$

where $\mathbf{a}(x,t) = A(x, \sigma u)$, $\mathbf{b}(x,t) = F^{(\sigma)}(x, u, Du)$, $\mathbf{g}(x,t) = f^{(\sigma)}(x,u)$, and $\mathbf{f}(x) = \hat{f}(x, 0, 0)$. The matrix $\mathbf{a}(u)$ is regular elliptic with uniform ellipticity constants by A), AR) if u is bounded, as we are assuming in (3.3.9) ((3.1.4)). We recall the following well-known result in [25, Chapter VII]. If there exist positive constants m and q, r such that (see the condition (1.5) in [25, Chapter VII]) $1/r + N/(2q) = 1$, $q > N/2$ and $r \ge 1$ and

$$\||\mathbf{b}(u)|^2\|_{q,r,Q} \,, \quad \|\mathbf{g}(u)\|_{q,r,Q} \,, \quad \|\mathbf{f}\|_{q,r,Q} \le m \,, \tag{3.3.11}$$

then the system (3.3.10) satisfies the assumptions of Theorem 1.1 in [25, Chapter VII], which asserts that (3.3.7) has a unique weak solution w.

Moreover, as the initial condition $w(\cdot, 0) = U_0(x)$ belongs to $W^{1,p_0}(\Omega)$ and then $C^{\beta_0}(\Omega)$ for $\beta_0 = 1 - n/p_0 > 0$, a combination of Theorems 2.1 and 3.1 in [25, Chapter VII] shows that w belongs to $C^{\alpha_0,\alpha_0/2}(\bar{Q}, \mathbb{R}^m)$ for some $\alpha_0 > 0$ depending only on β_0, $\|u\|_\infty$ and m.

Next, we will show that (3.3.11) holds by F) and (3.3.9). We consider the two cases f.1) and f.2). If f.1) holds then from the definition (3.3.5) there is a constant $C(|u|)$ such that

$$|\mathbf{b}(x,t)| = |F^{(\sigma)}(x,u,Du)| \le C(|u|)\,, \quad |\mathbf{g}(x,t)| = |f^{(\sigma)}(x,u)| \le C(|u|)\,.$$

From (3.3.9), we see that $\sup_{\tau\in(0,T_0)} \|u\|_\infty \le M_0$ and so there is a constant m depending on M_0 such that (3.3.11) holds for any q and n.

If f.2) holds then there is $\delta \in (0, 2/N)$ such that

$$|F^{(\sigma)}(x,u,Du)| \le C(|u|)|Du|^\delta\,, \quad |f^{(\sigma)}(x,u)| \le C(|u|)\,. \tag{3.3.12}$$

Therefore, a simple use of Hölder's inequality shows that $\||\mathbf{b}|^2\|_{L^q(Q)}$ is bounded by $C\|Du\|_{L^2(Q)}$ for $q = 1/\delta > N/2$. Then (3.3.9) implies the condition (3.3.11).

In both cases, (3.3.10) (or (3.3.7)) has a unique weak solution w. We then define $T_\sigma(u) = w$. Moreover, as we explained earlier, $w \in C^{\alpha_0,\alpha_0/2}(Q)$ for some $\alpha_0 > 0$ depending on M_0.

We now consider a fixed point u of T_σ. By Lemma 3.3.2 following this proof we see that u is a strong solution and we can use the assumption (3.3.8). The first bound in the assumption (3.3.8) implies u is Hölder continuous in x. This and the integrability of u_t in the second bound of the assumption (3.3.8) and [37, Lemma 4] provide positive constants α, M_1, with M_1 depending on M_*, such that any strong solution u of (3.3.2) satisfies $\|u\|_{C^{\alpha,\alpha/2}(\Omega)} \le M_1$. Also, the assumption AR) implies that $\lambda(u)$, ω are bounded from below, yielding that $\|Du\|_{L^2(Q)} \le C(M_*)$. Thus, there is a constant M_1, depending on M_* such that any strong solution u of (3.3.2) satisfies

$$\|u\|_{C^{\alpha,\alpha/2}(Q)} \le M_1\,, \quad \|Du\|_{L^2(Q)} \le M_1\,. \tag{3.3.13}$$

It is well known that there is a constant $c_0 > 1$, depending on α, T_0 and the diameter of Ω, such that $\|\cdot\|_{C^{\beta,\beta/2}(Q)} \le c_0\|\cdot\|_{C^{\alpha,\alpha/2}(Q)}$ for all $\beta \in (0,\alpha)$. We now let M_0, the constant in (3.3.9), be $M_0 = (c_0+1)M_1$, a constant depending on M_*.

Define $\mathbf{X} = C^{\beta,\beta/2}(Q) \cap V^{1,0}(Q)$ for some positive $\beta < \min\{\alpha, \alpha_0\}$, where α_0 is the Hölder continuity exponent for solutions of (3.3.10), and

$$V^{1,0}(Q) := \{u : Du \in L^2(Q)\}\,.$$

The space $\mathbf{X}$ is equipped with the norm

$$\|u\|_{\mathbf{X}} = \max\{\|u\|_{C^{\beta,\beta/2}(Q)}, \|Du\|_{L^2(Q)}\}$$

and consider the ball B_M in $\mathbf{X}$ centered at 0 with radius $M = M_0$.

We now see that T_σ is well defined and maps the ball $\bar{B}_M$ of $\mathbf{X}$ into $\mathbf{X}$. Moreover, from the definition $M = (c_0+1)M_1$, it is clear that T_σ has no fixed point on the boundary of B_M because such a fixed point u satisfies (3.3.13), which implies $\|u\|_{\mathbf{X}} \le c_0M_1 < M$.

Finally, we need only show that T_σ is compact. If u belongs to a bounded set K of $\bar{B}_M$ then $\|u\|_{\mathbf{X}} \le C(K)$ for some constant $C(K)$ and there is a constant $C_1(K)$ such that

$\|T_\sigma(u)\|_{C^{\alpha_0,\alpha_0/2}(Q)} = \|w\|_{C^{\alpha_0,\alpha_0/2}(Q)} \le C_1(K)$. Thus $T_\sigma(K)$ is compact in $C^{\beta,\beta/2}(Q)$ because $\beta < \alpha_0$. So, we need only show that $T(K)$ is precompact in $V^{1,0}(Q)$. We will discuss only the superlinear case where (3.3.12) in f.2) holds because the case $\hat{f}$ with linear growth in f.1) is similar and easier.

First of all, for $u \in K$ we easily see that $\|Dw\|_{L^2(Q)}$ is uniformly bounded by a constant depending on K. The argument is standard by testing the linear system (3.3.7) by w and using the boundedness of $\|w\|_{L^\infty}$ and $\|u\|_{L^\infty}$, (3.3.12), AR) and Young's inequality.

We will show that $T_\sigma(K)$ is sequentially compact in $V^{1,0}(Q)$. That is, every sequence $w_n = T_\sigma(u_n)$, $u_n \in K$, has a convergent subsequence.

Because $\{u_n\}$ is compact in $C^{\beta,\beta/2}(Q)$, passing to a subsequence, we can assume that $u_n \to u_*$ in $C^{\beta',\beta'/2}(Q)$ for some $\beta' \in (0,\beta)$. Clearly, Du_n weakly converges to Du_* in $L^2(Q)$. We then denote $w_* = T_\sigma(u_*)$. As we proved earlier that $T_\sigma(K)$ is compact in $C^{\beta,\beta/2}(Q)$, we can assume that $\{w_n\}$ converges. By uniqueness, $w_n \to w_*$ in $C^{\beta,\beta/2}(Q)$.

We have, writing $W = w_n - w_*$,

$$W_t - \operatorname{div}(A(x,\sigma u_n)DW) = \operatorname{div}(\alpha_n Dw_n) + \Psi_n\,,$$

where $\alpha_n = A(x,\sigma u_n) - A(x,\sigma u_*)$ and Ψ_n is defined by

$$F^{(\sigma)}(x,u_n,Du_n)Dw_n - F^{(\sigma)}(x,u_*,Du_*)Dw_* + f^{(\sigma)}(x,u_n)u_n - f^{(\sigma)}(x,u_*)u_*\,.$$

Testing the above system with W and using AR) and the fact that $W(x,0) = 0$, we have for $dz = dxdt$

$$\lambda_*(K)\mu_* \iint_Q |DW|^2\,dz \le \iint_Q [|\alpha_n||Dw_m||DW| + |\Psi_n||W|]\,dz\,.$$

By Young's inequality, we find a constant C depending on K and μ_* such that

$$\iint_Q |DW|^2\,dz \le C\iint_Q [(|\alpha_n||Dw_m|)^2\,dz + \sup_Q |W|\|\Psi_n\|_{L^1(Q)}\,.$$

By (3.3.12), it is clear that

$$|\Psi_n| \le C(K)[(|Du_n| + |Du_*|)(|Dw_n| + |Dw_*|) + 1]\,.$$

Using the fact that $\|Dw_n\|_{L^2(Q)}$ and $\|Du_n\|_{L^2(Q)}$ are uniformly bounded, we see that $\|\Psi_n\|_{L^1(Q)}$ is bounded. Hence,

$$\iint_Q |Dw_n - Dw_*|^2\,dz \le C(K)\max\{\sup_Q |A(x,\sigma u_n) - A(x,\sigma u_*)|, \sup_Q |w_n - w_*|\}\,.$$

As u_n, w_n converge in $C^0(Q)$ respectively to u_*, w_*, $\|A(x,\sigma u_n) - A(x,\sigma u_*)\|_\infty$, $\|w_n - w_*\|_\infty$ converge to 0. We then see from the above estimate that $Dw_n \to Dw_*$ converges in $L^2(Q)$. Thus, $T_\sigma(K)$ is precompact in $V^{1,0}(Q)$.

Hence, $T_\sigma : \mathbf{X} \to \mathbf{X}$ is a compact map. The proof is complete. □

3.3.2 A priori estimates

We now turn to the hardest part of the proof, and provide a uniform estimate for the fixed points of T_σ to justify the key assumption (3.3.8) of Lemma 3.3.1. The proof is long and will be divided into many lemmas and we will sketch the main ideas below.

- Lemma 3.3.2 is quite standard and shows that the fixed points of T_σ are strong solutions of (3.3.2).
- Lemma 3.3.3 establishes an energy estimate of Du for any strong solutions u of the system. Roughly speaking, we establish that

$$\sup_{(0,T_0)} \int_{\Omega_s} |Du|^{2p}\, dx + \iint_{Q_s} \lambda(u)|Du|^{2p-2}|D^2u|^2\, d\mu d\tau$$
$$\lesssim \iint_{Q_t} \Phi^2(u)|Du|^{2p+2}\, dz + \text{lower-order terms}$$

  where $\Omega_s = B_s \cap \Omega$, $Q_s = \Omega_s \times (0, T_0)$ and B_s, B_t are two concentric balls with radii $s < t$, and $\Phi(u) = (\sqrt{\lambda(u)})_u$.
- We wish to remove the integral of $\Phi^2(u)|Du|^{2p+2}$ on the right-hand side of the above energy estimate. This is the key part of the proof and it can be done by using the local Gagliardo–Nirenberg inequalities in the previous chapter. Here, under the assumptions K), we present such an inequality in Lemma 3.3.7. With this and the smallness assumption on the BMO norms in M.1), we achieve the goal and obtain in Lemma 3.3.8 a better estimate

$$\sup_{(0,T_0)} \int_{\Omega_R} |Du|^{2p}\, dx + \iint_{Q_R} \lambda(u)|Du|^{2p-2}|D^2u|^2\, d\mu d\tau + \iint_{Q_R} \Phi^2(u)|Du|^{2p+2}\, d\mu d\tau$$
$$\lesssim \text{lower-order terms}$$

  if R is sufficiently small.
- In Lemma 3.3.9 and Lemma 3.3.11, we will show that the above inequality is *self-improving* so that we can iterate it finite times, starting with the integrability assumption M.0), to obtain the desired bound for Du as in (3.3.8) of Lemma 3.3.1.

Hence, we begin with the following lemma.

Lemma 3.3.2. *A fixed point of T_σ is also a strong solution of* (3.3.2).

Proof. If u is a fixed point of T_σ in $\mathbf{X}$ then it solves (3.3.2) weakly and is continuous by the definition of $\mathbf{X}$. Thus, u is bounded and belongs to VMO(Q). By AR), the system (3.3.2) is regular elliptic. We can adapt the proof in [14] to show that Du is Hölder continuous. If $\hat{f}$ satisfies a quadratic growth in Du then, because u is bounded, the condition [14, (0.4)] that $|\hat{f}| \le a|Du|^2 + b$ is satisfied here. The proof of [14, Theorems 2.1 and 3.2] assumed the 'smallness condition' [14, (0.6)] $2aM < \lambda_0$, where $M = \sup|u|$. This 'smallness condition' was needed in that paper because only weak bounded solutions, which are not necessarily continuous, were considered. In our case, u is continuous so that we do not require this 'smallness condition'. Indeed, a careful checking of the arguments of the proof in [14, Lemma 2.1 and page 445] shows that if R is small and one knows that the solution u is continuous then these argument still hold as long we can absorb the integrals involving $|Du|^2$, $|Dw|^2$ in the estimates after [14, (3.7)] on the right hand side to the left right hand sides of those estimates. Thus, [14, Theorems 2.1 and 3.2] apply to our case and yield that $u \in C^{a,a/2}(Q)$ for all $a \in (0,1)$ and that, since $A(x,u)$ is differentiable, Du is locally Hölder continuous in Q. Therefore, u is also a strong solution. □

Thanks to Lemma 3.3.2, we need only consider a strong solution u of (3.3.2) and establish (3.3.8) for some $p > N/2$. Because the data of (3.3.7) satisfy the structural conditions A), F) with the same set of constants and the assumptions of the theorem are assumed to be uniform for all $\sigma \in [0,1]$, we will only present the proof for the case $\sigma = 1$ in the sequel.

Let u be a strong solution of (3.0.1) on Ω. We begin with an energy estimate for Du. For $p \geq 1$ and any ball B_s with center $x_0 \in \bar{\Omega}$ we denote $\Omega_s = B_s \cap \Omega$, $Q_s = \Omega_s \times (0, T_0)$ and

$$\mathcal{A}_p(s) = \sup_{\tau\in(0,T_0)} \int_{\Omega_s\times\{\tau\}} |Du|^{2p}\,dx\,, \tag{3.3.14}$$

$$\mathcal{H}_p(s) := \iint_{Q_s} \lambda(u)|Du|^{2p-2}|D^2u|^2\,d\mu d\tau\,, \tag{3.3.15}$$

$$\mathcal{B}_p(s) := \iint_{Q_s} \frac{|\lambda_u(u)|^2}{\lambda(u)}|Du|^{2p+2}\,d\mu d\tau\,, \tag{3.3.16}$$

$$\mathcal{C}_p(s) := \iint_{Q_s} (|f_u(u)| + \lambda(u))|Du|^{2p}\,d\mu d\tau\,, \tag{3.3.17}$$

and

$$\mathcal{F}_p(s) := \iint_{Q_s} |f(u)||Du|^{2p-1}(|D\omega_0|\omega_0)\,dz\,. \tag{3.3.18}$$

The following lemma establishes an energy estimate for Du.

Lemma 3.3.3. *Assume A), F). Let u be any strong solution of* (3.3.1) *on Ω and $p \geq 1$ satisfying the spectral gap condition (C_* is the constant in SG)*

$$1 - 1/p < C_*^{-1}\,. \tag{3.3.19}$$

There is a constant C, which depends only on the parameters in A) and F), such that for any two concentric balls B_s, B_t with center $x_0 \in \bar{\Omega}$ and $s < t$,

$$\mathcal{A}_p(s) + \mathcal{H}_p(s) \leq C\mathcal{B}_p(t) + C(1 + (t-s)^{-2})[\mathcal{C}_p(t) + \mathcal{F}_p(t)] + \||DU_0|^{2p}\|_{L^1(\Omega_t)}\,. \tag{3.3.20}$$

Proof. Since u is a strong solution we can differentiate the system in x to see that u weakly solves

$$(Du)_t - \operatorname{div}(A(x,u)D^2u + A_u(x,u)DuDu + A_x(x,u)Du) = D\hat{f}(x,u,Du)\,. \tag{3.3.21}$$

For any two concentric balls B_s, B_t, with $s < t$, let ψ be a cutoff function for B_s, B_t. That is, ψ is a C^1 function satisfying $\psi \equiv 1$ in B_s and $\psi \equiv 0$ outside B_t and $|D\psi| \le 1/(t-s)$. Consider any given triple t_0, T, T' satisfying $0 < t_0 < T < T' \le T_0$ and let η be a cutoff function for $(T - t_0, T'), (T, T')$. That is, $\eta(s) \equiv 1$ if $s \ge T$, $\eta(s) \equiv 0$ if $s \le T - t_0$ and $|\eta_t| \le t_0^{-1}$.

Let $\mathbf{w} = |Du|^{2p-2}Du$. We test (3.3.21) with $\mathbf{w}\psi^2\eta$ and obtain, using integration by parts and rearranging,

$$\begin{aligned}
&\int_{T'-t_0}^{T'}\int_\Omega (|Du|^{2p}\psi^2\eta)_t + \langle A(x,u)D^2u, D(\mathbf{w}\psi^2)\rangle\eta \, dxd\tau \\
&= \int_{T'-t_0}^{T'}\int_\Omega \langle -A_u(x,u)DuDu - A_x(x,u)Du, D(\mathbf{w}\psi^2)\rangle\eta \, dxd\tau \\
&\quad + \int_{T'-t_0}^{T'}\int_\Omega \langle D\hat{f}(x,u,Du), \mathbf{w}\rangle\psi^2\eta \, dxd\tau + \int_{T'-t_0}^{T'}\int_\Omega |Du|^{2p}\psi^2\eta_t \, dxd\tau \,. \qquad (3.3.22)
\end{aligned}$$

At first, we derive from this the following inequality:

$$\begin{aligned}
&\sup_{t\in(T,T')}\int_{\Omega_s} |Du|^{2p}\,dx + \int_{T-t_0}^{T'}\int_{\Omega_t} \lambda(u)|Du|^{2p-2}|D^2u|^2\psi^2\eta\omega \, dxd\tau \\
&\le C\int_{T-t_0}^{T'}\int_{\Omega_t} |\langle A(x,u)D^2u, \mathbf{w}D(\psi^2)\rangle|\eta \, dxd\tau \\
&\quad + C\int_{T-t_0}^{T'}\int_{\Omega_t} [|A_u(x,u)||Du|^2 + |A_x(x,u)||Du|]|D(\mathbf{w}\psi^2)|\eta \, dxd\tau \\
&\quad + C\int_{T-t_0}^{T'}\int_{\Omega_t} |D\hat{f}(x,u,Du)||\mathbf{w}|\psi^2\eta \, dxd\tau \\
&\quad + Ct_0^{-1}\int_{T-t_0}^{T}\int_{\Omega_t} |Du|^{2p}\,dxd\tau \,. \qquad (3.3.23)
\end{aligned}$$

Here, the first and last integrals of (3.3.23) result from the integrals involving the temporal derivative in (3.3.22) and the fact that $|\eta_t| \le t_0^{-1}$.

The second integral in the first line of (3.3.23) comes from the integral of $\langle A(x, u)D^2u, D(\mathbf{w})\psi^2\eta\rangle$ in (3.3.22), which is estimated from below by using (B.1.9) in Lemma B.1.3 with $X = Du$, $\Lambda = C_*^{-1}\lambda(u)$. The spectral gap condition (3.3.19) and the lemma then provide a positive constant c_1 such that

$$\begin{aligned}
\langle A(x,u)D^2u, D(\mathbf{w})\rangle &= \langle A(x,u)D^2u, D(|Du|^{2p-2}Du)\rangle \\
&\ge c_1C_*^{-2}\lambda(u)|Du|^{2p-2}|D^2u|^2\omega \,.
\end{aligned}$$

The integrand in the second line is estimated by

$$\begin{aligned}
|\langle A(x,u)D^2u, \mathbf{w}D(\psi^2)\rangle| &\lesssim \lambda(u)|Du|^{2p-1}||D^2u|\psi|D\psi|\omega \\
&\lesssim \varepsilon\lambda(u)|Du|^{2p-2}|D^2u|^2\psi^2\omega + C(\varepsilon)\lambda(u)|Du|^{2p}|D\psi|^2\omega \,.
\end{aligned}$$

The integrands in the third line come from the integral involving A_u, A_x in (3.3.22). Using the condition A), $|A_u| \lesssim |\lambda_u|\omega$, and the facts that $|\mathbf{w}| \le |Du|^{2p-1}$,

$|D\mathbf{w}| \lesssim |Du|^{2p-2}|D^2u|$ and Young's inequality, we have

$$
\begin{aligned}
&|\langle A_u(x,u)DuDu, D(\mathbf{w}\psi^2)\rangle| \\
&\le |\lambda_u||Du|^{2p}|D^2u|\psi^2\omega + |\lambda_u||Du|^{2p+1}\psi|D\psi|\omega \\
&\le \varepsilon\lambda(u)|Du|^{2p-2}|D^2u|^2\psi^2\omega + C(\varepsilon)\left[\frac{|\lambda_u(u)|^2}{\lambda(u)}|Du|^{2p+2} + |D\psi|^2\lambda(u)|Du|^{2p}\right]\omega .
\end{aligned}
$$

Similarly, as $|A_x(x,u)| \le C_*\lambda(u)|D\omega|$ and $\omega = \omega_0^2$,

$$
\begin{aligned}
&|\langle A_x(x,u)Du, D(\mathbf{w}\psi^2)\rangle| \\
&\le |\lambda(u)||Du|^{2p-1}|D^2u|\psi^2\omega_0|D\omega_0| + |\lambda(u)||Du|^{2p}\psi|D\psi|\omega_0|D\omega_0| \\
&\le \varepsilon\lambda(u)|Du|^{2p-2}|D^2u|^2\psi^2\omega_0^2 + C(\varepsilon)(\psi^2 + |D\psi|^2)(\omega_0|D\omega_0|)\lambda(u)|Du|^{2p}] .
\end{aligned}
$$

We now consider the integral of $|D\hat{f}(x,u,Du)||\mathbf{w}|\psi^2$. First, if $\hat{f}$ has a linear growth in Du then by f.1) in F) with $p = Du$,

$$
\begin{aligned}
|D\hat{f}(x,u,Du)| \lesssim{}& (\lambda(u)|D^2u| + |\lambda_u(u)||Du|^2)\omega + \lambda(u)|Du||D\omega| \\
&+ |f_u(u)||Du|\omega + |f(u)||D\omega| .
\end{aligned}
$$

Therefore, using Young's inequality, we easily get

$$
\begin{aligned}
|D\hat{f}(x,u,Du)||Du|^{2p-1} \lesssim{}& \varepsilon\lambda(u)|Du|^{2p-2}|D^2u|^2\omega + C(\varepsilon)\lambda(u)|Du|^{2p}\omega \\
&+ \lambda(u)|Du|^{2p}|D\omega_0|^2 + \frac{|\lambda_u(u)|^2}{\lambda(u)}|Du|^{2p+2}\omega \\
&+ |f(u)||Du|^{2p-1}|D\omega_0|\omega_0 + |f_u(u)||Du|^{2p}\omega . \qquad (3.3.24)
\end{aligned}
$$

Similarly, if $\hat{f}$ has a superlinear growth in Du as in f.2) then, because $\delta < 1$, by a simple use of Young's inequality,

$$
\begin{aligned}
|D\hat{f}(x,u,Du)| \lesssim{}& (|\lambda_u(u)||Du||D^2u| + |\lambda_{uu}(u)||Du|^3)\omega + |\lambda_u||Du|^2|D\omega| \\
&+ |f_u(u)||Du|\omega + |f(u)||D\omega| .
\end{aligned}
$$

We then have to deal with three new terms, which can be handled by Young's inequality and the assumption (3.1.8), that $|\lambda_{uu}(u)|\lambda(u) \lesssim |\lambda_u(u)|^2$, as follows:

$$
\begin{aligned}
|\lambda_u(u)||Du||D^2u||Du|^{2p-1} &\le \varepsilon\lambda(u)|Du|^{2p-2}|D^2u|^2 + C(\varepsilon)\frac{|\lambda_u(u)|^2}{\lambda(u)}|Du|^{2p+2} , \\
|\lambda_u||Du|^2|D\omega||Du|^{2p-1} &\lesssim \frac{|\lambda_u(u)|^2}{\lambda(u)}|Du|^{2p+2}\omega + \lambda(u)|Du|^{2p}|D\omega|^2\omega^{-1} , \\
|\lambda_{uu}(u)||Du|^3|Du|^{2p-1} &\lesssim \frac{|\lambda_u(u)|^2}{\lambda(u)}|Du|^{2p+2} .
\end{aligned}
$$

We then get the same terms as in (3.3.24) for the linear growth case.

Finally, having established the above inequalities, we use the definitions (3.3.15)–(3.3.18) to see that the integrals on the right-hand side of (3.3.23) can be estimated by

$$\varepsilon \int_{T-t_0}^{T'} \int_{\Omega_t} \lambda(u)|Du|^{2p-2}|D^2u|^2\psi^2\, d\mu$$
$$+ C(\varepsilon)[\mathcal{B}_p(t) + (1+(t-s)^{-2})(\mathcal{C}_p(t) + \mathcal{F}_p(t))]$$
$$+ Ct_0^{-1}\int_{T-t_0}^{T}\int_{\Omega_t} |Du|^{2p}\, dx d\tau\,. \tag{3.3.25}$$

We now choose ε sufficiently small so that the first integral can be absorbed into the left-hand side of (3.3.23). We then obtain (3.4.14) for $T, t_0 > 0$. Finally, We formally let $T, t_0 \to 0$ in the last integral, which will be justified below, to obtain (3.4.14).

Using the difference quotient operator δ_h instead of D in (3.3.21), we obtain

$$(\delta_h u)_t = \operatorname{div}(A(x,u)D(\delta_h u) + \delta_h(A(x,u))Du) + \delta_h \hat{f}(x,u,Du)\,. \tag{3.3.26}$$

We test this with $|\delta_h u|^{2p-2}\delta_h u\psi^2\eta$ to obtain a similar version of (3.3.23) with the operator D being replaced by δ_h. We can integrate the result over $(0, T_0)$ and obtain

$$\sup_{t\in(0,T_0)} \int_{\Omega_s} |\delta_h u|^{2p}\, dx + \iint_{Q_s} \lambda(u)|\delta_h u|^{2p-2}|D\delta_h u|^2\, dz$$
$$\le C\iint_{Q_t} \left[\frac{|\lambda_u(u)|^2}{\lambda(u)}|Du|^2|\delta_h u|^{2p} + |D\psi|^2\lambda(u)|\delta_h u|^{2p}\right] dz + \cdots$$
$$+ C\int_{\Omega_t} |\delta_h u(x,0)|^{2p}\, dx\,.$$

Since $u \in C([0,T'), L^{2p}(\Omega))$, we can let h tend to 0 and obtain a similar energy estimate (3.4.14) for Du with $T = t_0 = 0$ and $\eta \equiv 1$. We complete the proof. □

Remark 3.3.4. We note that the presence of $\mathcal{F}_p$ in the energy estimate is due to the dependence on x of $\hat{f}$ (see the estimate (3.3.24)). Otherwise, i.e., $\omega_* \equiv 1$, we have $\mathcal{F}_p \equiv 0$.

Remark 3.3.5. We discuss the case when the centers of B_s, B_t are on the boundary $\partial\Omega$. We assume that u satisfies the Neumann boundary condition on $\partial\Omega$. By flattening the boundary we can assume that $B_t \cap \Omega$ is the set

$$B^+ = \{x : x = (x_1, \ldots, x_n) \text{ with } x_n \ge 0 \text{ and } |x| < R\}\,.$$

For any point $x = (x_1, \ldots, x_N)$ we denote by $\bar{x}$ its reflection across the plane $x_N = 0$, i.e., $\bar{x} = (x_1, \ldots, -x_n)$. Accordingly, we denote by B^- the reflection of B^+. For a function u given on $B_+ \times (0,T)$ we denote its even reflection by $\bar{u}(x,t) = u(\bar{x},t)$ for $x \in B^-$. We then consider the even extension of $\hat{u}$ in $B = B^+ \cup B^-$,

$$\hat{u}(x,t) = \begin{cases} u(x,t) & \text{if } x \in B^+\,, \\ \bar{u}(x,t) & \text{if } x \in B^-\,. \end{cases}$$

With these notations, for $x \in B^+$ we observe that $u_t = \bar{u}_t$, $\mathrm{div}_x(D_x u) = \mathrm{div}_{\bar{x}}(D_{\bar{x}}\bar{u})$ and $D_x u D_x u = D_{\bar{x}}\bar{u} D_{\bar{x}}\bar{u}$. Therefore, it is easy to see that $\hat{u}$ satisfies in B a system similar to the one for u in B^+. Thus, the proof can apply to $\hat{u}$ to obtain the same energy estimate near the boundary.

Remark 3.3.6. For Dirichlet boundary condition we make use of the odd reflection $\bar{u}(x,t) = -u(\bar{x},t)$ and then define $\hat{u}$ as in Remark 3.3.5. Since $D_{x_i}U = 0$ on $\partial\Omega$ if $i \neq N$, we can test the system (3.3.21), obtained by differentiating the system of u with respect to x_i, with $|D_{x_i}u|^{2q-2}D_{x_i}u\psi^2$ and the proof goes as before because no boundary integral term appears. We need only consider the case $i = N$. We observe that $D_{x_N}\hat{u}$ is the even extension of $D_{x_N}u$ in B; therefore $\hat{u}$ satisfies a system similar to (3.3.21). The proof then continues.

We wish to remove the integral $\mathcal{B}_p$ on the right-hand side of (3.4.14). This is the most difficult task and we will apply the local Gagliardo–Nirenberg inequality in Corollary 2.3.5 for Ω, Ω_* being concentric balls for this purpose.

First of all, ω satisfies AR) so that the measure $d\mu = \omega dx$ supports the Poincaré and Hardy inequalities in L.1) and L.2) of Chapter 2 with $n = N$ (Remark 2.1.2). In addition, let Ω, Ω_* be concentric balls B_t, B_s, $0 < s < t$. Accordingly, we let ω_* be a cutoff function for B_s, B_t: ω_* is a C^1 function satisfying $\omega_* \equiv 1$ in B_s and $\omega_* \equiv 0$ outside B_t and $|D\omega_*| \leq 1/(t-s)$.

Secondly, we will consider the case $\Lambda(u) = \sqrt{\lambda(u)}$ and $\Phi(u) \sim |\Lambda_u(u)|$. The assumption K) on the map K and that $|\lambda_{uu}|\lambda \lesssim |\lambda_u|^2$ then imply (Remark 2.3.6) that the conditions (2.3.16) and (2.3.17) of Corollary 2.3.5 are verified so that it is applicable here. In addition, we are assuming the condition AR) so that the measure $d\mu = \omega dx$ is regular and the Poincaré–Sobolev and Hardy inequalities are available with the constants C_{PS}, C_H depending on those in AR).

The integrals defined in (2.2.4), (2.2.11) and (2.2.12) with $\Lambda(u) = \sqrt{\lambda(u)}$ and $\Phi(u) = |\lambda_u(u)|/\lambda(u) \sim |\Lambda_u(u)|$ now are

$$I_0(t,x_0) = \int_{B_t(x_0)} \lambda(u)|Du|^{2p}\,d\mu\,, \quad I_1(t,x_0) = \int_{B_t(x_0)} \frac{|\lambda_u(u)|^2}{\lambda(u)}|Du|^{2p+2}\,d\mu\,, \tag{3.3.27}$$

$$I_2(t,x_0) = \int_{B_t(x_0)} \lambda(u)|Du|^{2p-2}|D^2u|^2\,d\mu\,. \tag{3.3.28}$$

We then have the following lemma from Corollary 2.3.5.

Lemma 3.3.7. *Suppose that AR) and K) hold. Define*

$$\mathbf{W}_p(x) := \lambda^{p+\frac{1}{2}}(u)|\lambda_u(u)|^{-p}$$

and let $B_t(x_0)$ be any ball in Ω and assume that $[\mathbf{W}_p^{\alpha}]_{\beta+1,B_t(x_0)}$ is finite for some $\alpha > 2/(p+2)$ and $\beta < p/(p+2)$.

Then, for any concentric balls $B_s(x_0)$ and $B_t(x_0)$, $0 < s < t$, and any $\varepsilon > 0$ there are constants C, $C([\mathbf{W}_p^\alpha]_{\beta+1,B_t(x_0)})$ such that the integrals defined by (3.3.27) *and* (3.3.28) *satisfy*

$$I_1(s, x_0) \le C_{\varepsilon,u,\mathbf{W}}[I_1(t, x_0) + I_2(t, x_0) + (t-s)^{-2} I_0(t, x_0)] . \tag{3.3.29}$$

Here,

$$C_{\varepsilon,u,\mathbf{W}} = \varepsilon + \varepsilon^{-1} C \|K(u)\|^2_{BMO(B_t(x_0),\mu)}[1 + C([\mathbf{W}_p^\alpha]_{\beta+1,B_t(x_0)})] .$$

We now combine this lemma and Lemma 3.3.3 to to remove the integral $\mathcal{B}_p$ on the right-hand side of the energy estimate (3.4.14). We then obtain a key inequality, which allows us to start the iteration argument.

Lemma 3.3.8. *In addition to the assumptions of Lemma 3.3.3, we suppose that M.1) holds for some $p \ge 1$. That is, there exists a constant C_0 such that*

$$\sup_{\tau\in(0,T_0)} \left[\mathbf{W}_p^{\frac{2}{3}}\right]_{\frac{4}{3},\Omega} \le C_0 , \tag{3.3.30}$$

and for any given $\mu_0 > 0$ there is a positive R_{μ_0} sufficiently small such that

$$\sup_{x_0\in\bar{\Omega},\tau\in(0,T_0)} \|K(u)\|^2_{BMO(\Omega_R(x_0),\mu)} \le \mu_0 . \tag{3.3.31}$$

Then for sufficiently small μ_0 there is a constant C depending only on C_0 and the constants in A) and F) such that for $2R < R_{\mu_0}$ we have

$$\mathcal{A}_p(R) + \mathcal{B}_p(R) + \mathcal{H}_p(R) \le C(1 + R^{-2})[\mathcal{C}_p(2R) + \mathcal{F}_p(2R)] + \||DU_0|^{2p}\|_{L^1(\Omega_{2R})} . \tag{3.3.32}$$

Proof. Recall the energy estimate (3.4.14) in Lemma 3.3.3

$$\mathcal{A}_p(s) + \mathcal{H}_p(s) \le C\mathcal{B}_p(t) + C(1 + (t-s)^{-2})[\mathcal{C}_p(t) + \mathcal{F}_p(t)] + \||DU_0|^{2p}\|_{L^1(\Omega_t)} , \tag{3.3.33}$$

for any $0 < s < t$.

We apply Lemma 3.3.7 to estimate $\mathcal{B}_p(t)$, the integral on the right-hand side of (3.3.33). We compare the definitions (3.3.27) and (3.3.28) with those in (3.3.15)–(3.3.17) to see that for u being $u(\cdot, \tau)$ with $\tau \in (0, T_0)$,

$$\mathcal{B}_p(t) = \int_0^{T_0} I_1(t, x_0)d\tau , \quad \mathcal{C}_p(t) = \int_0^{T_0} I_0(t, x_0)d\tau , \quad \mathcal{H}_p(t) = \int_0^{T_0} I_2(t, x_0)d\tau .$$

Hence, for any $\varepsilon > 0$ we can use (3.3.29) to get

$$I_1(s, x_0) \le C_{\varepsilon,U,\mathbf{W}}[I_1(t, x_0) + I_2(t, x_0) + (t-s)^{-2} I_0(t, x_0)] . \tag{3.3.34}$$

From Remark 3.2.2, we see that the assumption (3.3.30) implies the existence of some $\alpha > 2/(p+2)$ and $\beta < p/(p+2)$ such that for any $p \ge 1$ $[\mathbf{W}_p^\alpha]_{\beta+1,B_{R_{\mu_0}}(x_0)\cap\Omega} \le C(C_0)$. Hence, Lemma 3.3.7 is applicable here.

By the definitions of μ_0 in (3.3.31) and $C(\varepsilon, U, \mathbf{W})$ in Lemma 3.3.7, we can find a constant C_1 depending on C_0 such that $C_{\varepsilon,U,\mathbf{W}} \le \varepsilon_0 := \varepsilon + C_1\varepsilon^{-1}\mu_0$. We then integrate (3.3.34) over $(0, T_0)$ to arrive at

$$\mathcal{B}_p(s) \le \varepsilon_0[\mathcal{B}_p(t) + \mathcal{H}_p(t) + (t-s)^{-2}\mathcal{C}_p(t)] \quad 0 < s < t \le R_{\mu_0} .$$

Define $F(t) := \mathcal{B}_p(t)$, $G(t) := \mathcal{H}_p(t)$, and $g(t) := \mathcal{C}_p(t)$. The above yields

$$F(s) \le \varepsilon_0[F(t) + G(t)] + \varepsilon_0(t-s)^{-2}g(t) . \tag{3.3.35}$$

Now, for $h(t) := \mathcal{F}_p(t) + \||DU_0|^{2p}\|_{L^1(\Omega_t)}$ the energy estimate (3.3.33) provides a constant C_2 such that

$$G(s) \le C_2[F(t) + (1 + (t-s)^{-2})(g(t) + h(t))] . \tag{3.3.36}$$

We now apply Lemma B.1.7 to the above two inequalities to estimate $F(s)$ and $G(s)$ in terms of $g(t)$, $h(t)$.

Let $\varepsilon = \frac{1}{4C_2}$ in $\varepsilon_0 = \varepsilon + C_1\varepsilon^{-1}\mu_0$. We have $2C_2\varepsilon_0 = \frac{1}{2} + 4C_1C_2\mu_0$. Thus, if $\mu_0 < \frac{1}{8C_1C_2}$, a constant depends on C_0 of (3.3.30) and the constants in A), F), then $2C_2\varepsilon_0 < 1$. So, we can apply Lemma B.1.7 to the two inequalities (3.3.35) and (3.3.36) and find a constant C_3 such that

$$F(s) + G(s) \le C_3(1 + (t-s)^{-2})[g(t) + h(t)] , \quad 0 < s < t \le R_{\mu_0} .$$

For any $R < R_{\mu_0}/2$ we take $t = 2R$ and $s = \frac{3}{2}R$ in the above to obtain

$$\mathcal{B}_p\left(\frac{3}{2}R\right) + \mathcal{H}_p\left(\frac{3}{2}R\right) \le C_4(1 + R^{-2})[\mathcal{C}_p(2R) + \mathcal{F}_p(2R) + \||DU_0|^{2p}\|_{L^1(\Omega_t)}] .$$

It follows that

$$\mathcal{B}_p(R) \le \mathcal{B}_p\left(\frac{3}{2}R\right) \le C_4(1 + R^{-2})[\mathcal{C}_p(2R) + \mathcal{F}_p(2R) + \||DU_0|^{2p}\|_{L^1(\Omega_t)}] .$$

Combining this and (3.3.33) with $s = R$ and $t = \frac{3}{2}R$, we see that

$$\mathcal{A}_p(R) + \mathcal{B}_p(R) + \mathcal{H}_p(R) \le C_4(1 + R^{-2})[\mathcal{C}_p(2R) + \mathcal{F}_p(2R) + \||DU_0|^{2p}\|_{L^1(\Omega_{2R})}] .$$

This is (3.3.32) and the proof is complete. □

Finally, we have the following lemma giving our desired uniform bound for strong solutions.

Lemma 3.3.9. *Assume as in Lemma 3.3.8. We assume also the integrability condition M.0). Then there exist $p > N/2$, $q_0 > 1$ and a constant M_* depending only on the parameters of A) and F), μ_0, R_{μ_0}, C_0 and the geometry of Ω such that*

$$\sup_{\tau \in (0,T_0)} \int_\Omega |Du(\cdot, \tau)|^{2p}\, dx \le M_* , \tag{3.3.37}$$

$$\|u_t\|_{L^{q_0}(Q)} \le M_* . \tag{3.3.38}$$

Proof. First of all, by the condition AR), there is a constant C_ω such that $|D\omega_0| \le C_\omega \omega_0$ and therefore we have from the the definition (3.3.18) that

$$\mathcal{F}_p(s) \le C_\omega \iint_{Q_s} |f(u)||Du|^{2p-1}\, d\mu d\tau\,.$$

By Young's inequality,

$$|f(u)||Du|^{2p-1} \lesssim |f_u(u)||Du|^{2p} + (|f(u)||f_u(u)|^{-1})^{2p}|f_u(u)|\,.$$

It follows from the assumption (3.2.11) that $|f(u)| \lesssim (1+|u|)|f_u(u)|$ so that

$$\mathcal{F}_p(s) \le C_\omega \iint_{Q_s} [|f_u||Du|^{2p} + |f_u(u)||u|^{2p} + |f_u|]\, d\mu d\tau\,.$$

We then derive from (3.3.32) and the integrability assumption (3.2.8) that there is $R_{\mu_0} > 0$ such that if $0 < R \le R_{\mu_0}$ then

$$\mathcal{A}_p(R) + \mathcal{B}_p(R) + \mathcal{H}_p(R) \le C(1+R^{-2})[\mathcal{C}_p(2R) + \mathcal{F}_{*,p}(2R) + C_0]\,, \tag{3.3.39}$$

where

$$\mathcal{F}_{*,p}(s) := \iint_{Q_s} |u|^{2p}|f_u(u)|\, d\mu d\tau\,. \tag{3.3.40}$$

Let $V = \lambda^{\frac{1}{2}}(u)|Du|^p$. We observe that

$$|DV|^2 \lesssim \frac{|\lambda_u(u)|^2}{\lambda(u)}|Du|^{2p+2} + \lambda(u)|Du|^{2p-2}|D^2u|^2\,.$$

From the definitions of $\mathcal{B}_p$, $\mathcal{H}_p$ and $\mathcal{C}_p$ we deduce from (3.3.39) that

$$\mathcal{A}_p(R) + \iint_{Q_R} [V^2 + |DV|^2]\, d\mu d\tau \le C(1+R^{-2})[\mathcal{C}_p(2R) + \mathcal{F}_{*,p}(2R) + C_0]\,. \tag{3.3.41}$$

Now, assume that for some $p \ge 1$ we can find a constant $C(C_0, R, p)$ such that

$$\mathcal{C}_p(2R) + \mathcal{F}_{*,p}(2R) \le C(C_0, R, p)\,. \tag{3.3.42}$$

Then, (3.3.41) and (3.3.42) yield that

$$\sup_{\tau\in(0,T_0)} \int_{\Omega_R} |Du|^{2p}\, dx + \iint_{Q_R} [V^2 + |DV|^2]\, d\mu d\tau \le C(C_0, R, p)\,. \tag{3.3.43}$$

Note that the assumptions (3.2.8) and (3.2.9) in M.0),

$$\iint_Q (|f_u(u)| + \lambda(u))|Du|^2\, d\mu d\tau\,, \quad \iint_Q |f_u(u)||u|^2\, d\mu d\tau \le C_0\,,$$

imply (3.3.42) holds for $p = 1$.

In the technical Lemma 3.3.11 following this proof, we will show that the assumption (3.3.42) is self-improving in the following sense: if (3.3.42) holds for some $p \ge 1$ then, together with its consequence (3.3.43), we find some *fixed* $\gamma_* > 1$ such that (3.3.42) holds again for the new exponent $\gamma_* p$ and R being $R/2$. Hence, (3.3.43) is valid if $\gamma_* p$ satisfies the gap condition (3.3.19) so that the energy estimate of Lemma 3.3.3) is available.

From the SG) condition and the elementary Lemma 3.3.10 following this proof, with $s_0 = 2$, we see that we can choose y_* obtained in the above argument smaller and close to 1 in order that there is an integer $k_0 \geq 1$ such that for $k = 0, \ldots, k_0$ the exponents $p_k = y_*^k$ verify the gap condition (3.3.19), $1 - /p_k < C_*^{-1}$, and we can iterate (3.3.43) k_0 times to find that (3.3.43) holds for $p_{k_0} = y_*^{k_0} > N/2$. It follows that there is a constant C depending only on the parameters of A) and F), μ_0, R_{μ_0} and k_0 such that

$$\sup_{(0,T_0)} \int_{\Omega_{R_0}} |Du|^{2p}\, d\mu \leq C \text{ for } R_0 = 2^{-k_0} R_{\mu_0},\ p = y_*^{k_0}\,. \tag{3.3.44}$$

Summing the above inequalities over a finite covering of balls B_{R_0} for Ω, we find a constant C, depending also on the geometry of Ω, and obtain the desired global estimate (3.3.37) because $p = p_{k_0} > N/2$.

Finally, we obtain from (3.3.43) with $p = 1$ that

$$\iint_Q \lambda(u)|D^2u|^2\, d\mu d\tau \leq C\,. \tag{3.3.45}$$

As u is a strong solution, we have $|u_t| \leq |\operatorname{div}(A(x,u)Du)| + |\hat{f}|$ a.e. in Q. Therefore,

$$\|u_t\|_{L^{q_0}(Q)} \lesssim \|\lambda(u)|D^2u|\|_{L^{q_0}(Q)} + \|\lambda_u(u)|Du|^2\|_{L^{q_0}(Q)} + \|f(u)\|_{L^{q_0}(Q)}\,.$$

We now know that u is bounded by (3.3.37) and Sobolev's embedding theorem. If $q_0 \in (1,2)$ then the first two norms on the right can be treated by Hölder's inequality and (3.3.45) and (3.3.37). Of course, the last norm is also bounded. Thus, there is $q_0 > 1$ such that (3.3.38) holds.

The lemma is proved. □

In the above proof, we assert the existence of the numbers y_* and k_0 in our iteration argument. Since this simple fact will be used later in several places with different settings, we will state and prove it in the following elementary lemma.

Lemma 3.3.10. *Suppose that N satisfies $1 - s_0/N < C_*^{-1}$ for some $s_0 > 0$ then there exist $y_0 > 1$ and an integer $k_0 \geq 1$ such that the numbers $p_k = y_0^k$, $k = 0, \ldots, k_0$, satisfy*

$$1 - 1/p_k < C_*^{-1} \text{ and } p_{k_0} > N/s_0\,. \tag{3.3.46}$$

Proof. From the condition $1 - s_0/N < C_*^{-1}$, we can find $N_1 > N$ such that N_1 still satisfies the same inequality. For some $y_0 > 1$ let $x_1 = \log(N/s_0)/\log(y_0)$ and $x_2 = \log(N_1/s_0)/\log(y_0)$. We see that we can choose $y_0 > 1$ smaller and close to 1 such that the difference between x_2, x_1 is $\log(N_1/N)/\log(y_0) > 2$. This means that there is an integer $k_0 \geq 1$ between the two positive numbers x_1, x_2. We observe that

$$\log(N/s_0)/\log(y_0) < k_0 < \log(N_1/s_0)/\log(y_0) \Leftrightarrow N < y_0^{k_0} s_0 < N_1\,.$$

The last inequality implies $p = y_0^{k_0} > N/s_0$. We also have $1 - 1/p_k < 1 - s_0/N_1 < C_*^{-1}$ because $p_k < N_1/s_0$. Hence (3.3.46) holds. □

Thus, to conclude the proof of the above lemma, we need to show that (3.3.42) is self-improving in the following lemma.

Lemma 3.3.11. *Assume as in Lemma 3.3.9. The quantities $\mathfrak{C}_p(R)$ and $\mathfrak{F}_{*,p}(R)$ are self-improving in the sense that: Suppose that for some $p \geq 1$ we can find a constant $C(C_0, R, p)$ such that*

$$\mathfrak{C}_p(2R) + \mathfrak{F}_{*,p}(2R) \leq C(C_0, R, p)\,, \tag{3.3.47}$$

then there exist a fixed $\gamma_* > 1$ and a new constant $C_1(C_0, R, p)$ such that

$$\mathfrak{C}_{\gamma_* p}(R) + \mathfrak{F}_{*,\gamma_* p}(R) \leq C_1(C_0, R, p)\,. \tag{3.3.48}$$

In the proof of this lemma, for any time interval I and any ball $B \subset R^N$ we will repeatedly make use of the following parabolic Sobolev inequality, denoting $\pi_* = N_* = 2N/(N-2)$, $q_* = 1 - \frac{2}{\pi_*}$ and $Q = B \times I$:

$$\iint_Q v^{2q_*}|V|^2 \, d\mu d\tau \lesssim \sup_I \left(\int_B v^2 \, d\mu\right)^{q_*} \iint_Q [|DV|^2 + V^2] \, d\mu d\tau\,. \tag{3.3.49}$$

To see this, we recall the inequality

$$\left(\int_B |V|^{\pi_*} \, d\mu\right)^{\frac{1}{\pi_*}} \lesssim \left(\int_B |DV|^2 \, d\mu\right)^{\frac{1}{2}} + \left(\int_B |V|^2 \, d\mu\right)^{\frac{1}{2}}\,,$$

which is just a simple consequence of the Poincaré–Sobolev inequality PS). For $q_* = (1 - \frac{2}{\pi_*})$ we use Hölder's inequality and the above inequality to get

$$\begin{aligned}\iint_Q v^{2q_*}|V|^2 \, d\mu d\tau &\leq \int_I \left(\int_B v^2 \, d\mu\right)^{1-\frac{2}{\pi_*}} \left(\int_B |V|^{\pi_*} \, d\mu\right)^{\frac{2}{\pi_*}} d\tau \\ &\lesssim \sup_I \left(\int_B v^2 \, d\mu\right)^{q_*} \int_I \left(\int_B |DV|^2 \, d\mu + \int_\Omega |V|^2 \, d\mu\right) d\tau\,.\end{aligned}$$

This is (3.3.49).

Proof of Lemma 3.3.11. We established in the proof of Lemma 3.3.9 that for $V = \lambda^{\frac{1}{2}}(u)|Du|^p$ (3.3.47) and (3.3.41) imply

$$\sup_{\tau \in (0,T_0)} \int_{\Omega_R} |Du|^{2p} \, d\mu + \iint_{Q_R} [V^2 + |DV|^2] \, d\mu d\tau \leq C(C_0, R, p)\,. \tag{3.3.50}$$

We consider first the quantity $\mathfrak{C}_{\gamma p}(R)$ for some $\gamma > 1$. Recalling the definition (3.3.17), $\mathfrak{C}_{\gamma p}(R)$ is the integral of $(|f_u(u)| + \lambda(u))|Du|^{2\gamma p}$.

For any $\gamma \in (1, q_* + 1)$ we write $|f_u(u)||Du|^{\gamma 2p} = v^{2q_*} V^2$ with

$$v = (|f_u(u)|\lambda^{-1}(u)|Du|^{2p(\gamma-1)})^{\frac{1}{2q_*}} \text{ and } V = \lambda^{\frac{1}{2}}(u)|Du|^p\,.$$

In order to apply (3.3.49), we need to estimate the integral of v^2 over Ω_R and the integral of $|DV|^2 + V^2$ over Q_R. A bound for the latter is already given in (3.3.47). Concerning v^2, we use Hölder's inequality with the exponent $q_1 = \frac{q_*}{\gamma-1}$ to get

$$\int_{\Omega_R} v^2\, d\mu \le \left(\int_{\Omega_R} (|f_u(u)|\lambda^{-1}(u))^{q_1'}\, d\mu\right)^{\frac{1}{q_1'}} \left(\int_{\Omega_R} |Du|^{2p}\, d\mu\right)^{\frac{1}{q_1}} . \tag{3.3.51}$$

We can find $\gamma = \gamma_*$ close to 1 such that $q_1' = \frac{q_*}{q_*-\gamma_*+1} \le r_0$, the exponent in (3.2.6) of the assumption M.0), which states: there exist C_0 and $r_0 > 1$ such that

$$\sup_{\tau\in(0,T_0)} \||f_u(u)|\lambda^{-1}(u)\|_{L^{r_0}(\Omega,\mu)} \le C_0 . \tag{3.3.52}$$

Hence, the first integral in (3.3.51) is bounded by the assumption (3.3.52). The second integral is bounded by (3.3.50). We obtain

$$\iint_{Q_R} |f_u(u)||Du|^{2\gamma_* p}\, d\mu d\tau \le C_1(C_0, R, p) .$$

In the same way, we replace $|f_u(u)|$ with $\lambda(u)$ in the above argument (and take r_0 in (3.3.52) to be ∞) to get

$$\iint_{Q_R} \lambda(u)|Du|^{2\gamma_* p}\, d\mu d\tau \le C_1(C_0, R, p) .$$

We then conclude that there is a constant $C_1(C_0, R, p)$ and some $\gamma_* > 1$ such that

$$\mathcal{C}_{\gamma_* p}(R) \le C_1(C_0, R, p) .$$

We now turn to $\mathcal{F}_{*,p}(R)$ defined by (3.3.40), the integral over Q_R of $|f_u(u)||u|^{2p}$. The treatment of this term is similar to the above argument of the integral of $|f_u(u)||Du|^{2p}$ in $\mathcal{C}_p$ with Du being replaced by $|u|$. Therefore, in order to use the inequality (3.3.49) again we need only estimate the following two integrals (compare with the last integral in (3.3.51)):

$$\int_{\Omega_R} |u|^{2p}\, d\mu , \quad \iint_{Q_R} [|DV|^2 + V^2]\, d\mu d\tau , \quad V = \lambda^{\frac{1}{2}}(u)|u|^p . \tag{3.3.53}$$

We consider the first integral and use Sobolev's inequality to get

$$\int_{\Omega_R} |u|^{2p}\, d\mu \lesssim \int_{\Omega_R} |D(|u|^p)|^2\, d\mu + \left(\int_{\Omega_R} |u|^{p\beta}\, d\mu\right)^{\frac{2}{\beta}} .$$

Because $|D(|u|^p)|^2 \sim |u|^{2p-2}|Du|^2 \le \varepsilon|u|^{2p} + C(\varepsilon)|Du|^{2p}$, we can choose ε sufficiently small to conclude from the above that

$$\int_{\Omega_R} |u|^{2p}\, d\mu \lesssim \int_{\Omega_R} |Du|^{2p}\, d\mu + \left(\int_{\Omega_R} |u|^{p\beta}\, d\mu\right)^{\frac{2}{\beta}} .$$

The first integral on the right-hand side is bounded by (3.3.50). Taking β such that $\beta p = \beta_0$, the exponent in the assumption (3.2.7), the second integral is bounded by the assumption (3.2.7), which reads

$$\sup_{\tau\in(0,T_0)} \|u^{\beta_0}\|_{L^1(\Omega,\mu)} , \qquad \sup_{\tau\in(0,T_0)} \|\lambda^{\beta_0}(u)\|_{L^1(\Omega,\mu)} \le C_0 . \tag{3.3.54}$$

Concerning the second integral in (3.3.53) with $V = \lambda^{\frac{1}{2}}(u)|u|^p$, we will show that there is a constant $C(C_0)$ such that

$$\iint_{Q_R} [|DV|^2 + V^2]\, d\mu d\tau \le C(C_0)\,. \tag{3.3.55}$$

We use the fact that $|\lambda_u(u)||u| \lesssim \lambda(u)$ and Young's inequality to see that

$$\begin{aligned} |DV|^2 &\lesssim \lambda(u)|D(|u|^p)|^2 + |\lambda_u(u)|^2\lambda^{-1}(u)|Du|^2|u|^{2p} \\ &\lesssim \lambda(u)|D(|u|^p)|^2 + \lambda(u)|Du|^2|u|^{2p-2} \\ &\lesssim C(\varepsilon)\lambda(u)|Du|^{2p} + \varepsilon\lambda(u)|u|^{2p}\,. \end{aligned} \tag{3.3.56}$$

Now, using Sobolev's inequality and integrating in t, we have

$$\iint_{Q_R} V^2\, d\mu d\tau \lesssim \iint_{Q_R} |DV|^2\, d\mu d\tau + \int_0^{T_0} \left(\int_{\Omega_R} V^\beta\, d\mu \right)^{\frac{2}{\beta}} d\tau\,.$$

Applying Hölder's inequality to $V^\beta = \lambda^{\beta/2}(u)|u|^{p\beta}$ and using (3.3.54), we will get

$$\|V^\beta\|_{L^1(\Omega_R)} \le \|\lambda^{\beta/2}(u)\|_{L^2(\Omega_R)} \||u|^{p\beta}\|_{L^2(\Omega_R)} \le C(C_0)$$

if β is sufficiently small such that $\beta \le \min\{\beta_0, \beta_0/2p\}$. So,

$$\iint_{Q_R} (|DV|^2 + V^2)\, d\mu d\tau \lesssim \iint_{Q_R} |DV|^2\, d\mu d\tau + C(C_0)\,.$$

Choosing ε sufficiently small in (3.3.56), we estimate the integral of $|DV|^2$ on the right-hand side of the above and arrive at

$$\iint_{Q_R} (|DV|^2 + V^2)\, d\mu d\tau \lesssim \iint_{Q_R} \lambda(u)|Du|^{2p}\, d\mu d\tau + C(C_0)\,.$$

By (3.3.50) the integral over Q_R of $\lambda(u)|Du|^{2p}$ is bounded by a constant $C(C_0, R, p)$. Therefore, the integral in (3.3.55) is bounded by a constant $C(C_0, R, p)$. We then find constants $\gamma_* > 1$ and $C_1(C_0, R, p)$ such that $\mathcal{F}_{*,\gamma_*p}(R) \le C_1(C_0, R, p)$ and thus $\mathcal{F}_{*,p}(R)$ is self-improving.

We complete the proof of the lemma. □

Remark 3.3.12. From Remark 3.3.4, $\mathcal{F}_p \equiv 0$ if $\hat{f}$ is independent of x. In this case, we do not have the term $\mathcal{F}_{*,p}$ in our consideration. Thus, the conditions (3.2.7) and (3.2.8) in M.0) can be dropped as we do not have to show that $\mathcal{F}_{*,p}$ is self-improving. In particular, the condition (3.1.9) that $|\lambda(u)||u| \lesssim \lambda(u)$ in F) was used only in (3.3.56) to estimate $\mathcal{F}_{*,p}$; it thus can be dropped too.

We are ready to provide the proof of the main theorem of this section.

Proof of Theorem 3.2.3. It is now clear that the assumptions M.0) and M.1) of our theorem allow us to apply Lemma 3.3.9 and obtain a priori uniform bounds for any continuous strong solution u of (3.3.2). The uniform estimate (3.3.37) shows that the assumption (3.3.8) of Lemma 3.3.1 holds true so that the map T_σ is well defined and compact on a ball $\bar{B}_M$ of $\mathbf{X}$ for some M depending on the bound M_* provided by Lemma 3.3.9.

Combining this with Lemma 3.3.2, the fixed points of T_σ are strong solutions of the system (3.3.2) so that T_σ does not have a fixed point on the boundary of $\bar{B}_M$. Thus, by the Leray–Schauder fixed-point theorem, T_1 has a fixed point in B_M, which is a strong solution to (3.3.1). This solution is unique because u, Du are bounded and (3.3.1) is now regular parabolic. The proof is complete. □

3.3.3 On the growth of *Du*

We assumed in F) that the reaction $\hat{f}$ has a superlinear growth in Du. That is $|\hat{f}(u, Du)$ can grow at most as $|Du|^r$ for some $r = 1 + \delta$ with $\delta < 2/N$, and thus $r < 2$. Although this assumption covers many real-life applications in physical science, the case $r = 2$ also plays an important role in mathematics and this case is very difficult to handle.

We conclude this section by discussing the interesting case when $N = 2$. That is, we will consider our system on Ω as a planar domain and the condition f.2) in F) holds with $\delta = 1$.

We will show that the proof in this section can be 'easily' modified to obtain the same conclusion in this case.

First of all, one can see that the energy estimate for Du in Lemma 3.3.3 continues to hold (without the need for some extra use of Young's inequality). Hence, the key estimates of Lemma 3.3.9 remain true and the necessary bound for fixed points of the map T_σ in the proof of the main Theorem 3.2.3 is available. The main issue is the setting of these fixed-point maps.

Recall the definition of the base space of these maps $\mathbf{X} = X_1 \cap X_2$, where $X_1 = C^{\beta,\beta/2}(Q)$ and $X_2 = L^1([0, T_0], W^{1,2}(\Omega))$ for some suitable $\beta > 0$. For each $u \in \mathbf{X}$ the map $w = T(u)$ is defined as the solution to the linear parabolic system (3.3.10)

$$w_t = \operatorname{div}(\mathbf{a}(u)Dw) + \mathbf{b}(u)Dw + \mathbf{g}(u)w + \mathbf{f}\,, \qquad (3.3.57)$$

where $|\mathbf{b}(u)| \lesssim |Du|^\delta$ due to f.2). This assumption is crucial in applying the theory in [25] to the existence of the solution w of (3.3.57). Indeed, we need that $|\mathbf{b}(u)|^2 \in L^1([0, T_0], L^q(Q))$ for some $q > N/2$ if $2\delta < 2$ and $u \in X_2$.

It is now clear that if $\delta = 1$ then the space $X_2 = L^1([0, T_0], W^{1,2}(\Omega))$ is no longer suitable for the definition of T. In the sequel, we will redefine

$$\mathbf{X} = X_1 \cap X_2\,, \quad \text{where } X_1 = C^{\beta,\beta/2}(Q)\,, \; X_2 = L^1([0, T_0], W^{1,2s}(\Omega))\,,$$

where s is some number greater than 1. We then see that $T(u)$ is well defined again with this new definition of X_2. Indeed, with $q = s > N/2 = 1$, $|\mathbf{b}(u)|^2 \sim |Du|^2 \in L^1([0, T_0], L^q(Q))$.

The remaining issue is to show that $T_\sigma\colon \mathbf{X} \to \mathbf{X}$ is compact. Let K be a bounded set in $\mathbf{X}$. The proof of Lemma 3.3.1 already showed that $T_\sigma(K)$ is compact in X_1 so that we need only establish its sequential compactness in X_2. That is, every sequence $w_n = T_\sigma(u_n)$, $u_n \in K$, has a convergent subsequence in X_2.

Because $\{u_n\}$ is compact in $X_1 = C^{\beta,\beta/2}(Q)$, passing to a subsequence, we can assume that $u_n \to u_*$ in $C^{\beta',\beta'/2}(Q)$ for some $\beta' \in (0,\beta)$. Clearly, Du_n weakly converges to Du_* in $L^2(Q)$. We then denote $w_* = T_\sigma(u_*)$. As we proved earlier that $T_\sigma(K)$ is compact in X_1, we can assume that $\{w_n\}$ converges. By uniqueness, $w_n \to w_*$ in X_1.

We have, writing $W = w_n - w_*$,

$$W_t - \operatorname{div}(A(x, \sigma u_n)DW) = \operatorname{div}(\alpha_n Dw_n) + \Psi_n\,, \tag{3.3.58}$$

where $\alpha_n = A(x,\sigma u_n) - A(x, \sigma u_*)$ and Ψ_n is defined by

$$F^{(\sigma)}(x, u_n, Du_n)Dw_n - F^{(\sigma)}(x, u_*, Du_*)Dw_* + f^{(\sigma)}(x, u_n)u_n - f^{(\sigma)}(x, u_*)u_*\,.$$

For any $(x_0, t_0) \in Q$ and $R > 0$ we denote below $B_R = B_R(x_0)$ and $Q_R = B_R(x_0) \times [t_0 - R^2, t_0]$ and

$$\bar{W}_R(t) = \fint_{B_R \times \{t\}} \bar{W}(x,t)\,dx\,.$$

Let ψ, η be cutoff functions for the pairs B_R, B_{2R} and $[t_0 - 4R^2, t_0]$, $[t_0 - R^2, t_0]$ respectively. Testing (3.3.58) with $(W - \bar{W}_{2R})\psi^2\eta$ we obtain a Caccioppoli type inequality

$$\iint_{Q_R} |DW|^2\,dz \le \frac{C}{R^2}\iint_{Q_{2R}} |W - \bar{W}_{2R}|^2\,dz + C\iint_{2R} g\,dz\,, \tag{3.3.59}$$

where $g := (|\alpha_n|^2|Dw_n|^2 + |W||\Psi_n|)$.

Similarly, we also have the following parabolic Poincaré inequality:

$$\sup_{t\in[t_0-R^2,t_0]} \int_{B_R} |W - \bar{W}_R|^2\,dx \le C\iint_{Q_{2R}} |DW|^2\,dz + C\iint_{2R} g\,dz\,. \tag{3.3.60}$$

Combining these two inequalities and arguing the same way as in the proof of [14, Proposition 1.3] we derive the following reverse Hölder inequality:

$$\fint\!\!\fint_{Q_R} |DW|^2\,dz \le \theta\fint\!\!\fint_{Q_R} |DW|^2\,dz + C(\theta)\left(\fint\!\!\fint_{Q_{2R}} |DW|^q\,dz\right)^{\frac{2}{q}} + C\fint\!\!\fint_{Q_{2R}} g\,dz, \tag{3.3.61}$$

for any $\theta > 0$ and $q \in (1,2)$ (as $N = 2$).

By (3.3.12), it is clear that

$$|\Psi_n| \le C(K)[(|Du_n| + |Du_*|)(|Dw_n| + |Dw_*|) + 1]\,.$$

Using the fact that $\{Dw_n\}$ and $\{Du_n\}$ are bounded in $L^{2s}(Q)$ and $\{w_n\}$ and $\{u_n\}$ are bounded in $C(Q)$, we see that $W\Psi_n$ and $|\alpha_n|^2|Dw_n|^2$ are in $L^s(Q)$. Hence, $g = |\alpha_n|^2|Dw_n|^2 + |W||\Psi_n| \in L^s(Q)$. The reverse Hölder inequality (3.3.61) then implies

$$\left(\fint\!\!\fint_{Q_R} |DW|^{2s}\,dz\right)^{\frac{1}{s}} \le C\left(\fint\!\!\fint_{Q_{2R}} |DW|^2\,dz\right)^{\frac{1}{2}} + C\left(\fint\!\!\fint_{Q_{2R}} |g|^s\,dz\right)^{\frac{1}{s}}\,. \tag{3.3.62}$$

We already showed that $DW \to 0$ in $L^2(Q)$ in Lemma 3.3.1. On the other hand, W and α_n converge to 0 in $C(Q)$ so that $g \to 0$ in $L^s(Q)$. We then see from the above that $DW \to 0$ in $L^{2s}(Q_R)$ for any $R > 0$. By a finite covering argument $DW \to 0$, or $Dw_n \to Dw_*$, in $L^{2s}(Q)$. This shows the desired sequential compactness of $T(K)$ in X_2. Thus, T is a compact map on $\mathbf{X}$.

3.4 The simpler case

We consider the case when the data of the system does not depend explicitly on x. In this case $\omega \equiv 1$ and $d\mu = dx$, the Lebesgue measure. In the sequel we write $dz = dxdt$.

We then consider in this section the following system:

$$\begin{cases} u_t - \operatorname{div}(A(u)Du) = \hat{f}(u, Du)\,, & x \in \Omega\,, \\ u = 0 \text{ or } \frac{\partial u}{\partial \nu} = 0 \text{ on } \partial\Omega\,. \end{cases} \tag{3.4.1}$$

We embed this system in the following family of systems:

$$\begin{cases} u_t - \operatorname{div}(A(\sigma u)Du) = \hat{f}(\sigma u, \sigma Du)\,, & x \in \Omega, \sigma \in [0,1]\,, \\ u = 0 \text{ or } \frac{\partial u}{\partial \nu} = 0 \text{ on } \partial\Omega\,. \end{cases} \tag{3.4.2}$$

The general result in the previous section relies on the existence of a general map K and the boundedness of the weight $\mathbf{W}$ to verify the main assumption M.1), so that some version of the local Gagliardo–Nirenberg inequalities in Chapter 2 is available. It is quite delicate to construct such a map and the checking of M.1) is complicated. Under slightly stronger structural assumptions of the systems, which often appear in applications, we will show in this section that we can take the map K to be the identity map and $\mathbf{W} = 1$, and hence the condition M.1) is greatly simplified.

If the spectral gap condition SG) holds then we can find $\alpha_* \in (0,1)$ such that:

SG') There is $\alpha_* \in (0,1)$ such that $(N - 2\alpha_*)/N < C_*^{-1}$.

Most importantly, we assume in this section the following:

A') In addition to A), we assume also that

$$\Lambda := \sup_{u \in \mathbb{R}^m} \frac{|\lambda_u(u)|}{\lambda(u)} < \infty\,. \tag{3.4.3}$$

It is easy to see that (3.4.3) is satisfied if $\lambda(u)$ is a polynomial in $|u|$ and $\lambda(u) \geq \lambda_0$ for some constant $\lambda_0 > 0$.

We consider the following integrability condition in place of M.0):

M.0') There exist C_0 and $r_0 > 1$, $\alpha_0 \in (0,1)$ such that

$$\sup_{\tau \in (0,T_0)} \| |f_u(u)| \lambda^{-1+\alpha_0}(u) \|_{L^{r_0}(\Omega)} \leq C_0\,, \tag{3.4.4}$$

$$\iint_Q (1 + |f_u(u)|\lambda^{-1}(u))|Du|^2 \, dz \leq C_0\,, \tag{3.4.5}$$

$$\sup_{\tau \in (0,T_0)} \int_{\Omega \times \{\tau\}} \lambda^{\frac{\alpha_*}{1-\alpha_*}}(u) \, dx \leq C_0\,. \tag{3.4.6}$$

Here, the number $\alpha_* \in (0,1)$ was introduced in the assumption SG').

Assume also that

$$\iint_Q (f(u)\lambda^{-1}(u))^{\frac{N_1}{2\alpha_*}} \, dz \leq C_0 \tag{3.4.7}$$

for some $N_1 > N$ and N_1 satisfying SG').

Theorem 3.4.1. *Assume that A'), F) and SG') hold and that any strong solution u to the family* (3.4.2) *has small BMO norm in small balls: for any given $\mu_0 > 0$ there is positive R_{μ_0} sufficiently small in terms of the constants in A) and F) such that*

$$\sup_{x_0 \in \bar{\Omega}, \tau \in (0,T_0)} \|u\|^2_{BMO(B_R(x_0)\cap\Omega)} \le \mu_0 . \tag{3.4.8}$$

Furthermore, u satisfies M.0') uniformly for $\sigma \in [0, 1]$.
Then there is a unique strong solution u to (3.4.1).

Remark 3.4.2. As we mentioned in Remark 3.2.1, the integrability condition M.0') is in fact a very mild one in conjunction with (3.4.8) in applications. In particular, if $f(u)$ and $\lambda(u)$ have polynomial growth in $|u|$ then so do the integrands $f_u(u)|\lambda^{-1+\alpha_0}(u)$, $\lambda^{\frac{\alpha_*}{1-\alpha_*}}(u)$ and $f(u)\lambda^{-1}(u)$. Thus, with the exception of (3.4.5), the condition M.0') reduces to only one condition that $\|u\|_{L^q(\Omega)}$ is uniformly bounded for sufficiently large q. This is guaranteed by (3.4.8), which implies u is BMO on Ω and therefore u is in $L^q(\Omega)$ for all $q \ge 1$ [16].

The proof of this theorem follows that of Theorem 3.2.3. The main difference here is a new energy estimate for Du. To this end, we introduce the following integrals:

$$\mathcal{A}_p(s) = \sup_{\tau \in (0,T_0)} \int_{\Omega_s \times \{\tau\}} \lambda^{-1}(u)|Du|^{2p}\, dx , \tag{3.4.9}$$

$$\mathcal{H}_p(s) := \iint_{Q_s} |Du|^{2p-2}|D^2u|^2\, dz , \tag{3.4.10}$$

$$\mathcal{B}_p(s) := \iint_{Q_s} |Du|^{2p+2}\, dz , \tag{3.4.11}$$

$$\mathcal{C}_p(s) := \iint_{Q_s} (1 + |f_u(u)|\lambda^{-1}(u))|Du|^{2p}\, dz , \tag{3.4.12}$$

$$\mathcal{F}_p(s) := \iint_{Q_s} (|f(u)|\lambda^{-1}(u))^{2p}\, dz . \tag{3.4.13}$$

Here and in the sequel, $\Omega_s = B(x_0, s) \cap \Omega$ for some ball $B(x_0, s)$ with center $x_0 \in \bar{\Omega}$ and $Q_s = \Omega_s \times (0, T_0)$.

We begin with the following lemma, which establishes another energy estimate for Du (compare to that of Lemma 3.3.3).

Lemma 3.4.3. *Assume A), F). Let u be any strong solution of* (3.3.1) *on Ω and p satisfying* (3.3.19), $1 - 1/p < C_*^{-1}$.

There is a positive constant C, which depends only on the parameters in A) and F) and Λ, such that for any two concentric balls B_s, B_t with center $x_0 \in \bar{\Omega}$ and $s < t$,

$$\begin{aligned} \mathcal{A}_p(s) + \mathcal{H}_p(s) \le C\mathcal{B}_p(t) + C(1 + (t-s)^{-2})\mathcal{C}_p(t) \\ + C\mathcal{F}_p(t) + \||DU_0|^{2p}\|_{L^1(\Omega_t)} . \end{aligned} \tag{3.4.14}$$

Proof. Let ψ be a cutoff function for B_s, B_t. That is, ψ is a C^1 function satisfying $\psi \equiv 1$ in B_s and $\psi \equiv 0$ outside B_t and $|D\psi| \le 1/(t-s)$. Consider any given triple t_0, T, T'

satisfying $0 < t_0 < T < T' \le T_0$ and let η be a cutoff function for $(T - t_0, T')$, (T, T'). That is, $\eta(s) \equiv 1$ if $s \ge T$, $\eta(s) \equiv 0$ if $s \le T - t_0$ and $|\eta_t| \le t_0^{-1}$.

Let $\mathbf{w} = \lambda^{-1}(u)|Du|^{2p-2}Du$. We revisit the proof of Lemma 3.3.3 and test (3.3.21) with $\mathbf{w}\psi^2\eta$ and obtain, using integration by parts,

$$\begin{aligned}
&\int_{T'-t_0}^{T'}\int_\Omega \mathbf{w}Du_t\psi^2\eta + \langle A(u)D^2u, D(\mathbf{w}\psi^2)\rangle\eta\, dxd\tau \\
&= \int_{T'-t_0}^{T'}\int_\Omega \langle -A_u(u)DuDu, D(\mathbf{w}\psi^2)\rangle\eta\, dxd\tau \\
&+ \int_{T'-t_0}^{T'}\int_\Omega \langle D\hat{f}(u, Du), \mathbf{w}\rangle\psi^2\eta\, dxd\tau\,.
\end{aligned} \tag{3.4.15}$$

We derive from this the following inequality:

$$\begin{aligned}
&\sup_{t\in(T,T')}\int_{\Omega_s} \lambda^{-1}(u)|Du|^{2p}\, dx + \int_{T-t_0}^{T'}\int_{\Omega_t} |Du|^{2p-2}|D^2u|^2\psi^2\eta\, dxd\tau \\
&\le C\int_{T-t_0}^{T'}\int_{\Omega_t} |\langle A(u)D^2u, |Du|^{2p-2}DuD(\lambda^{-1}(u)\psi^2)\rangle|\eta\, dxd\tau \\
&+ C\int_{T-t_0}^{T'}\int_{\Omega_t} (|A_u(u)||Du|^2|D(\mathbf{w}\psi^2)| + |D\hat{f}(u, Du)||\mathbf{w}|\psi^2)\eta\, dxd\tau \\
&+ Ct_0^{-1}\int_{T-t_0}^{T}\int_{\Omega_t} \lambda^{-1}(u)|Du|^{2p}\, dxd\tau \\
&+ \int_{T-t_0}^{T}\int_{\Omega_t} |(\lambda^{-1}(u))_u||u_t||Du|^{2p-1}\, dxd\tau\,.
\end{aligned} \tag{3.4.16}$$

Indeed, the first and last two integrals of (3.4.16) result from the integrals involving the temporal derivative in (3.4.15) and the fact that $|\eta_t| \le t_0^{-1}$ and that

$$2p\mathbf{w}Du_t\eta = (\lambda^{-1}(u)|Du|^{2p}\eta)_t - (\lambda^{-1}(u))_u u_t|Du|^{2p}\eta - \lambda^{-1}(u)|Du|^{2p}\eta_t\,.$$

The second integral in the first line of (3.4.16) comes from the integral of $\langle A(u)D^2u, \lambda^{-1}(u)D(|Du|^{2p-2}Du)\psi^2\eta\rangle$, which is again estimated from below by using Lemma B.1.3. The spectral gap condition (3.3.19), the ellipticity condition A) and the lemma then provide a positive constant c_0 such that

$$\langle A(u)D^2u, \lambda^{-1}(u)D(|Du|^{2p-2}Du)\rangle \ge c_0|Du|^{2p-2}|D^2u|^2\,.$$

We consider the integrand $|\langle A(u)D^2u, |Du|^{2p-2}DuD(\lambda^{-1}(u)\psi^2)\rangle|$ in the second line. By Young's inequality, for any $\varepsilon > 0$ we can find a constant $C(\varepsilon)$ such that

$$\begin{aligned}
|\langle A(u)D^2u, \mathbf{w}D(\psi^2)\rangle| &\lesssim |Du|^{2p-1}||D^2u|\psi|D\psi| \\
&\lesssim \varepsilon|Du|^{2p-2}|D^2u|^2\psi^2 + C(\varepsilon)|Du|^{2p}|D\psi|^2\,,
\end{aligned}$$

$$\begin{aligned}
|\langle A(u)D^2u, \frac{|\lambda_u|}{\lambda^2}|Du|^{2p}\psi^2\rangle| &\lesssim \frac{|\lambda_u|}{\lambda}|Du|^{2p}||D^2u|\psi^2 \\
&\lesssim \varepsilon|Du|^{2p-2}|D^2u|^2\psi^2 + C(\varepsilon)\frac{|\lambda_u(u)|^2}{\lambda^2(u)}|Du|^{2p+2}\psi^2\,.
\end{aligned}$$

The integrands in the third line come from the integral involving $A_u, D\hat{f}$ in (3.4.15). We note that

$$|\mathbf{w}| = \lambda^{-1}(u)|Du|^{2p-1}\,, \quad |D\mathbf{w}| \lesssim \lambda^{-1}(u)|Du|^{2p-2}|D^2u| + \frac{|\lambda_u|}{\lambda^2}|Du|^{2p}\,.$$

Using the condition A), $|A_u| \lesssim |\lambda_u|$, and Young's inequality, we have

$$\begin{aligned}
&|A_u(u)||Du|^2|D(\mathbf{w}\psi^2)| \\
&\leq \frac{|\lambda_u|}{\lambda(u)}|Du|^{2p}|D^2u|\psi^2 + \frac{|\lambda_u(u)|^2}{\lambda^2(u)}|Du|^{2p+2} + \frac{|\lambda_u|}{\lambda(u)}|Du|^{2p+1}\psi|D\psi| \\
&\leq \varepsilon\lambda(u)|Du|^{2p-2}|D^2u|^2\psi^2 + C(\varepsilon)\left[\frac{|\lambda_u(u)|^2}{\lambda^2(u)}|Du|^{2p+2} + |D\psi|^2|Du|^{2p}\right].
\end{aligned}$$

Next, we consider the integral of $|D\hat{f}(u, Du)||\mathbf{w}|\psi^2$. First, if $\hat{f}$ has a linear growth in Du then by f.1) in F) with $p = Du$,

$$|D\hat{f}(u, Du)| \lesssim \lambda(u)|D^2u| + |\lambda_u(u)||Du|^2 + |f_u(u)||Du|\,.$$

Therefore, using Young's inequality, we get

$$\begin{aligned}
|D\hat{f}(u, Du)|\mathbf{w} \lesssim\; & |Du|^{2p-1}|D^2u| + \frac{|\lambda_u(u)|}{\lambda(u)}|Du|^{2p+1} + |f_u(u)|\lambda^{-1}(u)|Du|^{2p} \\
& \varepsilon|Du|^{2p-2}|D^2u|^2 + C(\varepsilon)|Du|^{2p+2} + (C\varepsilon)\frac{|\lambda_u|^2}{\lambda^2} + |f_u(u)|\lambda^{-1})|Du|^{2p}\,. \qquad (3.4.17)
\end{aligned}$$

Similarly, if $\hat{f}$ has a quadratic growth in Du then by f.2) in F)

$$|D\hat{f}(u, Du)| \lesssim |\lambda_u(u)||Du||D^2u| + |\lambda_{uu}(u)||Du|^3 + |f_u(u)||Du|\,.$$

We then have to deal with two new terms, which can be handled by Young's inequality and the assumption (3.1.8) that $|\lambda_{uu}(u)|\lambda(u) \lesssim |\lambda_u(u)|^2$, as follows. Recall that $|\mathbf{w}| = \lambda^{-1}(u)|Du|^{2p-1}$.

$$|\lambda_u(u)||Du||D^2u||\mathbf{w}| \leq \varepsilon|Du|^{2p-2}|D^2u|^2 + C(\varepsilon)\frac{|\lambda_u(u)|^2}{\lambda^2(u)}|Du|^{2p+2}\,,$$

$$|\lambda_{uu}(u)||Du|^3|\mathbf{w}| \lesssim \frac{|\lambda_u(u)|^2}{\lambda^2(u)}|Du|^{2p+2}\,.$$

We then get the same terms as in (3.3.24) for the linear growth case.

We now look at the last integral of (3.4.16). As u is a strong solution, it satisfies the system a.e. in $\Omega \times (0, T_0)$ so that we can estimate u_t as follows:

$$|u_t| \leq |A(u)||D^2u| + |A_u(u)||Du|^2 + |\hat{f}(u, Du)|\,.$$

Hence,

$$|(\lambda^{-1}(u))_u||u_t||Du|^{2p} \lesssim \frac{|\lambda_u|}{\lambda}|D^2u||Du|^{2p} + \frac{|\lambda_u|^2}{\lambda^2}|Du|^{2p+2} + \frac{|\lambda_u|}{\lambda^2}|\hat{f}(u, Du)||Du|^{2p}\,.$$

Assuming either f.1) or f.2), the last term can be estimated by either

$$\frac{|\lambda_u|}{\lambda}|Du|^{2p+1} + \frac{|\lambda_u|}{\lambda(u)}f(u)\lambda^{-1}(u)|Du|^{2p} \text{ or } \frac{|\lambda_u|^2}{\lambda^2}|Du|^{2p+2} + \frac{|\lambda_u|}{\lambda}f(u)\lambda^{-1}(u)|Du|^{2p} .$$

Using Young's inequality, we can find a constant C such that

$$\frac{|\lambda_u|}{\lambda}f(u)\lambda^{-1}(u)|Du|^{2p} \le C|Du|^{2p+2} + C[\frac{|\lambda_u|}{\lambda}f(u)\lambda^{-1}(u)]^p .$$

Finally, because $|\lambda_u(u)|/\lambda(u)$ is bounded by the constant Λ defined in (3.4.3), we use the above inequalities and the definitions (3.4.10)–(3.4.12) to see that the integral on the right-hand side of (3.4.16) can be estimated by

$$\varepsilon\int_{T-t_0}^{T'}\int_{\Omega_t}|Du|^{2p-2}|D^2u|^2\psi^2\,dx + C(\varepsilon,\Lambda)[\mathcal{B}_p(t) + (1+(t-s)^{-2})\mathcal{C}_p(t) + C(\Lambda)\mathcal{F}_p(t) . \tag{3.4.18}$$

We then choose ε sufficiently small so that the first integral can be absorbed into the left-hand side of (3.4.16). The proof then continues as before to obtain (3.4.14). □

Proof of Theorem 3.4.1. From the energy estimate (3.4.14),

$$\begin{aligned}\mathcal{A}_p(s) + \mathcal{H}_p(s) \le{}& C\mathcal{B}_p(t) + C(1+(t-s)^{-2})\mathcal{C}_p(t)\\ &+ C\mathcal{F}_p(t) + \||DU_0|^{2p}\|_{L^1(\Omega_t)} .\end{aligned}$$

From the new definition (3.4.10) of $\mathcal{B}_p$, we can simply use the local Gagliardo–Nirenberg inequality in Corollary 2.3.2 instead of the more general version of Corollary 2.3.5 to estimate $\mathcal{B}_p$ in the above inequality, with $K(u) = u$ and $\mathbf{W} \equiv 1$. Following the proof of Lemma 3.3.8, for any given $\varepsilon_0 > 0$ if $\|u\|_{BMO(\Omega_{R_{\mu_0}})}$ is sufficiently small then

$$\mathcal{B}_p(s) \le \varepsilon_0[\mathcal{B}_p(t) + \mathcal{H}_p(t) + (t-s)^{-2}\mathcal{C}_p(t)] \quad 0 < s < t \le R_{\mu_0} .$$

Thus, we then use the assumption that u has small BMO norm in small balls to obtain a similar version of (3.3.32) in Lemma 3.3.8. Namely,

$$\mathcal{A}_p(R) + \mathcal{B}_p(R) + \mathcal{H}_p(R) \le C(1+R^{-2})\mathcal{C}_p(2R) + C\mathcal{F}_p(2R) + \||DU_0|^{2p}\|_{L^1(\Omega_{2R})} . \tag{3.4.19}$$

Hence, for sufficiently small $R < R_{\mu_0}/2$ if there is a constant $C(C_0, R, p)$ such that

$$\mathcal{C}_p(2R) \le C(C_0, R, p) , \tag{3.4.20}$$
$$\mathcal{F}_p(2R) \le C(C_0, R, p) , \tag{3.4.21}$$

then for $V = |Du|^p$, (3.4.19) implies (compare with (3.3.50))

$$\sup_{\tau\in(0,T_0)}\int_{\Omega_R}\lambda^{-1}(u)|Du|^{2p}\,dx + \iint_{Q_R}[V^2 + |DV|^2]\,dz \le C(C_0, R, p) . \tag{3.4.22}$$

This assertion holds as long as $1 - 1/p < C_*^{-1}$ for the energy estimate (3.4.14) is available.

Firstly, we will prove that (3.4.20) is self-improving so that we can iterate (3.4.22). That is, we will show that if (3.4.20) holds then, together with (3.4.22), they are true again with p, R being $\gamma_* p, R/2$ respectively for some $\gamma_* > 1$.

We recall the definition of $\mathfrak{C}_p$ in (3.4.12)

$$\mathfrak{C}_p(s) := \iint_{Q_s} (1 + |f_u(u)|\lambda^{-1}(u))|Du|^{2p}\, dz\,,$$

and the assumption (3.4.5) in the theorem

$$\iint_{Q} (1 + |f_u(u)|\lambda^{-1}(u))|Du|^{2}\, dz \le C_0\,,$$

which yields that (3.4.20) and (3.4.21) hold for $p = 1$.

To proceed, we recall our integrability assumptions (3.4.4) and (3.4.6): there exist C_0 and $r_0 > 1$, $\alpha_0 \in (0,1)$ such that

$$\sup_{\tau\in(0,T_0)} \||f_u(u)|\lambda^{-1+\alpha_0}(u)\|_{L^{r_0}(\Omega)} \le C_0\,, \tag{3.4.23}$$

$$\sup_{\tau\in(0,T_0)} \int_{\Omega\times\{\tau\}} \lambda^{\frac{\alpha_*}{1-\alpha_*}}(u)\, dx \le C_0\,. \tag{3.4.24}$$

Here, the number $\alpha_* \in (0,1)$ was introduced in the assumption SG').

Let $\gamma_1 \in (1,2)$. We write $|Du|^{2\gamma_1 p} = v^{2q_*}V^2$ with $v = |Du|^{\frac{p(\gamma_1-1)}{q_*}}$, $V = |Du|^p$ and apply (3.3.49) to get

$$\iint_{Q_R} |Du|^{2\gamma_1 p}\, dz \lesssim \sup_{(0,T_0)} \left(\int_{\Omega_R} |Du|^{2p(\gamma_1-1)}\, dx\right)^{q_*} \iint_{Q_R} [|DV|^2 + V^2]\, dz\,.$$

Concerning the first integral on the right-hand side, we apply Hölder's inequality, writing $|Du|^{2p(\gamma_1-1)} = (\lambda^{-1}|Du|^{2p})^{\gamma_1-1}\lambda^{\gamma_1-1}$,

$$\int_{\Omega_R} |Du|^{2p(\gamma_1-1)}\, dx \le \left(\int_{\Omega_R} \lambda^{-1}(u)|Du|^{2p}\, dx\right)^{\gamma_1-1} \left(\int_{\Omega_R} \lambda^{\frac{\gamma_1-1}{2-\gamma_1}}\, dx\right)^{2-\gamma_1}\,.$$

We now choose γ_1 close to 1 such that the last integral is bounded by (3.4.24). The first integral on the right-hand side is bounded by (3.4.22). Therefore,

$$\iint_{Q_R} |Du|^{2\gamma_1 p}\, dz \le C_1(C_0, R, p)\,.$$

Similarly, for any $\gamma_2 > 1$ we write $|f_u(u)|\lambda^{-1}(u)|Du|^{\gamma_2 p} = v^{2q_*}V^2$ with $v = (|f_u(u)|\lambda^{-1}(u)|Du|^{2p(\gamma_2-1)})^{\frac{1}{2q_*}}$ and $V = |Du|^p$. In order to apply (3.3.49) again here, we need to estimate the integral of v^2 over Ω_R. Assuming $\gamma_2 \in (1, q_*+1)$ and using Hölder's inequality with the exponent $q_1 = \frac{q_*}{\gamma_2-1}$, the integral of

$$v^2 = |f_u|\lambda^{-1+\frac{\gamma_2-1}{q_*}}(\lambda^{-1}|Du|^2p)^{\frac{\gamma_2-1}{q_*}}$$

is bounded by

$$\left(\int_{\Omega_R}(|f_u|\lambda^{-1+\frac{\gamma_2-1}{q_*}})^{q_1'}\,dx\right)^{\frac{1}{q_1'}}\left(\int_{\Omega_R}\lambda^{-1}(u)|Du|^{2p}\,dx\right)^{\frac{1}{q_1}}.$$

We can find γ_2 close to 1 such that $\frac{\gamma_2-1}{q_*} < \alpha_0$ and $q_1' = \frac{q_*}{q_*-\gamma_2+1} \le r_0$, the exponents in (3.4.23), so that the first integral is bounded by the assumption (3.4.23) and Hölder's inequality. The second integral is bounded by (3.4.22).

We then conclude that there is a constant $C_1(C_0, R, p)$ and some $\gamma_* = \min\{\gamma_1, \gamma_2\} > 1$ such that

$$\mathcal{C}_{\gamma_* p}(R) \le C_1(C_0, R, p).$$

Thus, $\mathcal{C}_p$ is self-improving.

Applying Lemma 3.3.10 with $s_0 = 2\alpha_*$, we can find an integer $k_0 \ge 1$ such that $N/(2\alpha_*) < \gamma_*^{k_0} \le N_1/(2\alpha_*)$, the exponent in the assumption (3.4.7) of M.0'). By Hölder's inequality, for $k = 0, \dots, k_0$ and $p_k = \gamma_*^k$, we see that $\mathcal{F}_{p_k}$ is bounded, thanks to (3.4.7).

Therefore, following the proof of Lemma 3.3.9 we can iterate the argument k_0 times, because p_k satisfies the spectral gap condition $1 - 1/p_k < C_*^{-1}$, to arrive at

$$\sup_{(0,T_0)}\int_{\Omega_R}\lambda^{-1}(u)|Du|^{2p}\,dx \le C(C_0, R, p, k_0), \quad p = \gamma_*^{k_0} \text{ and } R = 2^{-k_0}R_{\mu_0}.$$

Writing $|Du|^{\alpha_* p} = \lambda^{\alpha_*}\lambda^{-\alpha_*}|Du|^{\alpha_* p}$, by Hölder's inequality and (3.4.24) we have

$$\begin{aligned}\int_{\Omega_R}|Du|^{2\alpha_* p}\,dx &\le \left(\int_{\Omega_R}\lambda^{\frac{\alpha_*}{1-\alpha_*}}(u)\,dx\right)^{1-\alpha_*}\left(\int_{\Omega_R}\lambda^{-1}(u)|Du|^{2p}\,dx\right)^{\alpha_*}\\ &\le C(C_0, R, p).\end{aligned}$$

Summing these local estimates we obtain the global one

$$\sup_{(0,T_0)}\int_{\Omega}|Du|^{2\alpha_* p}\,dx \le C(C_0, R, p).$$

Because $2\alpha_* p = 2\alpha_* p_{k_0} > N$, the above shows that u is Hölder continuous in x and the proof can continue as before. □

One way, and perhaps the crudest one, to validate the key condition (3.4.8) on the smallest of the BMO norm of u is to use the Poincaré–Sobolev inequality as in the following corollary.

Corollary 3.4.4. *The condition* (3.4.8) *of Theorem 3.4.1 can be replaced by the following: there is a constant C_0 such that for any strong solution u to* (3.4.2),

$$\sup_{(0,T_0)}\int_{B_R}|Du(x,\tau)|^N\,dx \le C_0. \tag{3.4.25}$$

Then there exists a unique solution to (3.4.1).

Proof. We argue by contradiction. If (3.4.8) does not hold then there are sequences $\{x_n\} \subset \bar{\Omega}$, $\{\sigma_n\} \subset [0, 1]$, $\{t_n\} \subset (0, T_0)$, $\{r_n\}$, $r_n \to 0$ and a sequence of strong solutions $\{u_{\sigma_n}\}$ of (3.4.2) such that for $U_n(\cdot) = u_{\sigma_n}(\cdot, t_n)$,

$$\|U_n\|_{BMO(B_{r_n}(x_n)\cap\Omega)} > \varepsilon_0 \text{ for some } \varepsilon_0 > 0 . \tag{3.4.26}$$

By (3.4.25) we see that the sequence $\{U_n\}$ is bounded in $W^{1,N}(\Omega)$. We can then assume that U_n converges weakly to some U in $W^{1,2}(\Omega)$ and strongly in $L^2(\Omega)$. We then have $\|U_n\|_{BMO(B_R\cap\Omega)} \to \|U\|_{BMO(B_R\cap\Omega)}$ for any given ball B_R. It is easy to see that $U \in W^{1,N}(\Omega)$ and by the Poincaré–Sobolev inequality,

$$R^{-N}\int_{B_R} |U - U_R|^2\, dx \le C(N)\int_{B_R} |DU|^N\, dx ,$$

and the continuity of the integral of $|DU|^N$, that $\|U\|_{BMO(B_R\cap\Omega)} < \varepsilon_0/2$ if R is sufficiently small. Furthermore, we can assume also that x_n converges to some $x \in \bar{\Omega}$. Thus, for large n, we have $r_n < R/2$ and $x_n \in B_{R/2}(x)$. Then, for large n, $B_{r_n}(x_n) \subset B_R(x)$, and because $\|U_n\|_{BMO(B_R\cap\Omega)} \to \|U\|_{BMO(B_R\cap\Omega)}$, we have

$$\|U_n\|_{BMO(B_{r_n}(x_n)\cap\Omega)} \le \|U_n\|_{BMO(B_R(x)\cap\Omega)} \le \|U\|_{BMO(B_R(x)\cap\Omega)} + \varepsilon_0/2 < \varepsilon_0 .$$

We obtain a contradiction to (3.4.26). Thus, (3.4.8) holds and the proof is complete. □

3.5 Existence results for the general SKT system

We present an application of our main theorems in the previous sections. These examples concern cross-diffusion systems with polynomial growth data, which occur in many applications in mathematical biology and ecology that consider the effect of cross-diffusion.

For simplicity, we will discuss the case when the systems are independent of x. We consider the following system:

$$\begin{cases} u_t - \operatorname{div}(A(u)Du) = \hat{f}(u, Du) , & x \in \Omega , \\ u = 0 \text{ or } \frac{\partial u}{\partial \nu} = 0 \text{ on } \partial\Omega . \end{cases} \tag{3.5.1}$$

We embed this system in the following family of systems:

$$\begin{cases} u_t - \operatorname{div}(A(\sigma u)Du) = \hat{f}(\sigma u, \sigma Du) , & x \in \Omega, \sigma \in [0, 1] , \\ u = 0 \text{ or } \frac{\partial u}{\partial \nu} = 0 \text{ on } \partial\Omega . \end{cases} \tag{3.5.2}$$

We need to find an appropriate map $K\colon \mathbb{R}^m \to \mathbb{R}^m$ satisfying P.1) and establish two key assumptions that for *any strong solution* u of (3.5.2) $K(u)$ has small BMO norm in small balls and that $[\mathbf{W}^\alpha]_{\beta+1}$ is bounded.

We start with the simplest case $K(u) = u$ and apply Theorem 3.4.1, or more precisely, its Corollary 3.4.4. In this case, $\mathbf{W} \equiv 1$ so that we need only verify the condition (3.4.8) that any strong solution u to the family has small BMO norm in small balls: for any given $\mu_0 > 0$ there is positive R_{μ_0} sufficiently small in terms of the constants in A) and F) such that

$$\sup_{x_0 \in \bar{\Omega}, \tau \in (0,T_0)} \|u\|^2_{BMO(B_R(x_0)\cap\Omega,\mu)} \le \mu_0 . \tag{3.5.3}$$

It was proved in Corollary 3.4.4 that this condition can be verified if we can control the $L^N(\Omega)$ norm of Du.

We will assume the structural conditions A) and F) with $\lambda(u)$ have polynomial growth.

A") Assume A) and that λ satisfies the condition that there is $\lambda_0 > 0$ and $k \neq 0$ such that

$$\lambda(u) \sim (\lambda_0 + |u|)^k , \quad |\lambda_u(u)| \sim (\lambda_0 + |u|)^{k-1} .$$

From this, the condition (3.4.3) is clearly verified here as

$$\Lambda := \sup_{u \in \mathbb{R}^m} \frac{|\lambda_u(u)|}{\lambda(u)} \sim \frac{1}{\lambda_0 + |u|} < \infty . \tag{3.5.4}$$

The condition A") is inspired by the famous the SKT model [30, 44, 46] for two species with population densities u, v solving

$$\begin{cases} u_t = \Delta(u[d_1 + \alpha_{11}u + \alpha_{12}v]) + f_1(u,v) , \\ v_t = \Delta(v[d_2 + \alpha_{21}u + \alpha_{22}v]) + f_2(u,v) . \end{cases} \tag{3.5.5}$$

This system is a special case of (3.5.1) with $A(u) = P_u(u)$. In fact, it was proved in [46] that if d_i, α_{ij} are positive and

$$\alpha_{21}^2 < 8\alpha_{11}\alpha_{12} \text{ and } \alpha_{12}^2 < 8\alpha_{22}\alpha_{21}$$

then $A(u)$ satisfies the condition A) with $\lambda(u) = c + a|u| + |v|$ for some positive constants a, b, c depending on d_i, α_{ij}. Of course, A") holds with $k = 1$. In addition, the reaction terms f_i are of Lotka–Volterra type so that they are quadratics. We then consider the following condition:

F') $\bar{f}(u, Du)$ satisfies F) and for all $u \in \mathbb{R}^m$

$$f(u) \lesssim \lambda(u)|u| , \quad |f_u(u)| \lesssim \lambda(u) .$$

We then consider the following generalized SKT system with Dirichlet or Neumann boundary conditions on a bounded domain $\Omega \subset \mathbb{R}^N$ for vector-valued unknown $u\colon \Omega \times (0, T_0) \to \mathbb{R}^m$:

$$u_t = \Delta(P_i(u)) + B_i(u, Du) + f_i(u) , \quad i = 1, \dots, m . \tag{3.5.6}$$

Here, $P_i\colon \mathbb{R}^m \to \mathbb{R}$ are C^2 functions. The functions B_i, f_i are C^1 functions on $\mathbb{R}^m \times \mathbb{R}^{mn}$ and $\mathbb{R}^m$ respectively.

Of course, (3.5.6) is a special case of (3.5.1) with $A(u) = P_u(u)$. At the moment, for simplicity, we will assume that $B_i \equiv 0$ and show that Corollary 3.4.4 applies here for the system (3.5.6) under certain very mild integrability conditions of $f(u)$. The general case and much weaker integrability conditions will be discussed at length in Chapter 5.

We embed (3.5.6) in the following family of systems:

$$\begin{cases} u_t = \Delta(P(u)) + \sigma f(\sigma u)\,, & x \in \Omega, \sigma \in [0,1]\,, \\ u = 0 \text{ or } \frac{\partial u}{\partial \nu} = 0 \text{ on } \partial\Omega \\ u(x,0) = \sigma U_0(x) \text{ on } \Omega\,. \end{cases} \tag{3.5.7}$$

We first have the following a priori estimate for strong solutions to (3.5.7).

Lemma 3.5.1. *Let w be a strong solution u to* (3.5.7). *Then there is a constant C such that*

$$\sup_{\tau\in(0,T_0)} \int_{\Omega\times\{\tau\}} |\lambda(w)Dw|^2\, dx \le C\sigma^2 \int_0^{T_0}\int_\Omega |\lambda(u)\bar{f}(u)|^2\, dx\,, \tag{3.5.8}$$

where $\bar{f}(w) := [f_i(w)]_{i=1}^m$.

Proof. We test the i-th equation of the system with $(P_i(w))_t$ and sum the results to get

$$\int_\Omega \langle w_t, P_t(w)\rangle\, dx + \int_\Omega \langle DP(w), D(P_t(w))\rangle\, dx = \int_\Omega \langle \sigma\bar{f}(w), P_t(w)\rangle\, dx\,. \tag{3.5.9}$$

Because $|P_t(w)| = |A(w)w_t| \lesssim \lambda(w)|w_t|$, we have by Young's inequality

$$\langle \sigma\hat{f}(w), P_t(w)\rangle \le \varepsilon\lambda(w)|w_t|^2 + C(\varepsilon)\sigma^2\lambda(w)|\bar{f}(w)|^2\,.$$

As $\langle w_t, P_t(w)\rangle \ge \lambda(w)|w_t|^2$ and $\frac{d|DP(w)|^2}{dt} = \langle DP(w), D(P_t(w))\rangle$ (because u is a strong solution), we can choose ε to be small in the above and derive from (3.5.9) that

$$\frac{d}{dt}\int_\Omega |DP(w)|^2\, dx \le C\int_\Omega \lambda^2(w)|Dw|^2\, dx + C\sigma^2\int_\Omega \lambda(w)|\bar{f}(w)|^2\, dx\,. \tag{3.5.10}$$

By the ellipticity condition, $\lambda(w)|Dw|^2 \le \langle A(w)Dw, Dw\rangle = \langle DP(w), Dw\rangle$ so that a simple use of Young's inequality implies

$$\lambda(w)|Dw|^2 \le \frac{1}{2}\lambda^{-1}(w)|DP(w)|^2 + \frac{1}{2}\lambda(w)|Dw|^2\,,$$

and this shows that $|D(P(w))| \sim \lambda(w)|Dw|$. We then have from (3.5.10) the following Gronwall inequality $y'(t) \le Cy(t) + \alpha(t)$, where

$$y(t) = \int_{\Omega\times\{t\}} |DP(w)|^2\, dx\,, \quad \alpha(t) = \sigma^2\int_{\Omega\times\{t\}} |\lambda(w)\bar{f}(w)|^2\, dx\,.$$

We then have from the Gronwall inequality that

$$\sup_{\tau\in(0,T_0)} \int_{\Omega\times\{\tau\}} |\lambda(w)Dw|^2\, dx \le C\sigma^2 \iint_Q |\lambda(w)\bar{f}(w)|^2\, dz\,.$$

This is the desired inequality (3.5.8) and the proof is complete. □

Corollary 3.5.2. *Assume that $N = 2$, $A(u) = P_u(u)$, satisfies A") with $k > 0$ and f satisfies F). Suppose that there is a constant C_0 such that any strong solution u to* (3.5.7) *satisfies*

$$\int_0^{T_0} \int_\Omega |\lambda(u) f_i(u)|^2 \, dx \le C_0 \text{ for all } i = 1, \dots, m . \tag{3.5.11}$$

Then (3.5.6) *has a unique strong solution on $\Omega \times (0, T_0)$.*

Proof. We have to verify the condition (3.4.25) of Corollary 3.4.4 for (3.5.2), which is now

$$\begin{cases} u_t - \operatorname{div}(A(\sigma u) Du) = f(\sigma u) , & x \in \Omega, \sigma \in [0,1] , \\ u = 0 \text{ or } \frac{\partial u}{\partial \nu} = 0 \text{ on } \partial\Omega . \end{cases} \tag{3.5.12}$$

Multiplying the system by σ, we easily see that $w = \sigma u$ is the strong solution to (3.5.7). The inequality (3.5.8) then implies

$$\sup_{\tau \in (0, T_0)} \int_{\Omega \times \{\tau\}} |\lambda(w) Dw|^2 \, dx \le C\sigma^2 \iint_Q |\lambda(w) \bar{f}(w)|^2 \, dz .$$

Because $w = \sigma u$ is a strong solution to (3.5.7), the assumption (3.5.11) implies that the last integral is bounded by a constant $C(C_0)$. We then have

$$\sup_{\tau \in (0, T_0)} \int_{\Omega \times \{\tau\}} |\lambda(w) D(\sigma u)|^2 \, dx \le C\sigma^2 C(C_0) .$$

Dividing both sides by σ^2, we obtain from the above that

$$\sup_{\tau \in (0, T_0)} \int_{\Omega \times \{\tau\}} |\lambda(w) Du|^2 \, dx \le C(C_0) . \tag{3.5.13}$$

As $\lambda(w) \ge \lambda_0^k > 0$, because $k > 0$, we find

$$\sup_{\tau \in (0, T_0)} \int_{\Omega \times \{\tau\}} |Du|^2 \, dx \le C\lambda_0^{-k} C_0 , \tag{3.5.14}$$

which gives (3.4.25) of Corollary 3.4.4 when $N = 2$.

Next, we need to check the integrability assumptions in M.0'). We start with (3.4.4), (3.4.6) and (3.4.7): there exist C_0 and $r_0 > 1$, $\alpha_0 \in (0, 1)$ such that any strong solutions u of (3.5.12) satisfy

$$\sup_{\tau \in (0, T_0)} \| |f_u(u)| \lambda^{-1+\alpha_0}(u) \|_{L^{r_0}(\Omega)} \le C_0 , \tag{3.5.15}$$

$$\sup_{\tau \in (0, T_0)} \int_{\Omega \times \{\tau\}} \lambda^{\frac{\alpha_*}{1-\alpha_*}}(u) \, dx \le C_0 , \tag{3.5.16}$$

and for any p satisfying $(p-1)/p < C_*^{-1}$,

$$\iint_Q (f(u) \lambda^{-1}(u))^p \, dz \le C_0 . \tag{3.5.17}$$

By the assumption F'), $|f_u(u)|\lambda^{-1+\alpha_0}(u)$, $\lambda(u)$ and $f(u)\lambda^{-1}(u)$ can be bounded by polynomials in $|u|$. From (3.5.14) and Sobolev's embedding theorems, we see that $|u|$ belong to $L^q(\Omega)$ for any $q > 1$, because $N = 2$, so that the integrals in (3.5.15)–(3.5.17) are uniformly bounded.

Finally, we check the condition (3.4.5), which says

$$\iint_Q (1 + |f_u(u)|\lambda^{-1}(u))|Du|^2 \, dz \le C_0 .$$

By F'), this is just (3.5.14) because $|f_u(u)| \lesssim \lambda(u)$.

Thus, M.0') is verified and the proof is complete. □

The above result does not apply to the case $k < 0$ because the system becomes singular $\lambda(u) \to 0$ if $|u| \to \infty$ and (3.5.14) does not give the desired estimate (3.4.25), which implies the control of BMO norm of $K(u) = u$ in M.1'). We then consider another choice of K and apply Theorem 3.2.3. We need to check the two key assumptions in M.1). That is, for any given $\mu_0 > 0$ there is positive R_{μ_0} sufficiently small in terms of the constants in A) and F) such that

$$\sup_{x_0 \in \bar{\Omega}, \tau \in (0,T_0)} \|K(u)\|^2_{BMO(B_R(x_0)\cap\Omega,\mu)} \le \mu_0 . \tag{3.5.18}$$

Furthermore, for $\mathbf{W}_p(\sigma, x, \tau) := \lambda^{p+\frac{1}{2}}(\sigma u)|\lambda_u(\sigma u)|^{-p}$ and any p satisfying $(p-1)/p < C_*^{-1}$ there exist some $\alpha > 2/(p+2)$, $\beta < p/(p+2)$ and a constant C_0 such that

$$\sup_{\tau \in (0,T_0)} [\mathbf{W}_p^{\alpha}]_{\beta+1, B_{R_{\mu_0}}(x_0)\cap\Omega} \le C_0 .$$

To construct the map K we assume that:

L) There are m C^2 positive scalar functions λ_i, $i = 1, \dots, m$ on $\mathbb{R}^m$ satisfying

$$\lambda_i(u) \sim \lambda(u) , \quad |(\lambda_i(u))_{uu}| \lesssim |\lambda_{uu}(u)| , \tag{3.5.19}$$

$$\lambda(u)|\lambda_{uu}(u)| \lesssim |\lambda_u(u)|^2 . \tag{3.5.20}$$

We introduce the matrices

$$\mathbf{L} = \operatorname{diag}[\lambda_1(u), \dots, \lambda_m(u)] , \ \mathbf{L}_u = [D_u\lambda_i(u)]_{i=1}^m$$

and suppose that

$$|\mathbf{L}_u^{-1}| \lesssim |\lambda_u(u)|^{-1} . \tag{3.5.21}$$

We then define the map

$$K(u) = [K_i(u)]_{i=1}^m , \quad K_i(u) = \log(\lambda_i(u)) . \tag{3.5.22}$$

Remark 3.5.3. The existence of the functions λ_i in the condition L) is quite easy to verify when the system in consideration possesses some cone invariant property, which means its solutions stay in a cone $\mathbf{C}$ of $\mathbb{R}^m$. In many applications, we have $\mathbf{C} = \mathbb{R}^m_+ = \{x : x = [x_i]_i^m, \quad x_i \ge 0\}$ and $\lambda(u)$ is a polynomial satisfying $\lambda(u) \sim (1 + |u|)^k$ for some real number k. Such functions clearly satisfy (3.5.20).

Let $\alpha_1 = [1, \dots, 1]^T \in \mathbb{R}^m$. Of course, $\langle u, \alpha_1 \rangle \sim |u|$ for all $u \in \mathbf{C}$. It is clear that we can choose $m - 1$ vectors $\alpha_i = [\alpha_{ij}]_{j=1}^m$, $i = 2, \dots, m$ such that $\{\alpha_i : i = 1, \dots, m\}$ is linearly independent and $|\alpha_i - \alpha_1| < \varepsilon$ for any given $\varepsilon > 0$. Since $\langle u, \alpha_i \rangle = \langle u, \alpha_i - \alpha_1 \rangle + \langle u, \alpha_1 \rangle$ we easily see that $\langle u, \alpha_i \rangle \sim |u|$ if ε is sufficiently small.

For any $d_i > 0$ and $k \ne 0$ we define $\lambda_i(u) = d_i + \langle u, \alpha_i \rangle^k$. We have

$$\partial_{u_j}\lambda_i(u) = k(d_i + \langle u, \alpha_i \rangle)^{k-1}\alpha_{ij}\,, \qquad \partial_{u_j u_l}\lambda_i(u) = k(k-1)(d_i + \langle u, \alpha_i \rangle)^{k-2}\alpha_{ij}\alpha_{il}\,.$$

Thus, the condition (3.5.19) is satisfied with $\lambda(u) = (1 + |u|)^k$ because $d_i + \langle u, \alpha_i \rangle \sim 1 + |u|$. By the same reason, (3.5.21) holds because

$$\mathbf{L}_u^{-1} = (k\alpha)^{-1}\operatorname{diag}[(d_i + \langle u, \alpha_i \rangle)^{1-k}] \Rightarrow |\mathbf{L}_u^{-1}| \sim (1 + |u|)^{k-1}\,.$$

We see that the condition L) is validated. Note that this argument applies to any cone in $\mathbb{R}^m$ isomorphic to $\mathbb{R}^m_+$. Indeed, we consider $\mathbf{C} = A(\mathbb{R}^m_+)$ for some invertible linear transformation on $\mathbb{R}^m$. Let α_i be defined as above. For any $u \in \mathbf{C}$ we have $\langle u, A^T\alpha_i \rangle = \langle Au, \alpha_i \rangle \sim |Au| \sim |u|$ so that the vectors α_i's are now $A^T\alpha_i$.

We first have the following lemma showing that this map satisfies K) so that the local Gagliardo–Nirenberg inequality in Corollary 2.3.5, one of the key ingredients in the proof of Theorem 3.2.3, is available.

Lemma 3.5.4. *Assume L). Then the map K satisfies the condition K).*

Proof. Because $\partial_{u_j}K_i(u) = \lambda_i^{-1}(u)\partial_{u_j}\lambda_i(u)$, we have $K_u(u) = \mathbf{L}^{-1}\mathbf{L}_u$. Hence, by (3.5.19) and (3.5.21),

$$|\mathbb{K}(u)| = |K_u^{-1}(u)^T| = |K_u^{-1}(u)| \le |\mathbf{L}_u^{-1}||\mathbf{L}| \lesssim |\lambda_u(u)|^{-1}\lambda(u)\,.$$

Thus, the condition (3.2.3) in K) is verified. Also, as

$$\mathbb{K}_u^T = (K_u^{-1})_u = (\mathbf{L}_u^{-1}\mathbf{L})_u = \mathbf{L}_u^{-1}\mathbf{L}_{uu}\mathbf{L}_u^{-1}\mathbf{L} + \mathbf{L}_u^{-1}\mathbf{L}_u\,,$$

we have by the second assumption in (3.5.19),

$$|\mathbb{K}_u| = |\mathbb{K}_u^T| \le |\mathbf{L}_u^{-1}|^2|\mathbf{L}_{uu}||\mathbf{L}| + 1 \lesssim |\lambda_u|^{-2}|\lambda_{uu}|\lambda + 1\,.$$

We see that $|\mathbb{K}_u(u)|$ is bounded by a constant C_0 because of (3.5.20). Thus the condition K) is verified and the lemma is proved. □

By the assumption (3.5.20), $\lambda(u)|\lambda_{uu}(u)| \lesssim |\lambda_u(u)|^2$ and Remark 2.3.6, we see that (2.3.17) of Theorem 2.3.4, and then its local version Corollary 2.3.5 holds. We can now apply Theorem 3.2.3 to get the following result.

Corollary 3.5.5. *Assume L) and $N = 2$. Suppose that there is a constant C_0 such that*

$$|\lambda_u(u)|\lambda^{-2}(u) \le C_0 \qquad \text{for all } u \in \mathbb{R}^m\,, \tag{3.5.23}$$

$$\lambda(u) \le C_0\lambda(w) \qquad \text{if } u, w \in \mathbb{R}^m \text{ and } |w| \le |u|\,. \tag{3.5.24}$$

Also, suppose that any strong solution u to (3.5.7) *satisfies*

$$\sup_{\tau\in(0,T_0)} \|u\|_{L^1(\Omega,\mu)}\,, \quad \sup_{\tau\in(0,T_0)} \|\lambda(u)\|_{L^1(\Omega,\mu)} \le C_0\,, \tag{3.5.25}$$

$$\int_0^{T_0}\int_\Omega |\lambda(u)f_i(u)|^2\,dx \le C_0 \text{ for all } i = 1,\dots,m\,, \tag{3.5.26}$$

$$\int_0^{T_0}\int_\Omega u_i f_i(u)\,dx \le C_0 \text{ for all } i = 1,\dots,m\,. \tag{3.5.27}$$

Then (3.5.6) *has a unique strong solution on $\Omega \times (0, T_0)$.*

We note that if A") holds with $k > -1$ then (3.5.23) is valid. Indeed, we have $|\lambda_u(u)|\lambda^{-2}(u) \sim (\lambda_0 + |u|)^{-k-1}$, which is bounded by λ_0^{-k-1} if $k > -1$. If $k < 0$ and $|w| \le |u|$, we have $(\lambda_0 + |u|)^k \le (\lambda_0 + |w|)^k$, which implies $\lambda(u) \lesssim \lambda(w)$ so that (3.5.24) also holds.

Proof. By Lemma 3.5.4 the map K satisfies the condition K). We need to establish the main conditions M.1) and M.0) of Theorem 3.2.3.

We start with the first part of M.1) by showing that $K(u)$ has small BMO norm in small balls.

Again, multiplying the system (3.5.2) by σ, we see that $w = \sigma u$ is the strong solution to (3.5.7). The inequality (3.5.8) of Lemma 3.5.1 then implies

$$\sup_{\tau\in(0,T_0)} \int_{\Omega\times\{\tau\}} |\lambda(w)Dw|^2\,dx \le C\sigma^2 \iint_Q |\lambda(w)\bar{f}(w)|^2\,dz\,.$$

Because $w = \sigma u$ is a strong solution to (3.5.7), using the assumption (3.5.26) and then dividing the result by σ we obtain the following estimate:

$$\sup_{\tau\in(0,T_0)} \int_{\Omega\times\{\tau\}} |\lambda(w)Du|^2\,dx \le C(C_0)\,. \tag{3.5.28}$$

Using (3.5.21) in L), we get

$$|D(K(u))| = |K_u(u)Du| \lesssim |\mathbf{L}^{-1}\mathbf{L}_u|\lambda^{-1}(u)|\lambda(u)Du| \le \frac{|\lambda_u(u)|}{\lambda^2(u)}\lambda(u)|Du|$$

so that (3.5.23) implies $|DK(u)| \lesssim C_0\lambda(u)|Du|$. By (3.5.24), because $|w| \le |u|$, we have $\lambda(u) \lesssim \lambda(w)$ so that

$$|DK(u)| \lesssim \lambda(w)|Du|\,. \tag{3.5.29}$$

This and (3.5.28) then yield

$$\sup_{\tau\in(0,T_0)} \int_{\Omega\times\{\tau\}} |DK(u)|^2\,dx \le C(C_0)\,.$$

We then see that $K(u)$ has small BMO norm in small balls (see the proof of Corollary 3.4.4 when $N = 2$). The condition (3.5.18) is then verified.

Next, we need to show that $\mathbf{W}_p(x, \tau) := \lambda^{p+\frac{1}{2}}(u)|\lambda_u(u)|^{-p}$ is a weight. To this end, we will show that $\lambda(u)$ and $|\lambda_u(u)|$ are A_1 weights, $A_1 = \cap_{\gamma>1} A_\gamma$. For $w_1 = \log(\lambda(u))$ and $w_2 = \log(|\lambda_u(u)|^{-1})$ we have

$$|Dw_1| \le \frac{|\lambda_u(u)|}{\lambda(u)}|Du| \le \frac{|\lambda_u(u)|}{\lambda^2(u)}|\lambda(u)Du| ,$$
$$|Dw_2| \le \frac{|\lambda_{uu}(u)|}{|\lambda_u(u)|}|Du| \le \frac{|\lambda_{uu}(u)|\lambda(u)}{|\lambda_u(u)|\lambda^2(u)}|\lambda(u)Du| \le \frac{|\lambda_u(u)|}{\lambda^2(u)}|\lambda(u)Du| .$$

Using the facts that $|\lambda_u(u)|/\lambda^2(u)$ is bounded by the assumption (3.5.23) and that $\lambda(u) \lesssim \lambda(w)$, we find from the above that $|Dw_i| \lesssim \lambda(w)|Du|$ for $i = 1, 2$. This and (3.5.28) imply

$$\sup_{\tau\in(0,T_0)} \int_{\Omega\times\{\tau\}} |Dw_i|^2 \, dx \le C(C_0) .$$

Therefore, w_i's have small BMO norm in small balls. For any given $c > 0$, by Lemma 2.3.9, e^{cw_i}, and therefore $\lambda^c(u)$ and $|\lambda_u(u)|^{-c}$, are A_1 weights (in small balls). It is well known that the product of two A_1 weights is also an A_1 weight [17]. Hence, for each $\tau \in (0, T_0)$ and any power of $\mathbf{W}_p(x, \tau)$ is also an A_1 weight and the last condition in M.1) is then verified.

We now verify the condition M.0). The boundedness of the integral of $|f_u(\sigma u)|$ $\lambda^{-1}(\sigma u)$ in (3.2.6) is clear because $|f_u(u)| \lesssim \lambda(u)$. (3.2.7) holds for the exponent $\beta_0 = 1$ by our assumption (3.5.25). Since the system is independent of x, the condition (3.2.8) is not needed (Remark 4.2.9). What remains to check is the boundedness of the integral in (3.2.9), which is now (using $|f_u(u)| \lesssim \lambda(u)$ again)

$$\iint_Q \lambda(\sigma u)|Du|^2 \, d\mu d\tau .$$

To this end, we simply test the i-th equation in (3.5.2) by u_i and the ellipticity condition to get

$$\int_\Omega |u|^2 \, dx + \iint_Q \lambda(u)|Du|^2 \, d\mu d\tau \le \sum_i \iint_Q \langle u_i, f_i(u)\rangle \, dz + \int_\Omega |u(0)|^2 \, dx .$$

The right-hand side is bounded due to the assumption (3.5.27).

Theorem 3.2.3 applies here and the proof is complete. □

Remark 3.5.6. Inspired by the SKT system (3.5.5), we introduce the generalized SKT system (3.5.6). Here, we will consider the case $P_i = u_i\lambda_i(u)$ with $\lambda_i(u)$. The u_i's are population densities of the species in consideration so that $u_i \ge 0$. Hence, the system will often be positive invariant as we will see later. Let $\alpha_i = [\alpha_{ij}]_{j=1}^m$ be m linearly independent vectors in $\mathbb{R}^m$ described in Remark 3.5.3. We then have that $\langle u, \alpha_i\rangle \sim |u|$ for all i.

If $k > 0$ we define $\lambda_i(u) = d_i + \langle u, \alpha_i \rangle^k$ with $d_i > 0$. We see that the SKT system (3.5.5) is included in this case for $m = 2$, $k = 1$. In Chapter 5 we will show that $A(u) = P_u(u)$ satisfies A) with $\lambda(u) = (1 + |u|)^k$. The condition A") is then satisfied so that Corollary 3.5.2 applies to give the existence of strong solutions to the generalized SKT system (3.5.6). We note that the system (3.5.6) is degenerate when $|u| \to \infty$.

On the other hand, when $k < 0$, we define $\lambda_i(u) = (d_i + \langle u, \alpha_i \rangle)^k$. The system is now singular when $|u| \to \infty$ and Corollary 3.5.2 is no longer applicable. By Remark 3.5.3, we see that the condition L) is validated and we can use the map K defined in (3.5.22).

We consider the conditions (3.5.23) and (3.5.24) of Corollary 3.5.5. If $k > -1$ then $|\lambda_u(u)|\lambda^{-2}(u) \sim (1 + |u|)^{-k-1}$, which is bounded by λ_0^{-k-1}. Thus, (3.5.23) holds. If $k < 0$, we have $(1 + |u|)^k \le (1 + |w|)^k$, which implies $\lambda(u) \lesssim \lambda(w)$, so that (3.5.24) also holds. Therefore, we can apply Corollary 3.5.5 to obtain the existence of strong solutions to the generalized SKT system (3.5.6).

4 The elliptic systems

In this chapter, we consider the elliptic counterpart of the parabolic systems studied in the previous chapter:

$$\begin{cases} -\operatorname{div}(A(x,u)Du) = \hat{f}(x,u,Du)\,, & x \in \Omega\,, \\ u = 0 \text{ or } \frac{\partial u}{\partial \nu} = 0 \text{ on } \partial\Omega\,. \end{cases} \tag{4.0.1}$$

We will discuss the existence of strong solutions to (4.0.1). We say that u is a strong solution if u solves (4.0.1) a.e. on Ω with $Du \in L^{\infty}_{\text{loc}}(\Omega)$ and $D^2u \in L^2_{\text{loc}}(\Omega)$.

Again, without the boundedness assumption on the solutions of (4.0.1), we will consider the structural conditions A) and F) in Section 3.1 of the previous chapter.

If (4.0.1) is variational, i.e., it is the Euler–Lagrange system of some functional of vectorial function u, then the existence of its *weak* solutions has been investigated a great deal in the literature. The reader is referred to the book by Giusti [16] for an excellent historical account and variational methods in dealing with this problem.

Once a weak solution exists, the next question of its regularity immediately arises in the hope of showing that the solution is in fact classical. This is a fundamental and hard problem in the theory of regularity of elliptic systems. Unless (4.0.1) is a scalar equation, i.e, $m = 1$, the best answer so far for the case $m > 1$ is the *partial* regularity result: bounded weak solutions are Hölder continuous almost everywhere.

In general, the system (4.0.1) is not variational and it is well known that the best approach so far is the use of fixed-point theories to (4.0.1). This means we consider solutions to the system as fixed points of some map T defined on a base space $\mathbf{X}$. The main advantage of this approach is that we can obtain a classical/strong solution to (4.0.1) immediately by choosing an appropriate base space $\mathbf{X}$. However, the hardest part of this theory is the uniform estimate of the $\mathbf{X}$ norm of a given fixed point of T or, equivalently, a solution to (4.0.1).

This method was extensively developed in the literature and the reader can consult the survey by Amann [3] or the comprehensive book by Zeidler [47]. These two references discussed the existence problem via the fixed-point problems of monotone operators and the general Leray–Schauder theory. The first approach works well if certain comparison principles are available for (4.0.1). This is the case if the matrix A is *diagonal* and the right-hand side of the system satisfies some appropriate (and restrictive) structural conditions.

This chapter follows closely the same approach as in the preceding one. The absence of the temporal derivative makes the argument somewhat simpler. Since there are some important variations and our wish is to make it as independent as possible (for readers who are not interested in the evolution systems), we will present most of the needed details.

https://doi.org/10.1515/9783110608762-004

4.1 The main result

We consider the following system:

$$\begin{cases} -\operatorname{div}(A(x,u)Du) = \hat{f}(x,u,Du)\,, & x \in \Omega\,, \\ u = 0 \text{ or } \frac{\partial u}{\partial \nu} = 0 \text{ on } \partial\Omega\,. \end{cases} \tag{4.1.1}$$

First of all, we will assume that the data A and $\hat{f}$ satisfy the structural conditions A) and F).

The spectral gap condition SG) in the preceding chapter now takes a weaker form. We assume that:

ESG) There is $s_0 \geq 2$ such that $(N - s_0)/N < C_*^{-1}$.

In SG), we supposed that $s_0 = 2$. The compensation is the mild integrability condition (4.1.7) below.

We embed the system (4.1.1) in the following family of systems:

$$\begin{cases} -\operatorname{div}(A(x,\sigma u)Du) = \hat{f}(x,\sigma u,\sigma Du)\,, & x \in \Omega, \sigma \in [0,1]\,, \\ u = 0 \text{ or } \frac{\partial u}{\partial \nu} = 0 \text{ on } \partial\Omega\,. \end{cases} \tag{4.1.2}$$

For any strong solution u of (4.1.2) we will consider the following integrability hypotheses.

EM.0) There exists a constant C_0 such that for some $r_0 > 1$ and $\beta_0 \in (0,1)$

$$\||f_u(\sigma u)|\lambda^{-1}(\sigma u)\|_{L^{r_0}(\Omega,\mu)} \leq C_0\,, \tag{4.1.3}$$

$$\|u^{\beta_0}\|_{L^1(\Omega,\mu)},\ \|\lambda^{\beta_0}(u)\|_{L^1(\Omega,\mu)} \leq C_0\,, \tag{4.1.4}$$

$$\int_\Omega |f_u(\sigma u)|(1+|u|^2)\,d\mu \leq C_0\,, \tag{4.1.5}$$

$$\int_\Omega (|f_u(\sigma u)| + \lambda(\sigma u))|Du|^2\,d\mu \leq C_0\,. \tag{4.1.6}$$

In addition, we assume that

$$\int_\Omega \lambda^{-\alpha}(u)\,d\mu \leq C_0,\ \frac{1}{\alpha} = \frac{2}{s_0} - 1 + \frac{2}{N} = \frac{2(N_* - s_0)}{N_* s_0}\,, \tag{4.1.7}$$

where s_0 is the number introduced in ESG).

Remark 4.1.1. Here, we stated the above integrability conditions in the most general situation at the moment. In many applications of our main results, we will see that they can be greatly simplified. For example, if $\hat{f}$ is independent of x then (4.1.4)–(4.1.5) can be dropped. In particular, for the generalized SKT system we have $|f_u(u)| \lesssim \lambda(u)$ so that (4.1.3) is obvious with $r_0 = \infty$.

Remark 4.1.2. On the other hand, in many physical models, the ellipticity 'constant' $\lambda(u)$ in A) and the 'reaction' $f(u)$ have polynomial growths like $\lambda(u) \sim (\lambda_0 + |u|)^k$

and $f(u) \sim |u|^{k_0}\lambda(u)$ for some positive constants λ_0, k and k_0. In these cases, $\lambda(u)$ is bounded from below by a positive constants (i.e., the system is regularly uniform elliptic); hence (4.1.7) holds with $\alpha = \infty$ so that we can take $s_0 = N_*$ in the spectral gap condition ESG). We then see that ESG) is void when $N \le 4$ and reduced to $\frac{N-4}{N-2} < C_*^{-1}$ if $N > 4$. The condition (4.1.3) simply requires $\|u\|_{L^{l_0}(\Omega)} \le C_0$ for some $l_0 > k_0$. The conditions in (4.1.4) then come from this if we choose β_0 sufficiently small. In fact, we will show in Remark 4.2.10 that (4.1.4) and (4.1.5) imply (4.1.6) if $k_0 = 1$. The condition EM.0) then reduces to just a single condition (4.1.3).

Most importantly, as in Chapter 3, we assume the same condition K) on the existence of a map $K\colon \mathbb{R}^m \to \mathbb{R}^m$. Namely, there is a C^1 map $K\colon \mathbb{R}^m \to \mathbb{R}^m$ such that $\mathbb{K}(u) = (K_u(u)^{-1})^T$ exists and $\mathbb{K}_u \in L^\infty(\mathbb{R}^m)$. Furthermore, for all $u \in \mathbb{R}^m$,

$$|\mathbb{K}(u)| \lesssim \lambda(u)|\lambda_u(u)|^{-1} \,. \tag{4.1.8}$$

We consider the following condition, which relates to our use of the Gagliardo–Nirenberg inequalities in Chapter 2.

EM.1) For any given $\mu_0 > 0$ there is positive R_{μ_0} sufficiently small in terms of the constants in A) and F) such that

$$\sup_{x_0 \in \bar{\Omega}} \|K(u)\|^2_{BMO(B_R(x_0)\cap\Omega,\mu)} \le \mu_0 \,. \tag{4.1.9}$$

Furthermore, for $\mathbf{W}_p(\sigma, x) := \lambda^{p+\frac{1}{2}}(\sigma u)|\lambda_u(\sigma u)|^{-p}$ and any p such that $1 - 1/p < C_*^{-1}$ we have $[\mathbf{W}_p^{\frac{2}{3}}]_{\frac{4}{3}} \le C_0$.

The main theorem of this chapter is the following.

Theorem 4.1.3. *Assume A), F), AR) and K). Suppose that any strong solution u to* (4.1.2) *satisfies EM.0), EM.1) uniformly in* $\sigma \in [0, 1]$.

Then the system (4.1.1) *has a strong solution.*

4.2 Proof of the main theorem

As in Chapter 3, the proof of Theorem 4.1.3 relies on the Leray–Schauder fixed-point theorem. We obtain strong solutions u of (4.1.1) as fixed points of a nonlinear map defined on an appropriate Banach space **X**. The proof will be based on several lemmas and we will sketch the main steps of the proof below.

4.2.1 The setting of fixed-point theory

We will prove in the next section that there exist $p > N/2$ and a constant M_* depending only on the constants in A), F) and C_0 of EM.0) such that any strong solution u of (4.1.2) will satisfy

$$\|Du\|_{L^{2p}(\Omega,\mu)} \le M_* . \tag{4.2.1}$$

This implies that there are positive constants α, M_0 such that

$$\|u\|_{C^\alpha(\Omega)} \le M_0 . \tag{4.2.2}$$

For $\sigma \in [0,1]$ and any given $u \in W^{1,2}(\Omega, \mathbb{R}^m)$ satisfying (4.2.2) we consider the following linear systems:

$$\begin{cases} -\operatorname{div}(A(x,\sigma u)Dw) + Lw = \hat{\mathbf{f}}(\sigma, x, u, w) + Lu & x \in \Omega , \\ w = 0 \text{ or } \frac{\partial w}{\partial \nu} = 0 \text{ on } \partial\Omega , \end{cases} \tag{4.2.3}$$

where $\hat{\mathbf{f}}(\sigma, x, u, w)$ is linear in w, Dw and defined by

$$\hat{\mathbf{f}}(\sigma, x, u, w) = F^{(\sigma)}(x, u, Du)Dw + f^{(\sigma)}(x, u)w + \hat{f}(x, 0, 0) \tag{4.2.4}$$

with

$$F^{(\sigma)}(x, u, Du) := \int_0^1 \partial_\zeta F(\sigma, u, tDu)\, dt , \quad f^{(\sigma)}(x, u) := \int_0^1 \partial_u F(\sigma, x, tu, 0)\, dt . \tag{4.2.5}$$

Also, L in (4.2.3) is a suitable positive definite matrix depending on the constant M_0 such that the system (4.2.3) has a unique weak solution w if u satisfies (4.2.2). We then define $T_\sigma(u) = w$ and apply the Leray–Schauder fixed-point theorem to establish the existence of a fixed point of T_1. It is clear from (4.2.4) and (4.2.5) that $\hat{f}(x, \sigma u, \sigma Du) = \hat{\mathbf{f}}(\sigma, x, u, u)$. Therefore, from the definition of T_σ we see that a fixed point of T_σ is a weak solution of (4.1.2). By an appropriate choice of $\mathbf{X}$, we will show that these fixed points are strong solutions of (4.1.2), and so a fixed point of T_1 is a strong solution of (4.1.1).

From the proof of the Leray–Schauder fixed-point theorem in [15, Theorem 11.3], we need to find some ball B_M of radius M centered at 0 of $\mathbf{X}$ such that $T_\sigma : \bar{B}_M \to \mathbf{X}$ is compact and that T_σ has no fixed point on the boundary ∂B_M of B_M. The topological degree $\operatorname{ind}(T_\sigma, B_M)$ is then well defined and invariant by homotopy so that $\operatorname{ind}(T_1, B_M) = \operatorname{ind}(T_0, B_M)$. It is easy to see that the latter is nonzero because the corresponding (linear) system

$$\begin{cases} -\operatorname{div}(A(x,0)Du) = \hat{\mathbf{f}}(x, 0, 0) & x \in \Omega , \\ u = 0 \text{ or } \frac{\partial u}{\partial \nu} = 0 \text{ on } \partial\Omega , \end{cases}$$

has a unique solution in B_M. Hence, T_1 has a fixed point in B_M.

Therefore, the theorem is proved as we will establish the following claims.

Claim 1 There exist a Banach space $\mathbf{X}$ and $M > 0$ such that the map $T_\sigma : \bar{B}_M \to \mathbf{X}$ is well defined and compact.

Claim 2 T_σ has no fixed point on the boundary of $\bar{B}_M$. That is, $\|u\|_{\mathbf{X}} < M$ for any fixed points of $u = T_\sigma(u)$.

The following lemma establishes Claim 1.

Lemma 4.2.1. *Suppose that there exist $p^* > N$ and a constant M_* such that any strong solution u of* (4.1.2) *satisfies*

$$\|Du\|_{W^{1,p^*}(\Omega,\mu)} \le M_* \, . \tag{4.2.6}$$

Then, there exist $M, \beta > 0$ such that for $\mathbf{X} = C^{\beta}(\Omega) \cap W^{1,2}(\Omega)$ the map $T_\sigma \colon \bar{B}_M \to \mathbf{X}$ is well defined and compact for all $\sigma \in [0, 1]$. Moreover, T_σ has no fixed points on ∂B_M.

Proof. For some constant $M_0 > 0$ we consider $u \colon \Omega \to \mathbb{R}^m$ satisfying

$$\|u\|_{C(\Omega)} \le M_0, \ \|Du\|_{L^2(\Omega)} \le M_0 \, , \tag{4.2.7}$$

and write the system (4.2.3) as a linear elliptic system for w

$$-\operatorname{div}(\mathbf{a}(x)Dw) + \mathbf{b}(x)Dw + \mathbf{g}(x)w + Lw = \mathbf{f}(x) \, , \tag{4.2.8}$$

where $\mathbf{a}(x) = A(x, \sigma u)$, $\mathbf{b}(x) = F^{(\sigma)}(x, u, Du)$, $\mathbf{g}(x) = f^{(\sigma)}(x, u)$, and $\mathbf{f}(x) = \hat{f}(x, 0, 0) + Lu$.

The matrix $\mathbf{a}(x)$ is then regular elliptic with uniform ellipticity constants by A), AR) because u is bounded ((3.1.4)). From the theory of *linear* elliptic systems it is well known that if the operator

$$\mathcal{L}(w) = -\operatorname{div}(\mathbf{a}(x)Dw) + \mathbf{g}(x)w + Lw$$

is monotone and if there exist positive constants m and q such that

$$\||\mathbf{b}|^2\|_{L^q(\Omega)} \, , \quad \|\mathbf{g}\|_{L^q(\Omega)} \, , \quad \|\mathbf{f}\|_{L^q(\Omega)} \le m \, , \quad q > n/2 \, , \tag{4.2.9}$$

then the system (4.2.8) has a unique weak solution w.

It is easy to find a matrix L such that $\mathcal{L}(w)$ is monotone because the matrix $\mathbf{a}$ is regular elliptic and $\mathbf{g}$ is bounded (see below). We just need to choose a positive definite matrix L satisfying $\langle Lw, w\rangle \ge l_0|w|^2$ for some positive and sufficiently large l_0 in terms of M_0.

Next, we will show that (4.2.9) holds by F) and (4.2.7). We consider the two cases f.1) and f.2). If f.1) holds then from the definition (4.2.5) there is a constant $C(|u|)$ such that

$$|\mathbf{b}(x)| = |F^{(\sigma)}(x, u, \zeta)| \le C(|u|), \ |\mathbf{g}(x)| = |f^{(\sigma)}(x, u)| \le C(|u|) \, .$$

From (4.2.7), we see that $\|u\|_\infty \le M_0$ and so there is a constant m depending on M_0 such that (4.2.9) holds for any q and n.

If f.2) holds then

$$|F^{(\sigma)}(x, u, \zeta)| \le C(|u|)|\zeta|^{\delta}, \ |f^{(\sigma)}(x, u)| \le C(|u|) \, . \tag{4.2.10}$$

Therefore, $\||\mathbf{b}|^2\|_{L^q(\Omega)}$ is bounded by $C\|Du\|_{L^2(\Omega)}$ for $q = 1/\delta > N/2$. Thus, (4.2.7) implies the condition (4.2.9).

In both cases, (4.2.8) (or (4.2.3)) has a unique weak solution w. We then define $T_\sigma(u) = w$. Moreover, from the regularity theory of linear systems, $w \in C^{\alpha_0}(\Omega)$ for some $\alpha_0 > 0$ depending on M_0.

The bound in the assumption (4.2.6) implies that u is Hölder continuous and provides positive constants α, $C(M_*)$ such that $\|u\|_{C^\alpha(\Omega)} \le C(M_*)$. Also, the assumption (4.2.39) and AR), that $\lambda(u)$, ω are bounded from below, yields that $\|Du\|_{L^2(\Omega)} \le C(C_0)$. Thus, there is a constant M_1, depending on M_*, C_0 such that any strong solution u of (4.1.2) satisfies

$$\|u\|_{C^\alpha(\Omega)} \le M_1, \ \|Du\|_{L^2(\Omega)} \le M_1 . \tag{4.2.11}$$

It is well known that there is a constant $c_0 > 1$, depending on α and the diameter of Ω, such that $\|\cdot\|_{C^\beta(\Omega)} \le c_0\|\cdot\|_{C^\alpha(\Omega)}$ for all $\beta \in (0, \alpha)$. We now let M_0, the constant in (4.2.7), be $M = (c_0 + 1)M_1$.

Define $\mathbf{X} = C^\beta(\Omega) \cap W^{1,2}(\Omega)$ for some positive $\beta < \min\{\alpha, \alpha_0\}$. The space $\mathbf{X}$ is equipped with the norm

$$\|u\|_{\mathbf{X}} = \max\{\|u\|_{C^\beta(\Omega)}, \|Du\|_{L^2(\Omega)}\} .$$

We now see that T_σ is well defined and maps the ball $\bar{B}_M$ of $\mathbf{X}$ into $\mathbf{X}$. Moreover, from the definition $M = (c_0+1)M_1$, it is clear that T_σ has no fixed point on the boundary of B_M because such a fixed point u satisfies (4.2.11), which implies $\|u\|_{\mathbf{X}} \le c_0 M_1 < M$.

Finally, we need only show that T_σ is compact. If u belongs to a bounded set K of B_M then $\|u\|_{\mathbf{X}} \le C(K)$ for some constant $C(K)$ and there is a constant $C_1(K)$ such that $\|w\|_{C^{\alpha_0}(\Omega)} \le C_1(K)$. Thus $T_\sigma(K)$ is compact in $C^\beta(\Omega)$ because $\beta < \alpha_0$. So, we need only show that $T(K)$ is precompact in $W^{1,2}(\Omega)$. The proof is similar to the lines at the end of the proof of Lemma 3.3.1 and will be omitted. Hence, $T_\sigma : \mathbf{X} \to \mathbf{X}$ is a compact map. The proof is complete. □

4.2.2 A priori estimates

We now turn to the hardest part of the proof, and provide a uniform estimate for the fixed points of T_σ to justify the key assumption (4.2.6) of Lemma 4.2.1. The proof is long and will be divided into many lemmas and we will sketch the main ideas below, as we did in Chapter 3.

- Lemma 4.2.2 is quite standard and shows that the fixed points of T_σ are strong solutions of (4.1.2).
- Lemma 4.2.3 establishes an energy estimate of Du for any strong solutions u of the system. Roughly speaking, we establish that

$$\int_{\Omega_s} \lambda(u)|Du|^{2p-2}|D^2u|^2 \, d\mu \le \int_{\Omega_t} \Phi^2(u)|Du|^{2p+2} \, d\mu + \text{lower-order terms} ,$$

where $\Omega_s = B_s \cap \Omega$, B_s, B_t are two concentric balls with radii $s < t$, and $\Phi(u) = (\sqrt{\lambda(u)})_u$.

- We wish to remove the integral of $\Phi^2(u)|Du|^{2p+2}$ on the right-hand side of the above energy estimate. This is the key part of the proof and it can be done by using the local Gagliardo–Nirenberg inequalities in the previous chapter. Here, under the assumption K), we present such an inequality in Lemma 3.3.7. With this and the smallness assumption on the BMO norms in EM.1), we achieve the goal and obtain in Lemma 4.2.5 a better estimate

$$\int_{\Omega_R} \lambda(u)|Du|^{2p-2}|D^2u|^2\,d\mu + \int_{\Omega_R} \Phi^2(u)|Du|^{2p+2}\,d\mu \lesssim \text{lower-order terms}$$

  if R is sufficiently small.
- In Lemma 4.2.6 and Lemma 4.2.8, we will show that the above inequality is *self-improving* so that we can iterate it finite times, starting with the integrability assumption EM.0), to obtain the desired bound for Du assumed in (4.2.6) of Lemma 4.2.1. The proof of our main theorem then follows.

Hence, we begin with the following lemma.

Lemma 4.2.2. *A fixed point of T_σ is also a strong solution of* (4.1.2).

Proof. If u is a fixed point of T_σ in $\mathbf{X}$ then it solves (4.1.2) weakly and is continuous by the definition of $\mathbf{X}$. Thus, u is bounded and belongs to VMO(Ω). By AR), the system (4.1.2) is regular elliptic. The lemma can easily follow from the results in [16] on elliptic systems. Alternatively, we can adapt the argument for parabolic systems in Lemma 3.3.2 to our current situation to prove the lemma. □

Thanks to Lemma 4.2.2, we need only consider a strong solution u of (4.1.2) and establish (4.2.6) for some $p^* > N$. Because the data of (4.2.3) satisfy the structural conditions A), F) with the same set of constants and the assumptions of the theorem are assumed to be uniform for all $\sigma \in [0, 1]$, we will only present the proof for the case $\sigma = 1$ in the sequel.

Let u be a strong solution of (4.1.1) on Ω. We begin with an energy estimate for Du. For $p \geq 1$ and any ball B_s with center $x_0 \in \bar{\Omega}$ we denote $\Omega_s = B_s \cap \Omega$ and

$$\mathcal{H}_p(s) := \int_{\Omega_s} \lambda(u)|Du|^{2p-2}|D^2u|^2\,d\mu\,, \tag{4.2.12}$$

$$\mathcal{B}_p(s) := \int_{\Omega_s} \frac{|\lambda_u(u)|^2}{\lambda(u)}|Du|^{2p+2}\,d\mu\,, \tag{4.2.13}$$

$$\mathcal{C}_p(s) := \int_{\Omega_s} (|f_u(u)| + \lambda(u))|Du|^{2p}\,d\mu\,, \tag{4.2.14}$$

and

$$\mathcal{F}_p(s) := \int_{\Omega_s} |f(u)||Du|^{2p-1}(|D\omega_0|\omega_0)\,dx\,. \tag{4.2.15}$$

The following lemma establishes an energy estimate for Du.

Lemma 4.2.3. *Assume A), F). Let u be any strong solution of* (4.1.2) *on Ω and $p \geq 1$ satisfying the spectral gap condition*

$$1 - 1/p < C_*^{-1} . \tag{4.2.16}$$

There is a constant C, which depends only on the parameters in A) and F), such that for any two concentric balls B_s, B_t with center $x_0 \in \bar{\Omega}$ and $s < t$,

$$\mathcal{H}_p(s) \leq C\mathcal{B}_p(t) + C(1 + (t - s)^{-2})[\mathcal{C}_p(t) + \mathcal{F}_p(t)] . \tag{4.2.17}$$

Proof. Since u is a strong solution we can differentiate the system in x to see that u weakly solves

$$-\operatorname{div}(A(x, u)D^2u + A_u(x, u)DuDu + A_x(x, u)Du) = D\hat{f}(x, u, Du) . \tag{4.2.18}$$

For any two concentric balls B_s, B_t, with $s < t$, let ψ be a cutoff function for B_s, B_t. That is, ψ is a C^1 function satisfying $\psi \equiv 1$ in B_s and $\psi \equiv 0$ outside B_t and $|D\psi| \leq 1/(t - s)$.

Let $\mathbf{w} = |Du|^{2p-2}Du$. We test (4.2.18) with $\mathbf{w}\psi^2\eta$ and obtain, using integration by parts and rearranging,

$$\begin{aligned}
&\int_\Omega \langle A(x, u)D^2u, D(\mathbf{w}\psi^2)\rangle \, dx \\
&= \int_\Omega \langle -A_u(x, u)DuDu - A_x(x, u)Du, D(\mathbf{w}\psi^2)\rangle \, dx \\
&\quad + \int_\Omega \langle D\hat{f}(x, u, Du), \mathbf{w}\rangle \psi^2 \, dx .
\end{aligned} \tag{4.2.19}$$

The spectral gap condition (4.2.16) and Lemma B.1.3 then provide a positive constant c_0 such that

$$\begin{aligned}
\langle A(x, u)D^2u, D(\mathbf{w})\rangle &= \langle A(x, u)D^2u, D(|Du|^{2p-2}Du)\rangle \\
&\geq c_0\lambda(u)|Du|^{2p-2}|D^2u|^2\omega .
\end{aligned}$$

We derive from (4.2.19) the following inequality:

$$\begin{aligned}
&\int_{\Omega_t} \lambda(u)|Du|^{2p-2}|D^2u|^2\psi^2\omega \, dx \\
&\leq C\int_{\Omega_t} |\langle A(x, u)D^2u, \mathbf{w}D(\psi^2)\rangle| \, dx \\
&\quad + C\int_{\Omega_t} [|A_u(x, u)||Du|^2 + |A_x(x, u)||Du|]|D(\mathbf{w}\psi^2)| \, dx \\
&\quad + C\int_{\Omega_t} |D\hat{f}(x, u, Du)||\mathbf{w}|\psi^2 \, dx .
\end{aligned} \tag{4.2.20}$$

The integrands on the right-hand side are estimated exactly the same way as in Lemma 3.3.3. We see that the integrals on the right-hand side of (4.2.20) can be estimated by

$$\varepsilon\int_{\Omega_t} \lambda(u)|Du|^{2p-2}|D^2u|^2\psi^2 \, d\mu + C(\varepsilon)[\mathcal{B}_p(t) + (1 + (t - s)^{-2})(\mathcal{C}_p(t) + \mathcal{F}_p(t))] . \tag{4.2.21}$$

We now choose ε sufficiently small so that the first integral can be absorbed into the left-hand side of (4.2.20). We then obtain (4.3.10) and complete the proof. □

Remark 4.2.4. Again, as in Remark 3.3.4, we note that the presence of $\mathcal{F}_p$ in the energy estimate is due to the dependence on x of $\hat{f}$. Otherwise, i.e., $\omega \equiv 1$, we have $\mathcal{F}_p \equiv 0$.

We wish to remove the integral $\mathcal{B}_p$ on the right-hand side of (4.3.10). We achieve this goal by using Lemma 3.3.7 of Chapter 3 and get the following lemma, which provides a much better energy estimate.

Lemma 4.2.5. *In addition to the assumptions of Lemma 4.2.3, we suppose that EM.1) holds for some $p \ge 1$. That is, there exists a constant C_0 such that*

$$[\mathbf{W}_p^{\frac{2}{3}}]_{\frac{4}{3},\Omega} \le C_0\,, \tag{4.2.22}$$

and for any given $\mu_0 > 0$ there is a positive R_{μ_0} sufficiently small such that

$$\sup_{x_0\in\bar{\Omega},\tau\in(0,T_0)} \|K(u)\|^2_{BMO(\Omega_R(x_0),\mu)} \le \mu_0\,. \tag{4.2.23}$$

Then for sufficiently small μ_0 there is a constant C depending only on C_0 and the constants in A) and F) such that for $2R < R_{\mu_0}$ we have

$$\mathcal{B}_p(R) + \mathcal{H}_p(R) \le C(1 + R^{-2})[\mathcal{C}_p(2R) + \mathcal{F}_p(2R)]\,. \tag{4.2.24}$$

Proof. Recall the energy estimate (4.3.10) in Lemma 4.2.3:

$$\mathcal{H}_p(s) \le C\mathcal{B}_p(t) + C(1 + (t-s)^{-2})[\mathcal{C}_p(t) + \mathcal{F}_p(t)]\,, \tag{4.2.25}$$

for any $0 < s < t$.

We apply Lemma 3.3.7 to estimate $\mathcal{B}_p(t)$, the integral on the right-hand side of (4.2.25). We compare the definitions (3.3.27) and (3.3.28) with those in (4.2.12)–(4.2.14) to see that

$$\mathcal{B}_p(t) = I_1(t, x_0),\ \ \mathcal{C}_p(t) = I_0(t, x_0),\ \ \mathcal{H}_p(t) = I_2(t, x_0)\,.$$

Hence, for any $\varepsilon > 0$ we can use (3.3.29) of Lemma 3.3.7 to get

$$\mathcal{B}_p(s) \le C_{\varepsilon,U,\mathbf{W}}[\mathcal{B}_p(t) + \mathcal{H}_p(t) + (t-s)^{-2}\mathcal{C}_p(t)\,.$$

From this point, we can follow the proof of Lemma 3.3.8, using the local Gagliardo–Nirenberg inequality of Lemma 3.3.7, to assert that if R_{μ_0}, or μ_0, is sufficiently small then for any $R < R_{\mu_0}/2$,

$$\mathcal{B}_p(R) \le \mathcal{B}_p\left(\frac{3}{2}R\right) \le C_4(1 + R^{-2})[\mathcal{C}_p(2R) + \mathcal{F}_p(2R)]\,.$$

Combining this and the energy estimate (4.2.25) with $s = R$ and $t = \frac{3}{2}R$, we arrive at

$$\mathcal{B}_p(R) + \mathcal{H}_p(R) \le C_4(1 + R^{-2})[\mathcal{C}_p(2R) + \mathcal{F}_p(2R)]\,.$$

This is (4.2.24) and the proof is complete. □

We now have the following lemma giving our desired uniform bound for strong solutions.

Lemma 4.2.6. *Assume as in Lemma 4.2.5. We assume also the integrability condition EM.0). Then there exist $p > N/s_0$ and a constant M_* depending only on the parameters of A) and F), μ_0, R_{μ_0}, C_0 and the geometry of Ω such that*

$$\int_\Omega |Du|^{ps_0}\,dx \le M_* \,. \tag{4.2.26}$$

Proof. First of all, by the condition AR), there is a constant C_ω such that $|D\omega_0| \le C_\omega \omega_0$ and therefore we have from the the definition (4.2.15) that

$$\mathcal{F}_p(s) \le C_\omega \int_{\Omega_s} (\lambda(u)|Du|^{2p} + |f(u)||Du|^{2p-1})\,d\mu \,.$$

We write $f(u)|Du|^{2p-1} = f(u)f_u(u)|^{\frac{1-2p}{2p}}|f_u(u)|^{\frac{2p-1}{2p}}||Du|^{2p-1}$. By Young's inequality,

$$|f(u)||Du|^{2p-1} \lesssim |f_u(u)||Du|^{2p} + (|f(u)||f_u(u)|^{-1})^{2p}|f_u(u)| \,.$$

It follows from the assumption (3.2.11) in F) that $|f(u)| \lesssim (1+|u|)|f_u(u)|$ so that

$$\mathcal{F}_p(s) \le C_\omega \int_{\Omega_s} [(\lambda(u) + |f_u|)|Du|^{2p} + |f_u(u)||u|^{2p} + |f_u|]\,d\mu \,.$$

From the integrability assumption (4.2.38), we know that $\|f_u(u)\|_{L^1(\Omega)} \le C_0$. We then derive from (4.2.24) that there is $R_{\mu_0} > 0$ such that if $0 < R \le R_{\mu_0}/2$ then

$$\mathcal{B}_p(R) + \mathcal{H}_p(R) \le C(1 + R^{-2})[\mathcal{C}_p(2R) + \mathcal{F}_{*,p}(2R) + C_0] \,, \tag{4.2.27}$$

where

$$\mathcal{F}_{*,p}(s) := \int_{\Omega_s} |u|^{2p}|f_u(u)|\,d\mu \,. \tag{4.2.28}$$

Let $V = \lambda^{\frac{1}{2}}(u)|Du|^p$. We observe that

$$|DV|^2 \lesssim \frac{|\lambda_u(u)|^2}{\lambda(u)}|Du|^{2p+2} + \lambda(u)|Du|^{2p-2}|D^2u|^2 \,.$$

From the definitions of $\mathcal{B}_p$, $\mathcal{H}_p$ and $\mathcal{C}_p$ we deduce from (4.2.27) that

$$\int_{\Omega_R} [V^2 + |DV|^2]\,d\mu \le C(1 + R^{-2})[\mathcal{C}_p(2R) + \mathcal{F}_{*,p}(2R) + C_0] \,. \tag{4.2.29}$$

Now, assume that for some $p \ge 1$ we can find a constant $C(C_0, R, p)$ such that

$$\mathcal{C}_p(2R) + \mathcal{F}_{*,p}(2R) \le C(C_0, R, p) \,. \tag{4.2.30}$$

Then, (4.2.29) and (4.2.30) yield that

$$\int_{\Omega_R} [V^2 + |DV|^2]\,d\mu \le C(C_0, R, p), \quad V = \lambda^{\frac{1}{2}}(u)|Du|^p \,. \tag{4.2.31}$$

Note that the assumptions (4.2.38) and (4.2.39) in EM.0),

$$\int_\Omega (|f_u(u)| + \lambda(u))|Du|^2\, d\mu, \quad \int_\Omega |f_u(u)||u|^2\, d\mu \le C_0\,,$$

imply (4.2.30) holds for $p = 1$.

In the technical Lemma 4.2.8 following this proof, we will show that the assumption (4.2.30) is self-improving in the following sense: if (4.2.30) holds for some $p \ge 1$ then, together with its consequence (4.2.31), we find some *fixed* $\gamma_* > 1$ such that (4.2.30) holds again for the new exponent $\gamma_* p$ and R being $R/2$. Hence, (4.2.31) is valid if $\gamma_* p$ satisfies the gap condition (4.2.16) and the energy estimate of Lemma 4.2.3) continues to hold.

We now apply Lemma 3.3.10 to see that we can choose γ_* obtained in the above argument smaller and close to 1 and find an integer $k_0 \ge 1$ such that for $k = 0, \dots, k_0$ the exponents $p_k = \gamma_*^k$ verify the gap condition (4.2.16), $1 - 1/p_k < C_*^{-1}$, and $p_{k_0} s_0 > N$. Thus, we can iterate (4.2.31) k_0 times to see that (4.2.31) holds for $p = \gamma_*^{k_0}$. It follows that there is a constant C depending only on the parameters of A) and F), μ_0, R_{μ_0} and k_0 such that for $V = \lambda^{\frac{1}{2}}(u)|Du|^p$,

$$\int_{\Omega_{R_0}} (V^2 + |DV|^2)\, d\mu \le C\,, \quad R_0 = 2^{-k_0} R_{\mu_0}\,. \tag{4.2.32}$$

Summing the above inequalities over a finite covering of balls B_{R_0} for Ω, we find a constant C, depending also on the geometry of Ω, and obtain

$$\int_\Omega (V^2 + |DV|^2)\, d\mu \le C\,. \tag{4.2.33}$$

Sobolev's inequality then implies

$$\int_\Omega \lambda^{N_*/2}(u)|Du|^{N_* p}\, d\mu = \int_\Omega V^{N_*}\, d\mu \le C\,. \tag{4.2.34}$$

We now write $|Du|^{ps_0} = \lambda^{\frac{s_0}{2}}(u)|Du|^{ps_0}\lambda^{\frac{-s_0}{2}}(u)$ and use Hölder's inequality to get

$$\int_\Omega |Du|^{ps_0}\, d\mu \le \left(\int_\Omega \lambda^{\frac{N_*}{2}}(u)|Du|^{N_* p}\, d\mu\right)^{\frac{s_0}{N_*}} \left(\int_\Omega \lambda^{\frac{-s_0 N_*}{2(N_*-s_0)}}(u)\, d\mu\right)^{1-\frac{s_0}{N_*}}.$$

The first integral is bounded by (4.2.34). The second is bounded by the integrability assumption (4.2.40) on $\lambda(u)$.

$$\int_\Omega \lambda^{-\alpha}(u)\, d\mu \le C_0, \quad \frac{1}{\alpha} = \frac{2}{s_0} - 1 + \frac{2}{N} = \frac{2(N_* - s_0)}{N_* s_0}\,. \tag{4.2.35}$$

Thus, we find a constant C such that

$$\int_\Omega |Du|^{ps_0}\, d\mu \le C\,.$$

We then obtain the desired global estimate (4.2.26) as $ps_0 = \gamma_*^{k_0} s_0 > N$. The lemma is proved. □

Remark 4.2.7. If $\lambda(u) \geq \lambda_0$ for some $\lambda_0 > 0$ then we can take $\alpha = \infty$ in (4.2.40). That is, $s_0 = N_*$ in the spectral gap assumption ESG). In this case, ESG) reads $1 - N * /N = (N-4)/(N-2) < C_*^{-1}$, which is clearly weaker than SG) and void if $N \leq 4$.

Thus, to conclude the proof, we need to show that (4.2.30) is self-improving in the following lemma. For the convenience of the reader, we repeat the integrability conditions in EM.0) here: There exists a constant C_0 such that for some $r_0 > 1$ and $\beta_0 \in (0,1)$

$$\| |f_u(\sigma u)| \lambda^{-1}(\sigma u) \|_{L^{r_0}(\Omega,\mu)} \leq C_0 \,, \tag{4.2.36}$$

$$\|u^{\beta_0}\|_{L^1(\Omega,\mu)}, \ \|\lambda^{\beta_0}(u)\|_{L^1(\Omega,\mu)} \leq C_0 \,, \tag{4.2.37}$$

$$\int_\Omega |f_u(\sigma u)|(1+|u|^2)\, d\mu \leq C_0 \,, \tag{4.2.38}$$

$$\int_\Omega (|f_u(\sigma u)| + \lambda(\sigma u))|Du|^2\, d\mu \leq C_0 \,. \tag{4.2.39}$$

We also assume that

$$\int_\Omega \lambda^{-\alpha}(u)\, d\mu \leq C_0, \quad \frac{1}{\alpha} = \frac{2}{s_0} - 1 + \frac{2}{N} = \frac{2(N_* - s_0)}{N_* s_0}\,, \tag{4.2.40}$$

where s_0 is the number introduced in ESG).

Lemma 4.2.8. *Assume as in Lemma 4.2.6. The quantities $\mathfrak{C}_p(R)$ and $\mathfrak{F}_{*,p}(R)$ are self-improving in the sense that: Suppose that for some $p \geq 1$ we can find a constant $C(C_0, R, p)$ such that*

$$\mathfrak{C}_p(2R) + \mathfrak{F}_{*,p}(2R) \leq C(C_0, R, p)\,, \tag{4.2.41}$$

then there exist a fixed $\gamma_* > 1$ and a new constant $C_1(C_0, R, p)$ such that

$$\mathfrak{C}_{\gamma_* p}(R) + \mathfrak{F}_{*,\gamma_* p}(R) \leq C_1(C_0, R, p)\,. \tag{4.2.42}$$

In the proof of this lemma, we will repeatedly make use of $\pi_* = N_* = 2N/(N-2)$ and $q_* = 1 - \frac{2}{\pi_*}$, and the following inequality:

$$\int_{\Omega_R} v^{2q_*}|V|^2\, d\mu \lesssim \left(\int_{\Omega_R} v^2\, d\mu\right)^{q_*} \int_{\Omega_R} [|DV|^2 + V^2]\, d\mu\,. \tag{4.2.43}$$

To see this, we use Hölder's inequality

$$\int_\Omega v^{2q_*}|V|^2\, d\mu \leq \left(\int_\Omega v^2\, d\mu\right)^{1-\frac{2}{\pi_*}} \left(\int_\Omega |V|^{\pi_*}\, d\mu\right)^{\frac{2}{\pi_*}}.$$

The second factor on the right-hand side will be estimated by the Poincaré–Sobolev inequality PS)

$$\left(\int_\Omega |V|^{\pi_*}\, d\mu\right)^{\frac{2}{\pi_*}} \lesssim \int_\Omega |DV|^2\, d\mu + \int_\Omega |V|^2\, d\mu\,.$$

We arrive at (4.2.43).

Proof of Lemma 4.2.8. We established in the proof of Lemma 4.2.6 that (4.2.41) and (4.2.29) imply

$$\int_{\Omega_R} [V^2 + |DV|^2]\, d\mu \le C(C_0, R, p)\,, \quad V = \lambda^{\frac{1}{2}}(u)|Du|^p\,. \tag{4.2.44}$$

Sobolev's inequality then yields

$$\int_{\Omega_R} \lambda^{N_*/2}(u)|Du|^{N_* p}\, d\mu = \int_{\Omega_R} V^{N_*}\, d\mu \le C(C_0, R, p)\,. \tag{4.2.45}$$

We now write $|Du|^{2p} = \lambda(u)|Du|^{2p}\lambda^{-1}(u)$ and use Hölder's inequality to get, noting $1 - \frac{2}{N_*} = \frac{2}{N}$,

$$\int_\Omega |Du|^{2p}\, d\mu \le \left(\int_\Omega \lambda^{\frac{N_*}{2}}(u)|Du|^{N_* p}\, d\mu\right)^{\frac{2}{N_*}} \left(\int_\Omega \lambda^{\frac{-N}{2}}(u)\, d\mu\right)^{\frac{2}{N}}\,.$$

The first integral on the right-hand side is bounded by (4.2.45) and the second is bounded by a constant due to the assumption (4.2.40). Indeed, (4.2.40) states that $\lambda^{-1}(u) \in L^\alpha(\Omega)$ with $\frac{1}{\alpha} = \frac{2}{s_0} - 1 + \frac{2}{N}$. As $s_0 \ge 2$, we have $\alpha \ge N/2$ so that $\lambda^{-1}(u) \in L^{\frac{N}{2}}(\Omega)$. We conclude that

$$\int_{\Omega_R} |Du|^{2p}\, d\mu \le C(C_0, R, p)\,. \tag{4.2.46}$$

We now consider first the quantity $\mathfrak{C}_{\gamma p}(R)$ for some $\gamma > 1$. Recalling the definition (4.2.14), $\mathfrak{C}_{\gamma p}(R)$ is the integral of $(|f_u(u)| + \lambda(u))|Du|^{2\gamma p}$.

For any $\gamma \in (1, q_* + 1)$ we write $|f_u(u)||Du|^{\gamma 2p} = v^{2q_*} V^2$ with

$$v = (|f_u(u)|\lambda^{-1}(u)|Du|^{2p(\gamma-1)})^{\frac{1}{2q_*}} \text{ and } V = \lambda^{\frac{1}{2}}(u)|Du|^p\,.$$

In order to apply (4.2.43), we need to estimate the integrals of v^2 and $|DV|^2 + V^2$ over Ω_R. A bound for the latter is already given in (4.2.41). Concerning v^2, we use Hölder's inequality with the exponent $q_1 = \frac{q_*}{\gamma-1}$ to get

$$\int_{\Omega_R} v^2\, d\mu \le \left(\int_{\Omega_R} (|f_u(u)|\lambda^{-1}(u))^{q_1'}\, d\mu\right)^{\frac{1}{q_1'}} \left(\int_{\Omega_R} |Du|^{2p}\, d\mu\right)^{\frac{1}{q_1}}\,. \tag{4.2.47}$$

We can find $\gamma = \gamma_*$ close to 1 such that $q_1' = \frac{q_*}{q_* - \gamma_* + 1} \le r_0$, the exponent in (4.2.36) of the assumption EM.0), which states: there exist C_0 and $r_0 > 1$ such that

$$\||f_u(u)|\lambda^{-1}(u)\|_{L^{r_0}(\Omega,\mu)} \le C_0\,. \tag{4.2.48}$$

Hence, the first integral in (4.2.47) is bounded by the assumption (4.2.48). The second integral is bounded by (4.2.46). We obtain

$$\int_{\Omega_R} |f_u(u)||Du|^{2\gamma_* p}\, d\mu \le C_1(C_0, R, p)\,.$$

In the same way, we replace $|f_u(u)|$ with $\lambda(u)$ in the above argument (and of course take r_0 in (4.2.48) to be ∞) to get

$$\int_{\Omega_R} \lambda(u)|Du|^{2\gamma_* p}\, d\mu \le C_1(C_0, R, p)\,.$$

We then conclude that there is a constant $C_1(C_0, R, p)$ and some $\gamma_* > 1$ such that

$$\mathcal{C}_{\gamma_* p}(R) \le C_1(C_0, R, p)\,.$$

We now turn to $\mathcal{F}_{*,p}(R)$ defined by (4.2.28), the integral over Ω_R of $|f_u(u)||u|^{2p}$. The argument is similar to the above treatment of the integral of $|f_u(u)||Du|^{2p}$ in $\mathcal{C}_p$ with Du being replaced by $|u|$ as in the proof of Lemma 3.3.11. We present the details of the proof. In order to use (4.2.43) again we need only estimate the following two integrals:

$$\int_{\Omega_R} |u|^{2p}\, d\mu, \quad \int_{\Omega_R} [|DV|^2 + V^2]\, d\mu, \qquad V = \lambda^{\frac{1}{2}}(u)|u|^p\,. \tag{4.2.49}$$

We consider the first integral and use Sobolev's inequality to get

$$\int_{\Omega_R} |u|^{2p}\, d\mu \lesssim \int_{\Omega_R} |D(|u|^p)|^2\, d\mu + \left(\int_{\Omega_R} |u|^{p\beta}\, d\mu\right)^{\frac{2}{\beta}}\,.$$

Because $|D(|u|^p)|^2 \sim |u|^{2p-2}|Du|^2 \le \varepsilon|u|^{2p} + C(\varepsilon)|Du|^{2p}$, we can choose ε sufficiently small to conclude from the above that

$$\int_{\Omega_R} |u|^{2p}\, d\mu \lesssim \int_{\Omega_R} |Du|^{2p}\, d\mu + \left(\int_{\Omega_R} |u|^{p\beta}\, d\mu\right)^{\frac{2}{\beta}}\,.$$

The first integral on the right-hand side is bounded by (4.2.44). Taking β such that $\beta p = \beta_0$, the exponent in the assumption (4.2.37), the second integral is bounded by the assumption (4.2.37), which reads

$$\|u^{\beta_0}\|_{L^1(\Omega,\mu)},\ \|\lambda^{\beta_0}(u)\|_{L^1(\Omega,\mu)} \le C_0\,. \tag{4.2.50}$$

Concerning the second integral in (4.2.49) with $V = \lambda^{\frac{1}{2}}(u)|u|^p$, we will show that there is a constant $C(C_0)$ such that

$$\int_{\Omega_R} [|DV|^2 + V^2]\, d\mu \le C(C_0)\,. \tag{4.2.51}$$

We use the fact that $|\lambda_u(u)||u| \lesssim \lambda(u)$ and Young's inequality to see that

$$\begin{aligned} |DV|^2 &\lesssim \lambda(u)|D(|u|^p)|^2 + |\lambda_u(u)|^2\lambda^{-1}(u)|Du|^2|u|^{2p} \\ &\lesssim \lambda(u)|D(|u|^p)|^2 + \lambda(u)|Du|^2|u|^{2p-2} \\ &\lesssim C(\varepsilon)\lambda(u)|Du|^{2p} + \varepsilon\lambda(u)|u|^{2p}\,. \end{aligned} \tag{4.2.52}$$

Now, using Sobolev's inequality, we have

$$\int_{\Omega_R} V^2\, d\mu \lesssim \int_{\Omega_R} |DV|^2\, d\mu + \left(\int_{\Omega_R} V^\beta\, d\mu\right)^{\frac{2}{\beta}}\,.$$

Applying Hölder's inequality to $V^\beta = \lambda^{\beta/2}(u)|u|^{p\beta}$ and using (4.2.50), we will get

$$\|V^\beta\|_{L^1(\Omega_R)} \le \|\lambda^{\beta/2}(u)\|_{L^2(\Omega_R)}\||u|^{p\beta}\|_{L^2(\Omega_R)} \le C(C_0)$$

if β is sufficiently small such that $\beta \le \min\{\beta_0, \beta_0/2p\}$. So,

$$\int_{\Omega_R} (|DV|^2 + V^2)\, d\mu \lesssim \int_{\Omega_R} |DV|^2\, d\mu + C(C_0)\,.$$

Choosing ε sufficiently small in (4.2.52), we obtain from the above that

$$\int_{\Omega_R} (|DV|^2 + V^2)\, d\mu \lesssim \int_{\Omega_R} \lambda(u)|Du|^{2p}\, d\mu + C(C_0)\,.$$

By (4.2.44) the integral over Ω_R of $\lambda(u)|Du|^{2p}$ is bounded by a constant $C(C_0, R, p)$. Therefore, the integral in (4.2.51) is bounded by a constant $C(C_0, R, p)$. We then find constants $\gamma_* > 1$ and $C_1(C_0, R, p)$ such that $\mathcal{F}_{*,\gamma_* p}(R) \le C_1(C_0, R, p)$ and thus $\mathcal{F}_{*,p}(R)$ is self-improving.

We complete the proof of the lemma. □

Remark 4.2.9. From Remark 3.3.4, $\mathcal{F}_p \equiv 0$ if $\hat{f}$ is independent of x. In this case, we do not have the term $\mathcal{F}_{*,p}$ in our consideration. Thus, the conditions (4.2.37)–(4.2.38) in EM.0) can be dropped as they were used only for the estimation of $\mathcal{F}_{*,p}$.

Remark 4.2.10. On the other hand, we note that the condition EM.0) has been used only to show that the quantity $\mathcal{C}_p + \mathcal{F}_{*,p}$ is bounded for $p = 1$ so that we can start the arguments of Lemma 4.2.6 and Lemma 4.2.8 from $p = 1$. This condition can be simplified as we will see that (4.2.37) and (4.2.38) imply part of (4.2.39). Namely,

$$\int_\Omega \lambda(\sigma u)|Du|^2\, d\mu \le C_0\,.$$

We test the system with u to get

$$\int_\Omega \lambda(\sigma u)|Du|^2\, dx \le C\int_\Omega \lambda(\sigma u)|u|^2\, dx + \int_\Omega f(\sigma u)|u|^2\, dx\,. \tag{4.2.53}$$

We note that F) implies

$$f(\sigma u)|u| \lesssim |u|(1 + \sigma|u|)|f_u(\sigma u)| \lesssim (1 + |u|^2)|f_u(\sigma u)|$$

so that the last integral in (4.2.53) is bounded by some constant $C(C_0)$ thanks to the assumption.

Now, using the interpolation Sobolev's inequality for $V = \lambda^{\frac{1}{2}}(\sigma u)|u|$, we have

$$\int_{\Omega_R} V^2\, d\mu \lesssim \varepsilon\int_{\Omega_R} |DV|^2\, d\mu + C(\varepsilon, \beta)\left(\int_{\Omega_R} V^\beta\, d\mu\right)^{\frac{2}{\beta}}\,.$$

Because $|\lambda_u||u| \lesssim \lambda(u)$, we see that $|DV|^2 \lesssim \lambda(\sigma u)|Du|^2$. Using the above with ε, β sufficiently small in (4.2.53) and the hypothesis (4.2.37) we obtain

$$\int_\Omega \lambda(\sigma u)|Du|^2\, dx \le C(C_0)\,.$$

We are ready to provide the proof of the main theorem of this section.

Proof of Theorem 4.1.3. It is now clear that the assumptions EM.0) and EM.1) of the theorem allow us to apply Lemma 4.2.6 and obtain a priori uniform bounds for any continuous strong solution u of (4.1.2). The uniform estimate (4.2.26) shows that the assumption (4.2.6) of Lemma 4.2.1 holds true, with $p^* = ps_0 > N$, so that the map T_σ is well defined and compact on a ball $\bar{B}_M$ of $\mathbf{X}$ for some M depending on the bound M_* provided by Lemma 4.2.6. Combining this with Lemma 4.2.2, the fixed points of T_σ are strong solutions of the system (4.1.2) so that T_σ does not have a fixed point on the boundary of $\bar{B}_M$. Thus, by the Leray–Schauder fixed-point theorem, T_1 has a fixed point in B_M, which is a strong solution to (4.1.1). This solution is unique because u, Du are bounded and (4.1.1) is now regular elliptic. The proof is complete. □

4.3 The simpler case

We consider the case when the data of the system does not depend explicitly on x. In this case $\omega \equiv 1$ and $d\mu = dx$, the Lebesgue measure.

We then consider in this section the following system:

$$\begin{cases} -\operatorname{div}(A(u)Du) = \hat{f}(u, Du)\,, & x \in \Omega\,, \\ u = 0 \text{ or } \frac{\partial u}{\partial \nu} = 0 \text{ on } \partial\Omega\,. \end{cases} \tag{4.3.1}$$

We embed this system in the following family of systems:

$$\begin{cases} -\operatorname{div}(A(\sigma u)Du) = \hat{f}(\sigma u, \sigma Du)\,, & x \in \Omega, \sigma \in [0,1]\,, \\ u = 0 \text{ or } \frac{\partial u}{\partial \nu} = 0 \text{ on } \partial\Omega\,. \end{cases} \tag{4.3.2}$$

The general result in the previous section relies on the existence of a general map K and the boundedness of the weight $\mathbf{W}$ to verify the main assumption M.1), so that some version of the local Gagliardo–Nirenberg inequalities in Chapter 2 is available. It is quite delicate to construct such a map and the checking of M.1) is complicated. Under slightly stronger structural assumptions of the systems, which often appear in applications, we will show in this section that we can take the map K to be the identity map and $\mathbf{W} = 1$, and hence the condition M.1) is greatly simplified.

We assume the following spectral gap condition:

SG') If $N > 4$ then $1 - N_*/N = (N-4)/(N-2) < C_*^{-1}$.

Most importantly, we assume in this section the following:

A') In addition to A), we assume also that

$$\Lambda := \sup_{u \in \mathbb{R}^m} \frac{|\lambda_u(u)|}{\lambda(u)} < \infty\,. \tag{4.3.3}$$

It is easy to see that (4.3.3) is satisfied if $\lambda(u)$ is a polynomial in $|u|$ and $\lambda(u) \geq \lambda_0$ for some constant $\lambda_0 > 0$.

We consider the following integrability condition in place of M.0):

EM.0') There exist C_0 and $r_0 > 2N/(N+2)$ such that

$$\| |f_u(u)|\lambda^{-1}(u) \|_{L^{r_0}(\Omega)} \le C_0 \,, \tag{4.3.4}$$

$$\int_\Omega (1 + |f_u(u)|\lambda^{-1}(u))|Du|^2 \, dx \le C_0 \,. \tag{4.3.5}$$

Theorem 4.3.1. *Assume that A'), F) and SG') hold and that any strong solution u to the family (4.3.2) has small BMO norm in small balls: for any given $\mu_0 > 0$ there is positive R_{μ_0} sufficiently small in terms of the constants in A) and F) such that*

$$\sup_{x_0 \in \bar{\Omega}, \tau \in (0, T_0)} \|u\|^2_{BMO(B_R(x_0)\cap\Omega)} \le \mu_0 \,. \tag{4.3.6}$$

Furthermore, u satisfies EM.0') uniformly for $\sigma \in [0, 1]$.

Then there is a unique strong solution u to (4.3.1).

The proof of this theorem follows that of Theorem 4.1.3. The main difference here is a new energy estimate for Du. We introduce the following integrals:

$$\mathcal{H}_p(s) := \int_{\Omega_s} |Du|^{2p-2}|D^2u|^2 \, dx \,, \tag{4.3.7}$$

$$\mathcal{B}_p(s) := \int_{\Omega_s} |Du|^{2p+2} \, dx \,, \tag{4.3.8}$$

$$\mathcal{C}_p(s) := \int_{\Omega_s} (1 + |f_u(u)|\lambda^{-1}(u))|Du|^{2p} \, dx \,. \tag{4.3.9}$$

We begin with the following lemma (compare to that of Lemma 4.2.3).

Lemma 4.3.2. *Assume A), F). Let u be any strong solution of* (4.1.1) *on Ω and p satisfying* (4.2.16), $1 - 1/p < C_*^{-1}$.

For any $\varepsilon_0 > 0$, there are positive constants C, $C(\varepsilon_0)$, which depend only on the parameters in A) and F) and Λ, such that for any two concentric balls B_s, B_t with center $x_0 \in \bar{\Omega}$ and $s < t$,

$$\mathcal{H}_p(s) \le C\mathcal{B}_p(t) + C(1 + (t-s)^{-2})\mathcal{C}_p(t) \,. \tag{4.3.10}$$

Proof. The proof is similar to Lemma 4.2.3. Let $\mathbf{w} = \lambda^{-1}(u)|Du|^{2p-2}Du$. We revisit the proof of Lemma 4.2.3 and test (4.2.18) with $\mathbf{w}\psi^2\eta$ and obtain

$$\begin{aligned} &\int_{\Omega_t} |Du|^{2p-2}|D^2u|^2\psi^2 \, dx \\ &\le C\int_{\Omega_t} |\langle A(u)D^2u, |Du|^{2p-2}DuD(\lambda^{-1}(u)\psi^2)\rangle| \, dx \\ &+ C\int_{\Omega_t} (|A_u(u)||Du|^2|D(\mathbf{w}\psi^2)| + |D\hat{f}(u, Du)||\mathbf{w}|\psi^2) \, dx \,. \end{aligned} \tag{4.3.11}$$

The integrands on the right-hand side will be treated by Young's inequality in the same way as in the proof of Lemma 3.3.3. We do not have the term u_t involved here so that the calculation will be much simpler. Because $|\Lambda_u(u)|/\Lambda(u)$ is bounded by the constant Λ defined in (4.3.3), we use the definitions (4.3.7)–(4.3.9) to see that the integral on the right-hand side of (4.3.11) can be estimated by

$$\varepsilon\int_{\Omega_t} |Du|^{2p-2}|D^2u|^2\psi^2\,dx + C(\varepsilon,\Lambda)[\mathcal{B}_p(t) + (1+(t-s)^{-2})\mathcal{C}_p(t)\,. \tag{4.3.12}$$

We then choose ε sufficiently small so that the first integral can be absorbed into the left-hand side of (4.3.11) and we obtain (4.3.10). □

Proof of Theorem 4.3.1. We now see that a simpler version of Lemma 4.2.5 is available under our current setting with $K(u) = u$ and $\mathbf{W} \equiv 1$. Indeed, from the definitions (4.3.7) and (4.3.7), we can simply use the local Gagliardo–Nirenberg inequality in Corollary 2.3.2 to estimate $\mathcal{B}_p$. The assumption that u has small BMO norm in small balls then yields a similar version of (4.2.24) as in Lemma 4.2.5. Namely,

$$\mathcal{B}_p(R) + \mathcal{H}_p(R) \le C(1+R^{-2})\mathcal{C}_p(2R)\,. \tag{4.3.13}$$

Let $V = |Du|^p$. From the definitions of $\mathcal{B}_p$ and $\mathcal{C}_p$, we see that $|DV|^2 + V^2 \sim \mathcal{B}_p + \mathcal{C}_p$. Therefore, (4.3.13) yields

$$\int_{\Omega} (|DV|^2 + V^2)\,dx \le C(1+R^{-2})\mathcal{C}_p(2R)\,. \tag{4.3.14}$$

Hence, for sufficiently small $R < R_{\mu_0}/2$ if there is a constant $C(C_0, R, p)$ such that

$$\mathcal{C}_p(2R) \le C(C_0, R, p)\,, \tag{4.3.15}$$

then (4.3.14) implies (compare with (4.2.44))

$$\int_{\Omega_R} [V^2 + |DV|^2]\,dx \le C(C_0, R, p)\,. \tag{4.3.16}$$

Again, we will prove that (4.3.15) is self-improving so that we can iterate (4.3.16). We recall the definition of $\mathcal{C}_p$ in (4.3.9):

$$\mathcal{C}_p(s) := \int_{\Omega_s} (1 + |f_u(u)|\lambda^{-1}(u))|Du|^{2p}\,dx\,,$$

and the assumption (4.3.5) in the theorem

$$\int_{\Omega} (1 + |f_u(u)|\lambda^{-1}(u))|Du|^2\,dx \le C_0\,,$$

which shows that (4.3.15) holds for $p = 1$.

Let $\gamma_1 = N_*/2 > 1$. By Sobolev's inequality and (4.3.16), we have

$$\int_{\Omega_R} |Du|^{2\gamma_1 p}\,dx = \int_{\Omega_R} V^{N_*}\,dx \le C_1(C_0, R, p)\,. \tag{4.3.17}$$

On the other hand, for any $\gamma_2 \in (2\gamma_1, 1)$ we use Hölder's inequality to see that for $q = 2\gamma_1/(2\gamma_1 - \gamma_2)$,

$$\int_{\Omega_R} |f_u|\lambda^{-1}(u)|Du|^{2\gamma_2}\,dx \le \left(\int_{\Omega_R} (|f_u|\lambda^{-1})^q\,dx\right)^{\frac{1}{q}} \left(\int_{\Omega_R} |Du|^{2\gamma_1 p}\,dx\right)^{\frac{2\gamma_1}{\gamma_2}}.$$

The last integral is bounded by (4.3.17). The first integral on the right is also bounded by our integrability assumption (4.3.4) if $q \le r_0$. Indeed, we assumed that there exists $r_0 > 2N/(N+2)$ such that

$$\| |f_u(u)|\lambda^{-1}(u)\|_{L^{r_0}(\Omega)} \le C_0\,.$$

As $q = 2\gamma_1/(2\gamma_1 - \gamma_2)$ decreases to r_0 when $\gamma_2 \downarrow 1$, we see that there is $\gamma_2 > 1$ such that $q < r_0$.

We then conclude that there is a constant $C_1(C_0, R, p)$ and some $\gamma_* = \min\{\gamma_1, \gamma_2\} > 1$ such that

$$\mathcal{C}_{\gamma_* p}(R) \le C_1(C_0, R, p)\,.$$

Thus, $\mathcal{C}_p$ is self-improving and (4.3.16) holds for $V = |Du|^{p_1}$ if $p_1 = \gamma_* p$ and $1 - 1/p_1 < C_*^{-1}$.

If $N \le 4$ then with $p = 1$ the above argument provides $\gamma_* > 1$ and close to 1 such that $1 - 1/\gamma_* < C_*^{-1}$ and thus (4.3.16) holds with $V = |Du|^{\gamma_*}$. We then have from (4.3.17) that

$$\int_{\Omega_R} |Du|^{2\gamma_1\gamma_*}\,dx \le C_1(C_0, R),\ \ \gamma_1 = N_*/2\,. \tag{4.3.18}$$

The proof is done in this case as $2\gamma_1\gamma_* > N_* \ge N$.

If $N > 4$, the condition SG') reads $1 - N_*/N < C_*^{-1}$. Arguing as in the proof of Lemma 3.3.10 with $s_0 = N_*$, we can choose γ_* in the above argument smaller and close to 1 and an integer $k_0 \ge 1$ such that $N < \gamma_*^{k_0} N_*$. We then iterate the argument k_0 times to arrive at

$$\int_{\Omega_R} |Du|^{N_* p}\,dx \le C(C_0, R, p, k_0),\ \ p = \gamma_*^{k_0} \text{ and } R = 2^{-k_0} R_{\mu_0}\,.$$

Summing these local estimates we obtain the global one

$$\int_{\Omega} |Du|^{N_* p}\,dx \le C(C_0, R, p)\,.$$

Because $N_* p > N$, the above shows that u is Hölder continuous and the proof can continue as before. □

Remark 4.3.3. The coefficient C of $\mathcal{B}_p$ in the energy estimate (4.3.10) is proportional to Λ^2 so that the smallness condition in (4.3.6) of the theorem can be replaced by

$$\Lambda^2 \sup_{x_0\in\bar{\Omega},\tau\in(0,T_0)} \|u\|^2_{BMO(B_R(x_0)\cap\Omega)} \le \mu_0\,. \tag{4.3.19}$$

This observation is helpful in some situations where Λ is already small.

To validate the key condition (4.3.6) on the smallest of the BMO norm of u, we have the following version of Corollary 3.4.4.

Corollary 4.3.4. *The condition (4.3.6) of Theorem 4.3.1 can be replaced by the following: There is a constant C_0 such that for any strong solution u to (4.3.2),*

$$\int_{B_R} |Du(x)|^N \, dx \le C_0 \,. \tag{4.3.20}$$

Then there exists a unique solution to (4.3.1).

4.4 Proof of the theorem on the general SKT system

In this section, we present some applications of our technical main theorems in the previous sections. We need introduce appropriate maps K satisfying the condition K) and validate the condition EM.1).

For simplicity, we will discuss the case when the systems are independent of x. We consider the following system:

$$\begin{cases} -\operatorname{div}(A(u)Du) = \hat{f}(u, Du)\,, & x \in \Omega\,, \\ u = 0 \text{ or } \frac{\partial u}{\partial \nu} = 0 \text{ on } \partial\Omega\,. \end{cases} \tag{4.4.1}$$

We embed this system in the following family of systems:

$$\begin{cases} -\operatorname{div}(A(\sigma u)Du) = \hat{f}(\sigma u, \sigma Du)\,, & x \in \Omega, \sigma \in [0,1]\,, \\ u = 0 \text{ or } \frac{\partial u}{\partial \nu} = 0 \text{ on } \partial\Omega\,. \end{cases} \tag{4.4.2}$$

We need to find an appropriate map $K\colon \mathbb{R}^m \to \mathbb{R}^m$ satisfying P.1) and establish two key assumptions that for *any strong solution* u of (4.4.2) $K(u)$ has small BMO norm in small balls and that $[\mathbf{W}^\alpha]_{\beta+1}$ is bounded.

We start with the simplest case $K(u) = u$ and apply Theorem 4.3.1, or more precisely, its Corollary 4.3.4. In this case, $\mathbf{W} \equiv 1$ so that we need only verify the condition (4.3.6) that any strong solution u to the family has small BMO norm in small balls: for any given $\mu_0 > 0$ there is positive R_{μ_0} sufficiently small in terms of the constants in A) and F) such that

$$\sup_{x_0 \in \bar{\Omega}, \tau \in (0, T_0)} \|u\|^2_{BMO(B_R(x_0)\cap\Omega,\mu)} \le \mu_0\,. \tag{4.4.3}$$

It was proved in Corollary 3.4.4 that this condition can be verified if we can control the $L^N(\Omega)$ norm of Du.

We will assume the structural conditions A) and F) with $\lambda(u)$ has polynomial growth.

A") Assume A) and that λ satisfies the condition that there is $\lambda_0 > 0$ and $k \neq 0$ such that

$$\lambda(u) \sim (\lambda_0 + |u|)^k, \quad |\lambda_u(u)| \sim (\lambda_0 + |u|)^{k-1}\,.$$

From this, the condition (4.3.3) is clearly verified here as

$$\Lambda := \sup_{u \in \mathbb{R}^m} \frac{|\lambda_u(u)|}{\lambda(u)} \sim \frac{1}{\lambda_0 + |u|} < \infty\,. \tag{4.4.4}$$

We also consider the following condition:

F') $\bar{f}(u, Du)$ satisfies F) and for all $u \in \mathbb{R}^m$

$$f(u) \lesssim \lambda(u)|u|, \quad |f_u(u)| \lesssim \lambda(u)\,.$$

Our first result is just a restatement of Corollary 4.3.4.

Corollary 4.4.1. *Assume A"), F') and ESG'). Assume that there is a constant C_0 such that any strong solution u of* (4.4.2) *satisfies*

$$\|u\|_{L^1(\Omega)} \le C_0\,, \tag{4.4.5}$$

$$\|Du\|_{L^N(\Omega)} \le C_0\,. \tag{4.4.6}$$

Then there is a strong solution to (4.4.1).

Proof. The integrability condition (4.3.4) in EM.0') is verified by (4.4.5) because $|f_u|\lambda^{-1}(u) \in L^\infty(\Omega)$. By the same reason, (4.3.5) reduces to the boundedness of $\|Du\|_{L^2(\Omega)}$, which is given by that of $\|Du\|_{L^N(\Omega)}$ in (4.4.6) because $N \ge 2$. The last bound also yields (4.3.20) Corollary 4.3.4. The proof is complete. □

To continue, for $q \ge 0$ and any $U \in C^1(\Omega)$ we apply the interpolation inequality

$$\|w\|^2_{L^2(\Omega,\mu)} \le \varepsilon\|Dw\|^2_{L^2(\Omega,\mu)} + C(\varepsilon, \alpha_0)\|w^{\alpha_0}\|^{1/\alpha_0}_{L^2(\Omega,\mu)}\,,$$

with $\alpha_0 \in (0, 1)$, $q \ge 0$, and $w = U^{q/2}$, to obtain

$$\int_\Omega |U|^{q+2}\,dx \le \varepsilon\int_\Omega |U|^q|DU|^2\,dx + C(\varepsilon, \alpha_0)\||U|^{\alpha_0}\|^{1/\alpha_0}_{L^2(\Omega,\mu)}\,. \tag{4.4.7}$$

In the sequel, if $\hat{f}$ has a superlinear growth in Du as in (3.1.7) of f.2) then we use Young's inequality to write it in the following form for some $\varepsilon_0 > 0$:

$$|\hat{f}(x, u, p)| \le \varepsilon_0|\lambda_u(u)||p|^2 + f(u)\,. \tag{4.4.8}$$

We then have the following elementary lemma.

Lemma 4.4.2. *Suppose A), F) and* (4.4.8) *with ε_0 being sufficiently small. For any s satisfying*

$$s > -1 \text{ and } C_*^{-1} > s/(s+2) \tag{4.4.9}$$

and any $\alpha_0 \in (0, 1)$ we have for $U := \lambda_0 + |u|$ that

$$\int_\Omega U^{k+s}|Du|^2\,d\mu \le C\left(\int_\Omega U^{\alpha_0(k+s+2)}\,d\mu\right)^{\frac{1}{\alpha_0}} + C\int_\Omega U^s f(u)|u|\,d\mu\,. \tag{4.4.10}$$

Proof. Let $X = [\lambda_0 + |u_i|]_{i=1}^m$ and test the system with $|X|^s u$ to get

$$\int_\Omega \langle A(u)Du, D(|X|^s u)\rangle\,d\mu \le \int_\Omega \langle\hat{f}(u, Du), |X|^s u\rangle\,d\mu\,. \tag{4.4.11}$$

We note that $\langle A(u)Du, D(|X|^s u)\rangle = \langle A(u)DX, D(|X|^s X)\rangle$ so that, by the assumption (4.4.9) on s and (B.1.8) (in Section B.1 in Appendix B), there is $c_0 > 0$ such that

$$\langle A(u)Du, D(|X|^s u)\rangle \geq c_0\lambda(u)|X|^s|DX|^2 .$$

Because $|DX| = |Du|$ and $|X| \sim U$, the above yields

$$\int_\Omega \lambda(u)U^s|Du|^2\,d\mu \leq C\int_\Omega U^s\langle\hat{f}(u,Du),u\rangle\,d\mu . \tag{4.4.12}$$

If $\hat{f}$ satisfies f.1) then a simple use of Young's inequality gives

$$|\langle\hat{f}(u,Du),u\rangle| \leq \varepsilon\lambda(u)|Du|^2 + C(\varepsilon)\lambda(u)|u|^2 + f(u)|u| .$$

If f.2) holds with (4.4.8) then $|\langle\hat{f}(u,Du),u\rangle| \leq C\varepsilon_0|\lambda_u(u)||u||Du|^2 + f(u)|u|$. Because $|\lambda_u(u)||u| \lesssim \lambda(u)$, we obtain the above inequality again with $\varepsilon = C\varepsilon_0$. Therefore, if ε_0 is sufficiently small then (4.4.12) and the fact that $\lambda(u) \sim U^k$ imply

$$\int_\Omega U^{k+s}|Du|^2\,d\mu \leq C\int_\Omega U^{k+s+2}\,d\mu + C\int_\Omega U^s f(u)|u|\,d\mu . \tag{4.4.13}$$

We apply the interpolation inequality (4.4.7) with $q = k+s$ to the first integral on the right-hand side of (4.4.13), noting also that $|DU|^2 \sim |Du|^2$. For sufficiently small ε, we derive (4.4.10) from the above estimates and complete the proof. □

We now consider the case $N = 2$ and show that the condition (4.4.6) can be reduced to the boundedness for L^1 norm of u only.

Proof of Corollary 4.4.3. By F') and (4.4.10) of Lemma 4.4.2 with $s = 0$ and $\alpha_0 = 1/(k+2)$, we have

$$\int_\Omega (\lambda_0 + |u|)^k|Du|^2\,dx \lesssim \int_\Omega (\lambda_0 + |u|)^{k+2}\,dx + C(\|u\|_{L^1(\Omega)}) .$$

Applying (4.4.7) to the integral on the right-hand side with sufficiently small ε, $U = \lambda_0 + |u|$ and $\alpha_0 = 1/(k+2)$, we get

$$\int_\Omega (\lambda_0 + |u|)^k|Du|^2\,dx \lesssim C(\|u\|_{L^1(\Omega)}) . \tag{4.4.14}$$

This gives the bound for $\|Du\|_{L^2(\Omega)}$ because $k > 0$. As $N = 2$, the corollary follows from Corollary 4.4.1. □

Corollary 4.4.3. *Let $N = 2$. Suppose A"), F') and $k > 0$. If $\hat{f}$ has a quadratic growth in Du as in (3.1.7) of f.2) then we assume further that (4.4.8) holds with ε_0 being sufficiently small.*

If there is a constant C_0 such that for any strong solution u of (4.4.2),

$$\|u\|_{L^1(\Omega,\mu)} \leq C_0 , \tag{4.4.15}$$

then (4.0.1) has a strong solution u.

The above result does not apply to the case $k < 0$ because the system becomes singular $\lambda(u) \to 0$ if $|u| \to \infty$. Thus, from (4.4.14) we do not get the desired estimate for $\|Du\|_{L^2(\Omega)}$, which implies the control of BMO norm of $K(u) = u$. We then consider another choice of K and apply Theorem 4.1.3. We need to check the two key assumptions in M.1). That is, for any given $\mu_0 > 0$ there is positive R_{μ_0} sufficiently small in terms of the constants in A) and F) such that

$$\sup_{x_0 \in \bar{\Omega}, \tau \in (0,T_0)} \|K(u)\|^2_{BMO(B_R(x_0)\cap\Omega,\mu)} \le \mu_0 \,. \tag{4.4.16}$$

Furthermore, for $\mathbf{W}_p(\sigma, x) := \lambda^{p+\frac{1}{2}}(\sigma u)|\lambda_u(\sigma u)|^{-p}$ and any p such that $1 - 1/p < C_*^{-1}$ we have $[\mathbf{W}_p^{\frac{2}{3}}]_{\frac{4}{3}} \le C_0$.

To construct the map K we recall the condition L) in Section 5.5.

L) There are m C^2 positive scalar functions λ_i, $i = 1, \dots, m$ on $\mathbb{R}^m$ satisfying

$$\lambda_i(u) \sim \lambda(u), \;\; |(\lambda_i(u))_{uu}| \lesssim |\lambda_{uu}(u)| \,, \tag{4.4.17}$$

$$\lambda(u)|\lambda_{uu}(u)| \lesssim |\lambda_u(u)|^2 \,. \tag{4.4.18}$$

We introduce the matrices

$$\mathbf{L} = \mathrm{diag}[\lambda_1(u), \dots, \lambda_m(u)] \,, \; \mathbf{L}_u = [D_u\lambda_i(u)]_{i=1}^m$$

and suppose that

$$|\mathbf{L}_u^{-1}| \lesssim |\lambda_u(u)|^{-1} \,. \tag{4.4.19}$$

We then define the map

$$K(u) = [K_i(u)]_{i=1}^m, \; K_i(u) = \log(\lambda_i(u)) \,. \tag{4.4.20}$$

The existence of the function λ_i in the condition L) is quite easy to verify when the system in consideration possesses some cone invariant property as described in Remark 3.5.3. Lemma 3.5.4 shows that this map satisfies K) so that the local Gagliardo–Nirenberg inequality in Corollary 2.3.5, used in the proof of Theorem 4.1.3, is available.

Lemma 4.4.4. *Assume A") and L). The condition EM.1) holds for all strong solutions of* (4.4.2) *if they satisfy*

$$\int_\Omega (\lambda_0 + |u|)^{-N}|Du|^N \, dx \le C_0 \,. \tag{4.4.21}$$

In Corollary 4.4.1, we consider the case $\lambda(u) \sim (\lambda_0 + |u|)^k$ with $k > 0$ and assume that u has small BMO norm in small balls, which can be verified by establishing that $\|Du\|_{L^N(\Omega)}$ is bounded. The assumption (4.4.21) is of course much weaker, especially when $|u|$ is large, and can apply to the case $k < 0$.

Proof. To see that $K(u)$ has small BMO norm in small balls, we note that $|K_u| \lesssim |\lambda_u(u)|/\lambda(u)$ (Lemma 3.5.4) so that

$$|D(K(u))| \le |K_u(u)||Du| \lesssim \frac{|\lambda_u(u)|}{\lambda(u)}|Du| \lesssim (\lambda_0 + |u|)^{-1}|Du| \,.$$

Therefore, (4.4.21) yields that $D(K(u)) \in L^N(\Omega)$. A use of the Poincaré–Sobolev inequality as in Lemma B.1.4 shows that $K(u)$, as well as $\pm\log(\lambda_0 + |u|)$, has small BMO norm in small balls.

Next, we need to show that $\mathbf{W}_p(x) := \lambda^{p+\frac{1}{2}}(u)|\lambda_u(u)|^{-p}$ is a weight. By A"), $\mathbf{W}_p(x) \sim (\lambda_0 + |u|)^{k_p}$ for some exponent k_p. From the above argument, we see that $w = \pm\log(\lambda_0 + |u|)$ have small BMO norm in small balls. For any given $c, \gamma > 0$, by Lemma 2.3.9, e^{cw}, and therefore $(\lambda_0 + |u|)^c$ and $(\lambda_0 + |u|)^{-c}$, are $A_{1+\gamma}$ weights (in small balls). Thus, any power of $\mathbf{W}_p$ is an $A_{1+\gamma}$ weight and the last condition in M.1) is then verified. □

Remark 4.4.5. Alternatively, we can also consider the map K defined by

$$K_{\varepsilon_0}(u) = (|\log(|U|)| + \varepsilon_0)|U|^{-1}U\,, \quad U = [\lambda_0 + |u_i|]_{i=1}^m\,. \tag{4.4.22}$$

This map satisfies for any $\varepsilon_0 > 0$ the condition K) and Lemma 4.4.4 also applies to this case. However, the calculation is straightforward but not elegant; see Lemma B.1.4 and Remark B.1.5 in the Appendix.

We now consider the condition EM.0). By Remark 4.2.9, the conditions (4.2.37)–(4.2.38) in EM.0) can be dropped because we are considering $\hat{f}$ independent of x. Thus, EM.0) is now reduced to the following set of integrability conditions.

EM.0") There exists a constant C_0 such that for some $r_0 > 1$

$$\||f_u(\sigma u)|\lambda^{-1}(\sigma u)\|_{L^{r_0}(\Omega)} \le C_0\,, \tag{4.4.23}$$

$$\int_\Omega (|f_u(\sigma u)| + \lambda(\sigma u))|Du|^2\,dx \le C_0\,. \tag{4.4.24}$$

In addition,

$$\int_\Omega \lambda^{-\alpha}(u)\,dx \le C_0\,, \quad \frac{1}{\alpha} = \frac{2}{s_0} - 1 + \frac{2}{N} = \frac{2(N_* - s_0)}{N_* s_0}\,, \tag{4.4.25}$$

where s_0 is the number introduced in ESG).

Using Lemma 4.4.4, we immediately have the following consequence of Theorem 4.1.3.

Corollary 4.4.6. *Assume A"), ESG) and L). Suppose that any strong solution u of* (4.4.2) *satisfy EM0") and*

$$\int_\Omega (\lambda_0 + |u|)^{-N}|Du|^N\,dx \le C_0\,. \tag{4.4.26}$$

Then there is a strong solution to (4.4.1).

We revisit the case $k > 0$ first and present a result that is comparable to Corollary 4.4.1.

Corollary 4.4.7. *The conclusion of Corollary 4.4.1 still holds if L) is assumed and the condition* (4.4.6) *is replaced by its weaker version*

$$\int_\Omega (\lambda_0 + |u|)^{-N}|Du|^N\,dx \le C_0\,. \tag{4.4.27}$$

Proof. We apply Corollary 4.4.6. Since (4.4.26) is already stated in (4.4.27), we need only check its assumption EM.0"). Because $|f_u| \sim \lambda(u) \sim (\lambda_0 + |u|)^k$ by F'), (4.4.23) holds with $r_0 = \infty$. The estimate (4.4.14) in the proof of Corollary 4.4.3,

$$\int_\Omega (\lambda_0 + |u|)^k |Du|^2 \, dx \lesssim C(\|u\|_{L^1(\Omega)}) ,$$

still holds in this case. Again, as $|f_u| \sim \lambda(u) \sim (\lambda_0 + |u|)^k$, the above implies (4.4.24). Since $\lambda_0, k > 0$, (4.2.40) is true with $\alpha = \infty$. That is, we let $s_0 = N_*$ in ESG) and see that this is ESG') assumed here. Thus, EM.0") is validated and the proof is complete. □

We present an application of Corollary 4.4.7. This example concerns cross-diffusion systems with polynomial growth data on planar domains. This type of system occurs in many applications in mathematical biology and ecology. We will see that the assumptions of the corollary can be verified by a very simple integrability assumption on the solutions.

Corollary 4.4.8. *Let $n = 2$. Suppose A), F) and $f(u) \lesssim (\lambda_0 + |u|)^l$ and $\lambda(u) \sim (\lambda_0 + |u|)^k$ for some k, l satisfying*

$$k > \frac{-2C_*}{C_* - 1} \text{ and } l - k < \frac{C_* + 1}{C_* - 1} . \tag{4.4.28}$$

If $\hat{f}$ has a quadratic growth in Du as in (3.1.7) *of f.2) then we assume further that* (4.4.8) *holds with ε_0 being sufficiently small.*

If there is a constant C_0 such that for any strong solution u of (4.4.2),

$$\|u\|_{L^{l_0}(\Omega,\mu)} \le C_0 \quad \text{for some } l_0 > \max\{l, l - k - 1\} , \tag{4.4.29}$$

then (3.0.1) *has a strong solution u.*

The assumption (4.4.29) is a very weak one. For example, if $k \ge -1$ then we see from the growth condition $|f(u)| \lesssim (\lambda_0 + |u|)^l$ that (4.4.29) simply requires that $l_0 > l$, or equivalently $f(u) \in L^r(\Omega, \mu)$ for some $r > 1$.

This result greatly generalizes [30, Corollary 3.9] in many aspects. Besides the fact that we allow quadratic growth in Du for $\hat{f}(x, u, Du)$ and $k < 0$, we also consider a much more general relation between the growths of $f(u)$ and $\lambda(u)$ in (4.4.28), while we assume in F') that $f(u) \lesssim \lambda(u)|u|$ (i.e., $l - k = 1$).

Proof of Corollary 4.4.8. We apply Corollary 4.4.6 here. We will verify first the condition (4.4.26) and then the integrability assumptions in EM.0").

To begin we show that there is some $s > s_0 := \max\{-1, -k - 2\}$ and s is close to s_0 such that the condition (4.4.9) of Lemma 4.4.2 holds. That is,

$$s > -1 \text{ and } C_*^{-1} > s/(s + 2) .$$

This clearly holds if $k \ge -1$ because $s_0 = -1$. Otherwise, one can see that the assumption $k > -2C_*/(C_* - 1)$ in (4.4.28) and $s_0 = -k - 2$ yield that $C_*^{-1} > s_0/(s_0 + 2)$ so that if

s is close to s_0 then (4.4.9) holds. Hence, from (4.4.10) of Lemma 4.4.2 for $U = \lambda_0 + |u|$ and $\alpha_0 \in (0, 1)$,

$$\int_\Omega U^{k+s}|Du|^2\, d\mu \le C\left(\int_\Omega U^{\alpha_0(k+s+2)}\, d\mu\right)^{\frac{1}{\alpha_0}} + C\int_\Omega U^s f(u)|u|\, d\mu\,.$$

Choosing α_0 such that $\alpha_0(k+s+2) \le s+l+1$ and using the assumption $f(u) \lesssim U^l$ we see that

$$\int_\Omega U^{k+s}|Du|^2\, d\mu \le C\left(\int_\Omega U^{s+l+1}\, d\mu\right)$$

as $s \ge \max\{-1, -k-2\}$, $s+l+1 \le l_0 = \max\{l, l-k-1\}$. The assumption (4.4.29) then implies

$$\int_\Omega U^{k+s}|Du|^2\, dx \le C(C_0)\,. \tag{4.4.30}$$

We are now ready to verify the conditions of Corollary 4.4.6 with $N = 2$. As $k+s > -2 = -N$, (4.4.30) yields (4.4.26) of Corollary 4.4.6 because

$$\int_\Omega (\lambda_0 + |u|)^{-2}|Du|^2\, dx \le C(\lambda_0)\int_\Omega (\lambda_0 + |u|)^{k+s}|Du|^2\, dx \le C(\lambda_0, C_0)\,.$$

We now check the integrability conditions in EM.0”), which read

$$\||f_u(\sigma u)|\lambda^{-1}(\sigma u)\|_{L^{r_0}(\Omega)} \le C_0\,, \tag{4.4.31}$$

$$\int_\Omega (|f_u(\sigma u)| + \lambda(\sigma u))|Du|^2\, dx \le C_0\,. \tag{4.4.32}$$

In addition, for α such that $\frac{1}{\alpha} = \frac{2}{s_0} - 1 + \frac{2}{N} = \frac{2(N_*-s_0)}{N_* s_0}$,

$$\int_\Omega \lambda^{-\alpha}(u)\, dx \le C_0\,. \tag{4.4.33}$$

Because $N = 2$, we have $\|w\|_{L^q(\Omega)} \le C\|Dw\|_{L^2(\Omega)} + C(\beta)\|w^\beta\|^{\frac{1}{\beta}}_{L^1(\Omega)}$ for all $q \ge 1$ and $\beta \in (0, 1)$. Applying this to $w = |U|^{(k+s)/2+1}$ and sufficiently small β and using (4.4.30) and the assumption (4.4.29), we see that $\|U^q\|_{L^1(\Omega)} \le C(C_0)$ for all $q \ge 1$. By Hölder's inequality this is also true for $q \ge 0$. It is also true for $q < 0$ because $U \ge \lambda_0 > 0$. We then have

$$\|U^q\|_{L^1(\Omega)} \le C(C_0, \lambda_0, q) \quad \text{for all } q\,. \tag{4.4.34}$$

The above then immediately implies the integrability conditions (4.4.31) and (4.4.33) because $\lambda(u)$ and $|f_u(u)|$ are powers of U.

Next, (4.4.34) implies that (4.4.30) holds and is valid as long as $s \ge -1$ and $C_*^{-1} > s/(s+2)$. To verify (4.4.32) we need to find a constant $C(C_0)$ such that

$$\int_\Omega (\lambda_0 + |u|)^{l-1}|Du|^2\, dx + \int_\Omega (\lambda_0 + |u|)^k|Du|^2\, dx \le C(C_0)\,. \tag{4.4.35}$$

Letting $s = 0$ in (4.4.30), we see that the second integral on the left-hand side is bounded by a constant $C(C_0)$. If $l \le 1 + k$ then the first integral in (4.4.35) is bounded by the second one and we obtain the desired bound. If $l > k+1$ we let $s = l-k-1$ in (4.4.30). The condition on s in (4.4.9) holds because

$$\frac{s}{s+2} < C_*^{-1} \Leftrightarrow \frac{l-k-1}{l-k+1} < C_*^{-1} \Leftrightarrow l-k < \frac{C_*+1}{C_*-1},$$

which is assumed in (4.4.28). Hence, (4.4.35) holds. We have verified all assumptions of EM.0") and completed the proof. □

Up to this point, the above results in this section have not required that $A(u) = D_u P(u)$ for some map $P\colon \mathbb{R}^m \to \mathbb{R}^m$ as in Section 5.5. However, we will have much more general results in Chapter 5.

Nevertheless, we consider the following generalized SKT system [30, 44, 46] with Dirichlet or Neumann boundary conditions on a bounded domain $\Omega \subset \mathbb{R}^n$ with $n \le 4$:

$$-\Delta(P_i(u)) = B_i(u, Du) + f_i(u), \quad i = 1, \dots, m. \tag{4.4.36}$$

Here, $P_i\colon \mathbb{R}^m \to \mathbb{R}$ are C^2 functions. The functions B_i, f_i are C^1 functions on $\mathbb{R}^m \times \mathbb{R}^{mn}$ and $\mathbb{R}^m$ respectively. We will assume that $B_i(u, Du)$ has linear growth in Du.

By a different choice of the map K in the main technical theorem, we have the following.

Theorem 4.4.9. *Assume that the matrix $A(u) = \mathrm{diag}[(P_i)_u(u)]$ satisfies the condition A) with $\lambda(u) \sim (\lambda_0 + |u|)^k$ for some $k \ge -1$. Moreover, $\hat{f}(u, Du) = \mathrm{diag}[B_i(u, Du) + f_i(u)]$ satisfies the following special version f.1) of F):*

$$|B_i(u, Du)| \le C\lambda(u)|Du|, \quad |f_i(u)| \le f(u).$$

Thus, (3.5.6) can be written as (3.0.1). Assume that there exist $r_0 > n/2$ and a constant C_0 such that for any strong solution u of (4.4.2),

$$\|f_u(u)\lambda^{-1}(u)\|_{L^{r_0}(\Omega)} \le C_0, \tag{4.4.37}$$

and the following conditions:

i) *If $k \ge 0$ then $\|u\|_{L^1(\Omega)} \le C_0$.*
ii) *If $k \in [-1, 0)$ then $\|u\|_{L^{-kn/2}(\Omega)} \le C_0$. Furthermore,*

$$\|\lambda^{-2}(u)f_u(u)\|_{L^{\frac{n}{2}}(\Omega)} \le C_0. \tag{4.4.38}$$

Then (3.5.6) has a strong solution for $n = 2, 3, 4$.

The above result generalizes [30, Corollary 3.10] where we assumed that $\hat{f}$ is independent of Du, $k > 0$ and $f(u) \lesssim \lambda(u)|u|$. In this case, it is natural to assume that $|f_u(u)| \lesssim \lambda(u)$ so that (4.4.37) obviously holds. We also have $|\lambda^{-2}(u)f_u(u)| \lesssim \lambda^{-1}(u)$ so that (4.4.38) is in fact a consequence of the assumption $\|u\|_{L^{-kn/2}(\Omega)} \le C_0$ in ii).

5 Cross-diffusion systems of porous media type

In this chapter, we apply the general theories developed in the preceding chapters to study the solvability of the following parabolic system of m equations ($m \geq 2$):

$$u_t - \Delta(P(u)) = B(u, Du) + f(u)\,, \quad (x,t) \in \Omega \times (0, T_0)\,. \tag{5.0.1}$$

The system is equipped with the boundary and initial conditions

$$\begin{cases} u = 0 \text{ or } \frac{\partial u}{\partial \nu} = 0 \text{ on } \partial\Omega \times (0, T_0)\,, \\ u(x,0) = u_0(x)\,, \quad x \in \Omega\,. \end{cases}$$

Here, $P\colon \mathbb{R}^m \to \mathbb{R}^m$ is a C^2 map. $B\colon \mathbb{R}^m \times \mathbb{R}^{mN} \to \mathbb{R}^m$ and $f\colon \mathbb{R}^m \to \mathbb{R}^m$ are C^1 maps.

We also consider the elliptic counterpart of (5.0.1),

$$-\Delta(P(u)) = B(u, Du) + f(u)\,, \quad x \in \Omega\,. \tag{5.0.2}$$

The consideration of (5.0.1) is motivated by the extensively studied porous media equation for a scalar unknown $u\colon \Omega \times (0, T_0) \to \mathbb{R}$

$$u_t - \Delta(|u|^k u) = f(u)\,, \quad (x,t) \in \Omega \times (0, T_0) \text{ and some } k > 0\,.$$

The vectorial version of this is the system of m equations for the vector $u = [u_i]_{i=1}^m$, writing $|u| = \sqrt{u_1^2 + \cdots + u_m^2}$,

$$u_t - \Delta(P(u)) = f(u)\,, \quad P(u) = |u|^k u\,.$$

Let $A(u) = P_u(u)$, the Jacobian of P. We easily see that (5.0.1) is just a special case of the parabolic system (3.0.1) considered in Chapter 3. Indeed, for $\lambda(u) = |u|^k$ we have

$$\lambda(u) \leq \langle A(u)\zeta, \zeta\rangle \text{ and } |A(u)| \leq (1+k)\lambda(u) \quad \forall u \in \mathbb{R}^m, \zeta \in \mathbb{R}^{mN}\,.$$

In comparison with the structural condition A) of Chapter 3, the above leads us to the consideration of the following main condition of this chapter for the system (5.0.1).

P) $P\colon \mathbb{R}^m \to \mathbb{R}^m$ is a C^2 map. The Jacobian $A(u) = P_u(u)$ satisfies the condition that there are a constant $C_* > 0$ and a *nonnegative* scalar C^1 function $\lambda(u)$ on $\mathbb{R}^m$ such that for all $u \in \mathbb{R}^m$, $\zeta \in \mathbb{R}^{mn}$

$$\lambda(u)|\zeta|^2 \leq \langle A(u)\zeta, \zeta\rangle \text{ and } |A(u)| \leq C_*\lambda(u)\,. \tag{5.0.3}$$

In addition, $A(u)$ and $\lambda(u)$ have a polynomial growth in $|u|$. That is, $|A(u)|, \lambda(u) \sim |u|^k$ for some $k > 0$.

Of course, the polynomial growth of $A(u), \lambda(u)$ implies the assumption in the condition A) that $|A_u(u)| \leq C|\lambda_u(u)|$.

https://doi.org/10.1515/9783110608762-005

We note that the ellipticity $\lambda(u)$ in A) was assumed to be positive for any $u \in \mathbb{R}^m$, i.e., $\lambda(u) \ge \lambda_0 > 0$ ((3.1.4)). Here, we assume only that $\lambda(u) \ge 0$ so that $A(u)$ can be singular, i.e., $\lambda(u) \equiv 0$, in a bounded set of $\mathbb{R}^m$, say $\{0\}$. Therefore, in order to discuss the existence of strong solutions to the system (5.0.1) (and (4.0.1)) we need also to consider the following hypothesis:

PR) P) holds and there is some $\lambda_0 > 0$ such that $\lambda(u) \ge \lambda_0$ for all $u \in \mathbb{R}^m$.

For $\lambda(u)$ with polynomial growth this condition is equivalent to the assumption that $\lambda(u) \sim (\lambda_0 + |u|)^k$ for some $\lambda_0 > 0$.

Concerning the reaction term f, we assume the following condition:

f) $f : \mathbb{R}^m \to \mathbb{R}^m$ is a C^1 map with polynomial in u components and there exists a constant C such that

$$|f_u(u)| \le C(1 + \lambda(u)) . \tag{5.0.4}$$

We note that f) also implies

$$|f(u)| \le C(1 + |u|)(1 + \lambda(u)) . \tag{5.0.5}$$

If the drift term $B(u, Du)$ is included in the consideration then we assume:

B) For any differentiable vector-valued functions $u : \Omega \to \mathbb{R}^m$ and $\zeta : \Omega \to \mathbb{R}^{mn}$ we assume that there is a constant C such that

$$|B(u, \zeta)| \le C\lambda(u)|\zeta| ,$$
$$|DB(u, \zeta)| \le C(\lambda(u)|D\zeta| + |\lambda_u(u)||Du||\zeta|) .$$

The structural conditions P), PR) and f) are also motivated by the well-known SKT model introduced by Shigesada *et al.* in [44]:

$$\begin{cases} u_t = \Delta(d_1 u + \alpha_{11} u^2 + \alpha_{12} uv) + \operatorname{div}[b_1 u \nabla \Phi(x)] + f_1(u, v) , \\ v_t = \Delta(d_2 v + \alpha_{21} uv + \alpha_{22} v^2) + \operatorname{div}[b_2 v \nabla \Phi(x)] + f_2(u, v) , \end{cases} \tag{5.0.6}$$

where $f_i(u, v)$ are reaction terms of Lotka–Volterra type and quadratic in u, v. Dirichlet or Neumann boundary conditions were usually assumed for (5.0.6). This model was used to describe the population dynamics of the species densities u, v that move under the influence of population pressures and the environmental potential $\Phi(x)$.

Under suitable assumptions on the constant parameters α_{ij}'s and that Ω is a planar domain ($N = 2$), Yagi proved in [46] the global existence of positive solutions, with positive initial data. In this chapter, we will extend this result and others in the literature by considering much more general structural conditions like P) and f).

We will show in Section 5.1 that the SKT system (5.0.6) is a special case of (5.0.1) with $P : \mathbb{R}^2 \to \mathbb{R}^2$ being a quadratic map that satisfies PR) for $k = 1$. Because f_i's in (5.0.6) are quadratic in u, v, it is clear that (5.0.4) of the condition f) is satisfied here. The drift term $B(u, Du)$ has a *linear growth* in Du and includes the potential $\Phi(x)$ in (5.0.6). In fact, we will introduce in Section 5.1 a much more general multispecies version of the SKT system (5.0.6) and allow P, f to have *polynomial growths of any order* $k > 0$ in u. We will refer to it as the *generalized (SKT) system*.

We first discuss the existence of *strong solutions* of (5.0.1) and (5.0.2) when they are regular (i.e., PR) holds). We will see that these results are just simple consequences of the theories for general parabolic and elliptic systems in Chapter 3 and Chapter 4.

Once the existence of strong solutions for regular systems is proved, we will follow the standard approach to establish the existence of *weak solutions* to the degenerate systems. We will approximate the degenerate systems by a sequence of regular ones whose strong solutions can be estimated uniformly so that we can pass to the limit. Uniform estimates of strong solutions to (5.0.1) and (5.0.2) will be the most important matter of this chapter and they can be established under very mild uniform integrability assumptions on the strong solutions and data of the approximation systems. Again, we would like to emphasize that no boundedness of solutions will be assumed here because maximum or comparison principles are not available.

We organize our chapter as follows. We introduce the generalized SKT model in Section 5.1, which will serve as the main example/application of our general results in this chapter. In Section 5.2 and Section 5.5 we state our main existence results for the general evolutionary system (5.0.1) and its elliptic counterpart (5.0.2). The proofs of these results are presented in Section 5.3 and Section 5.6. The existence of weak solutions to the evolutionary systems is established by a standard approximation argument described above and the uniqueness of such solutions is discussed in Section 5.4. We end the chapter by showing that our theory applies to the generalized (SKT) system in Section 5.7.

5.1 The generalized SKT systems

We see that the general (5.0.1) includes the (SKT) system (5.0.6) if we write

$$P(u) = [P_i(u)]_{i=1}^2 , \quad P_i(u) = d_i u_i + u_i\langle u, \alpha_i\rangle ,$$

where $\alpha_i := [\alpha_{ji}]_{j=1}^2$ are two vectors in $\mathbb{R}_+^2 = \{(x_1, x_2) : x_i \geq 0\}$. Inspired by this, we introduce the *generalized (SKT) system* whose diffusions are 'polynomials' of any order $k > 0$.

Definition SKT). Consider m linearly independent vectors $\alpha_i = [\alpha_{ji}]_{j=1}^m \in \mathbb{R}^m$. We define the map P in (5.0.1) by

$$P(u) = [u_i\lambda_i(u)]_{i=1}^m, \; \lambda_i(u) = d_i + |\langle u, \alpha_i\rangle|^{k-1}\langle u, \alpha_i\rangle + \sum_{j=1}^{l} \beta_{ij}|\langle u, \alpha_i\rangle|^{\kappa_j - 1}\langle u, \alpha_i\rangle . \quad (5.1.1)$$

Here, the coefficients d_i, β_{ij} are nonnegative, the exponent k is positive and $\kappa_j \in (0, k)$.

Concerning the reaction terms, we consider $f_i(u) = k_i u_i + g_i(u)$ where g_i's are C^1 functions on $\mathbb{R}^m$ and $|(g_i)_u(u)| \leq C|u|^k$ for all $u \in \mathbb{R}^m$ and some constant C.

We first discuss the simplest case of the above definition: $d_i = 0$ and $\beta_{ij} = 0$ for all i, j. That is,

$$P(u) = [u_i\lambda_i(u)]_{i=1}^m\,, \qquad \text{where } \lambda_i(u) := |\langle u, \alpha_i\rangle|^{k-1}\langle u, \alpha_i\rangle\,. \tag{5.1.2}$$

We then have $P_u(u) = \mathrm{diag}[|\langle u, \alpha_i\rangle|^{k-1}\langle u, \alpha_i\rangle] + k\,\mathrm{diag}[|\langle u, \alpha_i\rangle|^{k-1}u_i]\alpha^T$, where $\alpha = [\alpha_{ji}]$. We introduce the matrices

$$\mathbf{A}_{(\alpha)}(u) = \mathrm{diag}[\langle u, \alpha_i\rangle] + \mathrm{diag}[u_i]\alpha^T, \; d_{k,\alpha}(u) := \mathrm{diag}[|\langle u, \alpha_i\rangle|^{k-1}]\,.$$

Clearly, $\mathbf{A}_{(\alpha)}(u) = P_u(u)$ when $k = 1$ and

$$P_u(u) = d_{k,\alpha}(u)\,\mathrm{diag}[\langle u, \alpha_i\rangle] + k d_{k,\alpha}(u)\,\mathrm{diag}[u_i]\alpha^T\,. \tag{5.1.3}$$

For $k = 1$, $m = 2$ we see that (5.1.2) is the classical (SKT) system. The vector $u = [u_1, u_2]^T$ is the density vector and one usually considered nonnegative u_1, u_2. The quadratic $\langle\mathbf{A}_{(\alpha)}(u)\zeta, \zeta\rangle$ is generated by the symmetric matrix $S_\alpha(u) := (\mathbf{A}_{(\alpha)}(u) + \mathbf{A}^T_{(\alpha)}(u))/2$ whose trace and determinant are

$$\begin{aligned} T &:= 2\,u_1\alpha_{11} + u_2\alpha_{21} + u_1\alpha_{12} + 2\,u_2\alpha_{22}\,, \\ \Delta &:= \left(2\,\alpha_{11}\alpha_{12} - 1/4\,\alpha_{21}{}^2\right)u_1{}^2 + (1/2\,\alpha_{21}\alpha_{12} + 4\,\alpha_{11}\alpha_{22})\,u_1u_2 \\ &\quad + \left(-1/4\,\alpha_{12}{}^2 + 2\,\alpha_{21}\alpha_{22}\right)u_2{}^2\,. \end{aligned}$$

Thus, if $u_i \ge 0$, $\alpha_{ij} > 0$ and

$$\alpha_{21}{}^2 < 8\alpha_{11}\alpha_{12} \text{ and } \alpha_{12}{}^2 < 8\alpha_{21}\alpha_{22} \tag{5.1.4}$$

then there is a constant $c(\alpha) > 0$ such that $T, \sqrt{\Delta} \ge c(\alpha)(u_1 + u_2)$. The eigenvalues of $S_\alpha(u)$ are then greater $c(\alpha)(u_1+u_2)$ and thus $\langle\mathbf{A}_{(\alpha)}(u)\zeta, \zeta\rangle \ge c(\alpha)|u||\zeta|^2$ for any $\zeta \in \mathbb{R}^{2N}$. We note that the condition (5.1.4) was considered by Yagi, who assumed further that $\alpha_{21} = \alpha_{12}$.

Inspired by this example, we assume for general $m \ge$ the following hypotheses:

a.1) There is a constant $c(\alpha) > 0$ such that

$$\langle\mathbf{A}_{(\alpha)}(u)\zeta, \zeta\rangle \ge c(\alpha)|u||\zeta|^2 \text{ for any } \zeta \in \mathbb{R}^{Nm}\,. \tag{5.1.5}$$

a.2) For $u \in \mathbb{R}^m_+$ we have $\langle u, \alpha_i\rangle \sim \langle u, \alpha_j\rangle \sim |u|$, $i \neq j$.

a.3) If $k > 1$ then α^T is semipositive, i.e., $\langle\alpha^T\zeta, \zeta\rangle \ge 0$ for all $\zeta \in \mathbb{R}^{mN}$.

We then show that the condition P) is verified. First of all, we need that $A(u) = P_u(u)$ is positive definite. Namely, there exists $C > 0$ such that

$$\langle P_u(u)\zeta, \zeta\rangle \ge C|u|^k|\zeta|^2 \text{ for any } \zeta \in \mathbb{R}^{Nm}\,. \tag{5.1.6}$$

We recall a simple fact from linear algebra. Let B, A be matrices such that B is diagonal. If there are $c_1, c_2 \ge 0$ such that

$$\langle B\zeta, \zeta\rangle \ge c_1|\zeta|^2 \text{ and } \langle A\zeta, \zeta\rangle \ge c_2|\zeta|^2 \text{ then } \langle BA\zeta, \zeta\rangle \ge c_1c_2|\zeta|^2\,. \tag{5.1.7}$$

Indeed, as $c_1 \geq 0$, $B^{\frac{1}{2}}$ is well defined and $|B^{\frac{1}{2}}\zeta|^2 = \langle B\zeta, \zeta\rangle \geq c_1|\zeta|^2$. We then have $\langle B^{\frac{1}{2}}AB^{\frac{1}{2}}\zeta, \zeta\rangle = \langle AB^{\frac{1}{2}}\zeta, B^{\frac{1}{2}}\zeta\rangle \geq c_2|B^{\frac{1}{2}}\zeta|^2 \geq c_2c_1|\zeta|^2$. By [5, Corollary 4.4.10] we know that the (multi)spectrum sets of $BA = B^{\frac{1}{2}}B^{\frac{1}{2}}A$ and $B^{\frac{1}{2}}AB^{\frac{1}{2}}$ (B is diagonal) are the same, so that $\langle B^{\frac{1}{2}}AB^{\frac{1}{2}}\zeta, \zeta\rangle \geq c_2c_1|\zeta|^2$ implies $\langle BA\zeta, \zeta\rangle \geq c_2c_1|\zeta|^2$, which is (5.1.7).

Let $A = \mathbf{A}_{(\alpha)}(u)$, $B := d_{k,\alpha}(u)$. By a.2), the definition of $d_{k,\alpha}(u)$ and (5.1.5), $\langle B\zeta, \zeta\rangle \sim |u|^{k-1}|\zeta|^2$ and $\langle A\zeta, \zeta\rangle \geq c(\alpha)|u||\zeta|^2$. Using (5.1.7), we see that

$$\langle d_{k,\alpha}(u)\mathbf{A}_{(\alpha)}(u)\zeta, \zeta\rangle \geq c(\alpha)|u|^k|\zeta|^2 . \tag{5.1.8}$$

We now consider two cases, where either $k > 1$ or $k \in (0, 1]$. If $k > 1$ then we use a.3) and (5.1.7), with $A = \alpha^T$ and $B := d_{k,\alpha}(u)\,\mathrm{diag}[u_i]$, to see that $\langle d_{k,\alpha}(u)\,\mathrm{diag}[u_i]\alpha^T\zeta, \zeta\rangle$ is semipositive. We then write (5.1.3) as

$$P_u(u) = d_{k,\alpha}(u)\mathbf{A}_{(\alpha)}(u) + (k-1)d_{k,\alpha}(u)\,\mathrm{diag}[u_i]\alpha^T ,$$

and use (5.1.8) to see that $\langle P_u(u)\zeta, \zeta\rangle \geq c(\alpha)|u|^k|\zeta|^2$ so that (5.1.6) holds.

Similarly, for $k \in (0, 1]$ we write

$$P_u(u) = kd_{k,\alpha}(u)\mathbf{A}_{(\alpha)}(u) + (1-k)d_{k,\alpha}(u)\,\mathrm{diag}[\langle u, \alpha_i\rangle] .$$

Because $\mathrm{diag}[\langle u, \alpha_i\rangle]$ is semipositive, we need only consider the first matrix on the right. We use (5.1.8) again to see that $\langle P_u(u)\zeta, \zeta\rangle \geq kc(\alpha)|u|^k|\zeta|^2$ so that (5.1.6) holds.

We now see that $\langle P_u(u)\zeta, \zeta\rangle \geq \lambda(u)|\zeta|^2$, with $\lambda(u) = C|u|^k$ for some positive constant C. With this choice of λ, the conditions $|\Lambda(u)| \leq C_*\lambda(u)$, $|A_u(u)| \leq C|\lambda_u(u)|$ and $|\lambda_u(u)||u| \lesssim \lambda(u)$ are obvious. Therefore, under the assumptions a.1)–a.3), the map P satisfies P).

If $d_i > 0$, $i = 1, \ldots, m$, we define $\lambda_i(u) = d_i + |\langle u, \alpha_i\rangle|^{k-1}\langle u, \alpha_i\rangle$ in (5.1.2) and easily see that P satisfies PR) because

$$\langle P_u(u)\zeta, \zeta\rangle \geq C(\min\{d_i\} + |u|^k)|\zeta|^2 \quad \text{for any } \zeta \in \mathbb{R}^{Nm} . \tag{5.1.9}$$

Finally, we consider the generalized (SKT) system defined in (5.0.1) (SKT) with P is given by (5.1.1) and

$$\lambda_i(u) = d_i + |\langle u, \alpha_i\rangle|^{k-1}\langle u, \alpha_i\rangle + \sum_{j=1}^{l} \beta_{ij}|\langle u, \alpha_i\rangle|^{\kappa_j - 1}\langle u, \alpha_i\rangle . \tag{5.1.10}$$

We apply the above argument to the exponents k and κ_j's, which are in $(0, k)$. One then see easily that $\mathrm{diag}[\sum_j \beta_{ij}|\langle u, \alpha_i\rangle|^{\kappa_j-1}]_{i=1}^m$ is semipositive, and the previous argument continues to hold and shows that the map P satisfies (5.1.9) and thus PR). The ellipticity function $\lambda(u) \sim \min\{d_i\} + |u|^k$.

Concerning the reaction terms $f_i(u) = k_iu_i + g_i(u)$, the assumption that $|(g_i)_u(u)| \leq C|u|^k$ clearly shows that f) is verified.

Thus, under the conditions a.1)–a.3), if $d_i > 0$ then the generalized (SKT) system satisfies PR) and f). Otherwise, i.e., $d_i = 0$ for all i, if $\beta_{ij} > 0$ then P) is satisfied.

5.2 Existence of strong/weak solutions to the evolutionary systems

We consider in this section the following boundary and initial condition parabolic system:

$$\begin{cases} u_t - \Delta(P(u)) = B(u, Du) + f(u)\,, & (x,t) \in \Omega \times (0, T_0)\,, \\ (\mathrm{BIC})\,, \end{cases} \tag{5.2.1}$$

where (BIC) denotes the boundary and initial conditions

$$(\mathbf{BIC}) \begin{cases} u = 0 \text{ or } \frac{\partial u}{\partial \nu} = 0 \text{ on } \partial\Omega \times (0, T_0)\,, \\ u(x, 0) = u_0(x)\,, \quad x \in \Omega\,. \end{cases} \tag{5.2.2}$$

Firstly, we will apply the theory in Chapter 3 to discuss the existence of strong solutions to this system when it is regular, i.e., PR) holds, with initial data u_0 in $W^{1,p_0}(\Omega)$ for some $p_0 > N$. We embed this system in the following family parameterized by $\sigma \in [0, 1]$:

$$\begin{cases} u_t - \Delta(P(u)) = \sigma^2 B(u, Du) + \sigma^2 f(u)\,, & (x,t) \in \Omega \times (0, T_0)\,, \\ u = 0 \text{ or } \frac{\partial u}{\partial \nu} = 0 \text{ on } \partial\Omega \times (0, T_0)\,, \\ u(x, 0) = \sigma u_0(x)\,, \quad x \in \Omega\,. \end{cases} \tag{5.2.3}$$

The existence of a strong solution to the regular system (5.2.1) will be established under the crucial assumption that the *strong* solutions to the family (5.2.3) *a priori* have small BMO norms in small balls (uniformly in $\sigma \in [0, 1]$). Namely, we consider the following property:

(SBMO) (Small BMO norm in small balls property) We say that a function $u\colon \Omega \times (0, T_0) \to \mathbb{R}^m$ satisfies (SBMO) if for any given $\mu_0 > 0$ there is $R > 0$ depending on the parameters in PR) and μ_0 such that for any ball B_R in $\mathbb{R}^N$ with $\Omega_R = B_R \cap \Omega \neq \emptyset$,

$$\sup_{t \in (0, T_0)} \|u(\cdot, t)\|_{BMO(\Omega_R)} \leq \mu_0\,.$$

Our first main result on the existence of strong solutions to the parabolic system (5.2.1) is the following theorem.

Theorem 5.2.1. *Assume that PR), SG), B) and f) hold. Suppose further that any strong solution u to the family* (5.2.3) *a priori satisfy the following conditions:*

a.1) $\sigma^{-1}u$ *satisfies (SBMO) uniformly in* $\sigma \in (0, 1]$.

a.2) *There is a constant* C_0 *such that*

$$\sup_{t \in (0, T_0)} \|u\|_{L^1(\Omega)} \leq C_0\,. \tag{5.2.4}$$

Then there exists a strong solution u to the system (5.2.1).

Next, we study the existence of a weak solution to the *degenerate* (5.2.1) where P is only assumed to satisfy the condition P), i.e., λ_0 can be 0.

We state the standard definition of weak solutions here.

Definition W). We say that u is a weak solution to (5.2.1) in $Q = \Omega\times(0, T_0)$ if $u \in L^1_{\mathrm{loc}}(Q)$ and $P(u) \in L^1_{\mathrm{loc}}(0, T_0 : W^{1,1}(\Omega))$; and for any $\eta \in C^1(\bar{Q})$, $\eta = 0$ on $\partial\Omega \times (0, T)$ and $\Omega \times \{T\}$ the following holds:

$$\iint_Q (-\langle u, \eta_t\rangle + \langle DP(u), D\eta\rangle)\, dz = \int_\Omega u_0\eta(x, 0)\, dx + \iint_Q \langle B(u, Du) + f(u), \eta\rangle\, dz\,.$$

Concerning the term $B(u, Du)$, we note that the growth condition B) implies $B(u, Du) = 0$ if $\lambda(u) = 0$. Otherwise, $A^{-1}(u)$ exists, and therefore the term Du in $B(u, Du)$ should be understood as $Du = A^{-1}(u)D(P(u))$.

We assume further that:

Ph) P^{-1} exists and is Hölder continuous for some $\alpha_P \in (0, 1]$: There is a constant $[P]_{\alpha_P} > 0$ such that $|P^{-1}(u)-P^{-1}(v)| \le [P]_{\alpha_P}|u-v|^{\alpha_P}$ for all $u, v \in \mathbb{R}^m$. Equivalently,

$$|u - v| \le [P]_{\alpha_P}|P(u) - P(v)|^{\alpha_P} \quad \text{for all}\;\; u, v \in \mathbb{R}^m\,. \tag{5.2.5}$$

An example of such P is $P(u) = |u|^k u$ for some $k > 0$. Then $P^{-1}(u) = |u|^{\frac{-k}{1+k}} u$, which is Hölder continuous with the exponent $\alpha_P = 1/(k + 1)$. The map P defined for the generalized (SKT) system in Section 5.1 also satisfies this condition. Indeed, away from the singular point $u = 0$, P is Lipschitz because P_u^{-1} exists and is bounded. At $u = 0$, it is clear that (5.2.5) holds because $|P(u)| \ge C|u|^{k+1}$ for some positive constant C.

The case $B(u, Du) \ne 0$ will be discussed later in Corollary 5.3.5. We state here the results without the drift term $B(u, Du)$ and consider the system

$$\begin{cases} u_t - \Delta(P(u)) = f(u)\,, & (x, t) \in \Omega \times (0, T_0)\,,\\ u = 0 \text{ or } \frac{\partial u}{\partial \nu} = 0 \text{ on } \partial\Omega \times (0, T_0)\,, & \\ u(x, 0) = u_0(x)\,, & x \in \Omega\,. \end{cases} \tag{5.2.6}$$

We will obtain a weak solution to this system as the limit of a sequence of strong solutions to regularized systems. To this end, let $\{\lambda_{0,n}\}$ be a sequence in $(0, 1)$ and $\lim_{n\to\infty} \lambda_{0,n} = 0$. We consider the following approximation systems for (5.2.6):

$$\begin{cases} u_t - \Delta(\lambda_{0,n}u + P(u)) = f(u)\,, & (x, t) \in \Omega \times (0, T_0)\,,\\ u = 0 \text{ or } \frac{\partial u}{\partial \nu} = 0 \text{ on } \partial\Omega \times (0, T_0)\,, & \\ u(x, 0) = u_{0,n}(x)\,, & x \in \Omega\,. \end{cases} \tag{5.2.7}$$

The system (5.2.7) is just a special case of that considered in Theorem 5.2.1 because $\lambda_{0,n} > 0$ and hence PR) is satisfied. Following Theorem 5.2.1, for each n we embed

(5.2.7) in the following family parameterized by $\sigma \in [0, 1]$:

$$\begin{cases} u_t - \Delta(\lambda_{0,n}u + P(u)) = \sigma^2 f(u)\,, & (x,t) \in \Omega \times (0, T_0)\,, \\ u = 0 \text{ or } \frac{\partial u}{\partial \nu} = 0 \text{ on } \partial\Omega \times (0, T_0)\,, & \\ u(x, 0) = \sigma u_{0,n}(x)\,, & x \in \Omega\,. \end{cases} \tag{5.2.8}$$

If strong solutions to the above system a priori satisfy the assumptions a.1) and a.2) of Theorem 5.2.1 then we obtain a sequence of strong solutions $\{u_n\}$ for (5.2.7). However, in order to pass to the limit to obtain the existence of a weak solution to our degenerate system (5.2.6), we have to assume that these strong solutions satisfy a stronger integrability condition than a.2) of Theorem 5.2.1 *uniformly in n* (see (5.2.10) below).

Concerning the initial condition of (5.2.6) and (5.2.7), we also assume that:

IC) There exists a sequence $\{u_{0,n}\}$ in $C^1(\Omega)$ that converges to u_0 in $L^1(\Omega)$. Furthermore, there is a constant C_0 such that for all n

$$\|(\lambda_{0,n} + \lambda(u_{0,n}))Du_{0,n}\|_{L^2(\Omega)} \le C_0\,. \tag{5.2.9}$$

Theorem 5.2.2. *Assume P), Ph), IC), f). Let $\{\lambda_{0,n}\}$ be a sequence in* $(0, 1)$ *and* $\lim_{n\to\infty} \lambda_{0,n} = 0$. *Consider the family* (5.2.8) *and assume that its strong solutions apriori satisfy the condition a.1) of Theorem 5.2.1 (with $B \equiv 0$) for each n.*

Assume also that there are constants $q_0 > N/2$, $q_1 > 2N/(N+2)$ and C_0, C_1 such that any strong solutions u_n of (5.2.7) *satisfy*

$$\sup_{t\in(0,T_0)} \|\lambda(u_n)\|_{L^{q_0}(\Omega)}, \quad \sup_{t\in(0,T_0)} \|u_n\|_{L^{q_1}(\Omega)} \le C_1\,. \tag{5.2.10}$$

Then there exists a weak solution u to the system (5.2.6).

We should emphasize that the condition (SBMO) is only assumed for each n in order to obtain the strong solutions to the regular systems (5.2.7) and this condition is uniform only in $\sigma \in (0, 1]$ but not in n. Meanwhile, the integrability condition (5.2.10) is assumed to be uniform in n. The condition (SBMO) seems to be the weakest one in the literature concerning Hölder regularity of solutions; yet it seems to be very hard to establish when we deal with general systems of more than one equation given on higher-dimension domains. Nevertheless, we can say more for systems given on planar domains below.

The planar case $N = 2$: In this case, thanks to the special structure of the systems, we will prove that (SBMO) can be verified under a very weak a priori integrability condition of strong solutions. On the other hand, as the Hölder continuity of the strong solutions u_n obtained by Theorem 5.2.1 is not uniform when $\lambda_{0,n} \to 0$ (see the proof of Theorem 5.2.3), this regularity cannot pass to that of the weak solution u found in Theorem 5.2.2. At least, we can show that this weak solution u is VMO (vanishing mean oscillation). That is,

$$\limsup_{R\to 0} \|u\|_{BMO(\Omega_R(x,t))} = 0\,, \quad \forall (x,t) \in \Omega_R \times (0, T_0)\,.$$

Theorem 5.2.3. *Let $N = 2$. Assume that P), Ph), f) and (5.0.5) hold. For any given $\lambda_{0,n} > 0$ assume that there are constants $q_0 > 1$ and C_1 such that strong solutions of (5.2.8) a priori satisfy*

$$\|\lambda(u)\|_{L^{q_0}(\Omega)}\,, \quad \|u\|_{L^{q_0}(\Omega)} \le C_1\,. \tag{5.2.11}$$

Then there exists a weak solution u to (5.2.6). Moreover, u is VMO.

5.3 The proofs

We present the proof of our theorems stated in the previous section concerning the boundary and initial condition problem.

$$\begin{cases} u_t - \Delta(P(u)) = B(u, Du) + f(u)\,, & (x,t) \in \Omega \times (0, T_0)\,, \\ (\text{BIC})\,. \end{cases} \tag{5.3.1}$$

We first give the proof of Theorem 5.2.1 on the existence of strong solutions to the boundary and initial condition problem (5.3.1) with initial data u_0 being in $W^{1,p_0}(\Omega)$ for some $p_0 > N$.

In the proof we will frequently make use of the following interpolation Sobolev inequality: For any $\varepsilon > 0$, $\beta \in (0, 1]$ and $W \in W^{1,2}(\Omega)$ we can find a constant $C(\varepsilon, \beta)$ such that

$$\|W\|_{L^q(\Omega)} \le \varepsilon \|DW\|_{L^p(\Omega)} + C(\varepsilon, \beta) \|W^\beta\|_{L^1(\Omega)}^{\frac{1}{\beta}} \quad \text{for any } q \in [1, p_*)\,. \tag{5.3.2}$$

We also note that the ellipticity condition (5.0.3) of P) and Young's inequality imply

$$\lambda(u)|Du|^2 \le \langle A(u)Du, Du\rangle = \langle DP(u), Du\rangle \le \frac{1}{2}\lambda^{-1}(u)|DP(u)|^2 + \frac{1}{2}\lambda(u)|Du|^2\,.$$

We then have $\lambda(u)|Du|^2 \le \lambda^{-1}(u)|DP(u)|^2$ so that $\lambda(u)|Du| \le |DP(u)|$. Of course, $|DP(u)| = |A(u)Du| \le C_*\lambda(u)|Du|$. Hence,

$$\lambda(u)|Du| \sim |DP(u)|\,. \tag{5.3.3}$$

Proof of Theorem 5.2.1. We apply Theorem 3.4.1 here by verifying its assumptions. First of all, we need to show that the number $\Lambda = \sup_{u \in \mathbb{R}^m} \Lambda(u)$, defined in (3.4.3) with $\Lambda(u) = |\lambda_u(u)|/\lambda(u)$, is finite. Since $\lambda(u) \ge \lambda_0 > 0$, if $|u|$ is bounded then so is $\Lambda(u)$. For large $|u|$ we use the assumption in P) that $|\lambda_u(u)| \lesssim \lambda(u)/|u|$ to see that $\Lambda(u) \lesssim 1/|u|$ is also bounded. Hence, the number Λ is finite.

Next, following Theorem 3.4.1, we consider the following family (which is not exactly (3.3.2) but we can see that the theory of Chapter 3 is still available):

$$\begin{cases} u_t - \operatorname{div}(A(\sigma u)Du) = \sigma B(\sigma u, \sigma Du) + \sigma f(\sigma u)\,, & (x,t) \in \Omega \times (0, T_0)\,, \\ \text{(BIC)}\,. \end{cases} \tag{5.3.4}$$

Multiplying $\sigma > 0$ to the equation in (5.3.4), we see easily that $w = \sigma u$ is a strong solution to

$$\begin{cases} u_t - \Delta(P(u)) = \sigma^2 B(u, Du) + \sigma^2 f(u)\,, & (x,t) \in \Omega \times (0, T_0)\,, \\ u = 0 \text{ or } \frac{\partial u}{\partial \nu} = 0 \text{ on } \partial\Omega \times (0, T_0)\,, & \\ u(x,0) = \sigma u_0(x)\,, & x \in \Omega\,. \end{cases} \tag{5.3.5}$$

Thus, the condition (3.4.8) in Theorem 3.4.1, that strong solutions to (3.3.2) have small BMO norms in small balls, is already assumed in a.1) where $u = \sigma^{-1}w$ satisfies (SBMO). We need only check the integrability condition M.0'). To proceed, we note the following consequence of (SBMO). Fixing $\mu_0 = 1$, we can find a fixed $R_1 > 0$ such that any strong solution u to (5.3.4) satisfies

$$\sup_{t\in(0,T_0)} \|u(\cdot,t)\|_{BMO(B_{R_1})} \le 1\,.$$

Since Ω is bounded, by using a finite covering of finitely many balls of radius R_1 we deduce from the above that

$$\sup_{t\in(0,T_0)} \|u(\cdot,t)\|_{BMO(\Omega)} \le C(R_1)\,. \tag{5.3.6}$$

For each $q \ge 1$ and $t \in (0, T_0)$ it is well known [16] that

$$\|u(\cdot,t)\|_{L^q(\Omega)} \le C(q, \|u(\cdot,t)\|_{BMO(\Omega)}, \|u(\cdot,t)\|_{L^1(\Omega)})\,.$$

We then conclude from (5.3.6) and the assumption (5.2.4) in a.2) of the theorem that there is a constant $C(q, C_0)$ that also depends on R_1 (or the geometry of Ω) such that

$$\sup_{t\in(0,C_0)} \|u(\cdot,t)\|_{L^q(\Omega)} \le C(q, T_0) \quad \forall q \ge 1\,. \tag{5.3.7}$$

From the polynomial growths of λ and f, we now see that the integrands $\lambda(u)$, $f(u)\lambda^{-1}(u)$ in M.0') are bounded by powers of $|u|$ so that the integrability conditions (3.4.6) and (3.4.7) are verified by (5.3.7). Therefore, in order to check the condition M.0') of Theorem 3.4.1 we need only show that

$$\int_0^{T_0}\int_\Omega |Du|^2\,dxdt \le C_0(T_0) \tag{5.3.8}$$

is verified. To prove this, we test the system with u and easily obtain

$$\int_0^{T_0}\int_\Omega \lambda(\sigma u)|Du|^2\,dxdt \le \int_\Omega \sigma|u_0|^2\,dx + \int_0^{T_0}\int_\Omega \sigma\langle B(\sigma u, \sigma Du) + f(\sigma u), u\rangle\,dxdt\,.$$

Using B) and Young's inequality,

$$|\langle B(\sigma u, \sigma Du), u\rangle| \le \varepsilon\lambda(\sigma u)|Du|^2 + C(\varepsilon)\lambda(\sigma u)|\sigma u|^2 .$$

In addition, by (5.0.5), $|f(\sigma u)| \le C(1 + |\sigma u|)(1 + \lambda(\sigma u))$ so that

$$\int_0^{T_0}\int_\Omega \lambda(\sigma u)|Du|^2\,dxdt \le C + C\int_0^{T_0}\int_\Omega (|\sigma u|^2 + \lambda(\sigma u)|\sigma u|^2)\,dxdt . \tag{5.3.9}$$

From the growth condition of $\lambda(u)$, the integrand on the right-hand side of the above is a polynomial in $|u|$. By (5.3.7), we conclude that the right-hand side of (5.3.9) is bounded uniformly in σ.

Because $\lambda(\sigma u)$ is bounded from below by $\lambda_0 > 0$, we obtain (5.3.8). The hypotheses of Theorem 3.4.1 are all verified and the proof is complete. □

Next, assuming $B \equiv 0$, we will prove Theorem 5.2.2 and establish the existence of a weak solution to the following boundary and initial condition problem:

$$\begin{cases} u_t - \Delta(P(u)) = f(u) , & (x,t) \in \Omega \times (0, T_0) , \\ \text{(BIC)} . \end{cases} \tag{5.3.10}$$

Let $\{\lambda_{0,n}\}$ be a sequence in $(0, 1)$ and $\lim_{n\to\infty} \lambda_{0,n} = 0$. We denote $P_n(u) = \lambda_{0,n}u + P(u)$ and consider the following approximation systems with initial data $u_{0,n}$ in $W^{1,p_0}(\Omega)$ for some $p_0 > N$:

$$\begin{cases} u_t - \Delta(P_n(u)) = f(u) , & (x,t) \in \Omega \times (0, T_0) , \\ u = 0 \text{ or } \frac{\partial u}{\partial \nu} = 0 \text{ on } \partial\Omega \times (0, T_0) , & \\ u(x, 0) = u_{0,n}(x) , & x \in \Omega , \end{cases} \tag{5.3.11}$$

which is embedded in the following family parameterized by $\sigma \in [0, 1]$:

$$\begin{cases} u_t - \Delta(P_n(u)) = \sigma^2 f(u) , & (x,t) \in \Omega \times (0, T_0) , \\ u = 0 \text{ or } \frac{\partial u}{\partial \nu} = 0 \text{ on } \partial\Omega \times (0, T_0) , & \\ u(x, 0) = \sigma u_{0,n}(x) , & x \in \Omega . \end{cases} \tag{5.3.12}$$

As we assume in Theorem 5.2.2 that for each n the strong solutions to (5.3.12) a priori satisfy the small BMO norm condition a.1) of Theorem 5.2.1 and the condition a.2) is also satisfied because the exponents q_0, q_1 in (5.2.10) are greater than 1, and there is a sequence of strong solutions $\{u_n\}$ of (5.3.11). In order to pass to the limit to obtain the existence of a weak solution to our degenerate system (5.3.10), we have to to establish *uniform* estimates for these strong solutions u_n under the integrability conditions (5.2.9) and (5.2.10) of Theorem 5.2.2.

The following proposition provides the needed uniform estimates.

Proposition 5.3.1. *Assume P) and f). Assume also that there are constants $q_0 > N/2$, $q_1 > 2N/(N+2)$ and C_0, C_1 such that the initial data $u_{0,n} \in C^1(\Omega)$ and the corresponding strong solutions u_n of* (5.3.11) *satisfy*

$$\|\Lambda(u_{0,n})Du_{0,n}\|_{L^2(\Omega)} \le C_0\,, \tag{5.3.13}$$

$$\sup_{t\in(0,T_0)} \|\lambda(u_n)\|_{L^{q_0}(\Omega)}, \quad \sup_{t\in(0,T_0)} \|u_n\|_{L^{q_1}(\Omega)} \le C_1\,. \tag{5.3.14}$$

Then there are constants $C(C_0, C_1)$ and $q_2 > 1$ such that for every n

$$\sup_{t\in[0,T_0]} \int_\Omega (\lambda_{0,n}^2 + \lambda^2(u_n))|Du_n|^2\,dx \le C(C_0, C_1)\,, \tag{5.3.15}$$

$$\iint_{\Omega\times[0,T_0]} \lambda(u_n)|(u_n)_t|^2\,dz \le C(C_0, C_1)\,, \tag{5.3.16}$$

and

$$\sup_{t\in(0,T_0)} \int_\Omega |f(u_n)|^{q_2}\,dx \le C(C_0, C_1)\,. \tag{5.3.17}$$

The uniform estimates (5.3.15)–(5.3.17) will come from the following two lemmas, which discuss the strong solutions of

$$\begin{cases} u_t - \Delta(\mathcal{P}(u)) = f(u)\,, & (x,t) \in \Omega \times (0, T_0)\,, \\ (\mathrm{BIC})\,. \end{cases} \tag{5.3.18}$$

Here, $\mathcal{P} = [\mathcal{P}_i]_{i=1}^m$ is a C^2 map on $\mathbb{R}^m$ and satisfies the condition P). In the lemmas and their proof, we will denote $\mathcal{A}(u) = \mathcal{P}_u(u)$ and the ellipticity function λ for $\mathcal{A}$ by $\lambda_{\mathcal{A}}$.

The first lemma provides a differential inequality for $\|\mathcal{A}(u)Du\|_{L^2(\Omega)}$.

Lemma 5.3.2. *Let u be a strong solution to* (5.3.18). *For any $t \in (0, T_0)$,*

$$\int_{\Omega\times\{t\}} \lambda_{\mathcal{A}}(u)|u_t|^2\,dx + \frac{d}{dt}\int_{\Omega\times\{t\}} |\mathcal{A}(u)Du|^2\,dx \le C\int_{\Omega\times\{t\}} \lambda_{\mathcal{A}}(u)|f(u)|^2\,dx\,. \tag{5.3.19}$$

Proof. Because u is a strong solution, for any $t \in (0, T_0)$ we can test the system with $\mathcal{P}(u)_t$. This means we multiply the i-th equation of the system by $(\mathcal{P}_i(u))_t$ and integrate by parts in x over Ω. Summing the results, we get, writing Ω for $\Omega \times \{t\}$,

$$\int_\Omega (\langle \mathcal{P}(u)_t, u_t\rangle + \langle D(\mathcal{P}(u)), D(\mathcal{P}(u)_t)\rangle)\,dx = \int_\Omega \langle f(u), \mathcal{P}(u)_t\rangle\,dx\,.$$

As $\mathcal{P}_u = \mathcal{A}(u)$ and $D(\mathcal{P}(u)_t) = (D\mathcal{P}(u))_t = D(\mathcal{A}(u)u_t)$, we have

$$\int_\Omega \left[\langle \mathcal{A}(u)u_t, u_t\rangle + \frac{1}{2}\frac{\partial}{\partial t}(|\mathcal{A}(u)Du|^2)\right]dx = \int_\Omega \langle f(u), \mathcal{A}(u)u_t\rangle\,dx\,, \tag{5.3.20}$$

because $\langle D(\mathcal{P}(u)), D(\mathcal{A}(u)u_t)\rangle = \frac{1}{2}\frac{\partial}{\partial t}(|D(\mathcal{A}(u)Du)|^2)$.

We now use the ellipticity of $\mathcal{A}(u)$ in the first integrand on the left-hand side of (5.3.20) to get $\langle \mathcal{A}(u)u_t, u_t\rangle \ge \lambda_{\mathcal{A}}(u)|u_t|^2$. Also, as $|\mathcal{A}(u)| \le C\lambda_{\mathcal{A}}(u)$, we use

Young's inequality to find a constant $C(\varepsilon)$ such that for any $\varepsilon > 0$ we can estimate the second integrand on the right-hand side as follows: $|\langle f(u), \mathcal{A}(u)u_t\rangle| \le \varepsilon\lambda_{\mathcal{A}}(u)|u_t|^2 + C(\varepsilon)\lambda_{\mathcal{A}}(u)|f(u)|^2$. Using these facts in (5.3.20) with sufficiently small ε, we get (5.3.19). □

In order to estimate the integral of $\lambda_{\mathcal{A}}(u)|f(u)|^2$ in (5.3.19) we need the following lemma.

Lemma 5.3.3. *Assume that $|u||(\lambda_{\mathcal{A}})_u(u) \lesssim \lambda_{\mathcal{A}}(u)$ and that there are constants $q_0 > N/2$, $q_1 > 2N/(N+2)$ and C_1 such that*

$$\|\lambda_{\mathcal{A}}(u)\|_{L^{q_0}(\Omega)},\ \|u\|_{L^{q_1}(\Omega)} \le C_1\,. \tag{5.3.21}$$

Then there are constants C, $C(C_1)$ and $q \in (2, 2_*)$ such that

$$\left(\int_\Omega (\lambda_{\mathcal{A}}(u)|u|)^q\,dx\right)^{\frac{2}{q}} \le C\int_\Omega |\lambda_{\mathcal{A}}(u)Du|^2\,dx + C(C_1)\,, \tag{5.3.22}$$

and

$$\int_\Omega (|u|^2\lambda_{\mathcal{A}}(u) + |u|^2\lambda_{\mathcal{A}}^3(u))\,dx \le C\int_\Omega |\lambda_{\mathcal{A}}(u)Du|^2\,dx + C(C_1)\,. \tag{5.3.23}$$

Proof. From the assumptions on q_0, q_1 it is clear that we can find $q \in (2, 2_*)$ such that $\frac{N}{2} < (\frac{q}{2})' = q/(q-2) \le q_0$ and $2N/(N+2) < q' \le q_1$. By the Hölder and Cauchy inequalities and the assumption (5.3.21), we have

$$\begin{aligned}\int_\Omega \lambda_{\mathcal{A}}^3(u)|u|^2\,dx &\le \left(\int_\Omega (\lambda_{\mathcal{A}}(u)|u|)^q\,dx\right)^{\frac{2}{q}} \|\lambda_{\mathcal{A}}(u)\|_{L^{(\frac{q}{2})'}(\Omega)}\\ &\le C_1\left(\int_\Omega (\lambda_{\mathcal{A}}(u)|u|)^q\,dx\right)^{\frac{2}{q}}\,,\end{aligned} \tag{5.3.24}$$

$$\begin{aligned}\int_\Omega \lambda_{\mathcal{A}}(u)|u|^2\,dx &\le \left(\int_\Omega (\lambda_{\mathcal{A}}(u)|u|)^q\,dx\right)^{\frac{1}{q}} \|u\|_{L^{q'}(\Omega)}\\ &\le \left(\int_\Omega (\lambda_{\mathcal{A}}(u)|u|)^q\,dx\right)^{\frac{2}{q}} + C_1^2\,.\end{aligned} \tag{5.3.25}$$

Thus, (5.3.23) follows from (5.3.22), which we will prove below.

Because $q < 2_*$, we can apply the interpolation inequality (5.3.2) to estimate the integral of $(\lambda_{\mathcal{A}}(u)|u|)^q$. Note that

$$|D(\lambda_{\mathcal{A}}(u)|u|)| \lesssim \lambda_{\mathcal{A}}(u)|Du| + |u||(\lambda_{\mathcal{A}})_u(u)||Du| \lesssim \lambda_{\mathcal{A}}(u)|Du|$$

thanks to the assumption $|u||(\lambda_{\mathcal{A}})_u(u) \lesssim \lambda_{\mathcal{A}}(u)$ of the lemma. We then have, by the interpolation inequality, for any given $\varepsilon, \beta > 0$,

$$\left(\int_\Omega (\lambda_{\mathcal{A}}(u)|u|)^q\,dx\right)^{\frac{2}{q}} \le \varepsilon\int_\Omega |\lambda_{\mathcal{A}}(u)Du|^2\,dx + C(\varepsilon,\beta)\left(\int_\Omega (\lambda_{\mathcal{A}}(u)|u|)^\beta\,dx\right)^{\frac{2}{\beta}}. \tag{5.3.26}$$

As $(\lambda_{\mathcal{A}}(u)|u|)^\beta = (\lambda_{\mathcal{A}}^3(u)|u|^2)^{\beta/3}|u|^{\beta/3}$, we have

$$\int_\Omega (\lambda_{\mathcal{A}}(u)|u|)^\beta\,dx \le C_1\left(\int_\Omega \lambda_{\mathcal{A}}^3(u)|u|^2\,dx\right)^{\frac{\beta}{3}}\left(\int_\Omega |u|^{\frac{\beta}{3-\beta}}\,dx\right)^{\frac{3-\beta}{3}}.$$

We choose β small enough such that $\beta/(3-\beta) \le q_1$. By (5.3.21) and Young's inequality, we obtain from the above that

$$\left(\int_\Omega (\lambda_{\mathcal{A}}(u)|u|)^\beta\,dx\right)^{\frac{2}{\beta}} \le C_1\left(\int_\Omega \lambda_{\mathcal{A}}^3(u)|u|^2\,dx\right)^{\frac{2}{3}} \le \varepsilon\int_\Omega \lambda_{\mathcal{A}}^3(u)|u|^2\,dx + C(\varepsilon, C_1).$$

We estimate the last integral by (5.3.24) and then use (5.3.26) to get

$$\left(\int_\Omega (\lambda_{\mathcal{A}}(u)|u|)^\beta\,dx\right)^{\frac{2}{\beta}} \le C_1\varepsilon\int_\Omega |\lambda_{\mathcal{A}}(u)Du|^2\,dx + C_1\varepsilon\left(\int_\Omega (\lambda_{\mathcal{A}}(u)|u|)^\beta\,dx\right)^{\frac{2}{\beta}} + C(\varepsilon, C_1).$$

For some sufficiently small ε, $C_1\varepsilon < 1/2$, the second integral on the right-hand side can be absorbed into the left. We then have

$$\left(\int_\Omega (\lambda_{\mathcal{A}}(u)|u|)^\beta\,dx\right)^{\frac{2}{\beta}} \le C\int_\Omega |\lambda_{\mathcal{A}}(u)Du|^2\,dx + C(C_1).$$

Using this in (5.3.26), we obtain (5.3.22) and complete the proof. □

We now combine the last two lemmas to prove Proposition 5.3.1.

Proof of Proposition 5.3.1. Let $\mathcal{A}(u) = \lambda_{0,n}I + A(u)$. The growth condition (5.0.5) gives $\lambda_{\mathcal{A}}(u)|f(u)|^2 \lesssim \lambda_{\mathcal{A}}(u) + |u|^2\lambda_{\mathcal{A}}(u) + |u|^2\lambda_{\mathcal{A}}^3(u)$. The integrability condition (5.3.14) implies (5.3.21) of Lemma 5.3.3 so that its conclusion (5.3.23) and the fact that $|\lambda_{\mathcal{A}}(u)Du| \sim |\mathcal{A}(u)Du|$ ((5.3.3)) yield

$$\int_\Omega \lambda_{\mathcal{A}}(u)|f(u)|^2\,dx \le C\int_\Omega |\mathcal{A}(u)Du|^2\,dx + C(C_1).$$

We can use this in (5.3.19) of Lemma 5.3.2 to obtain

$$\int_{\Omega\times\{t\}} \lambda_{\mathcal{A}}(u)|u_t|^2\,dx + \frac{d}{dt}\int_{\Omega\times\{t\}} |\mathcal{A}(u)Du|^2\,dx \le C\int_{\Omega\times\{t\}} |\mathcal{A}(u)Du|^2\,dx + C(C_1). \tag{5.3.27}$$

Define $y(t) := \|\mathcal{A}(u)Du\|^2_{L^2(\Omega\times\{t\})}$. We obtain from (5.3.27) that $y' \le Cy + C(C_1)$. By Gronwall's inequality and the assumption on the initial condition (5.3.13) (the initial data $u_{0,n}$ is C^1 smooth so that u_n is a classical solution in Q), we see that any strong solution u of (5.3.11) satisfies

$$\|\mathcal{A}(u)Du\|^2_{L^2(\Omega\times\{t\})} \le C\|\mathcal{A}(u)Du\|^2_{L^2(\Omega\times\{0\})} + C(C_1) \quad \text{for any } t \in (0, T_0).$$

As $\lambda_{\mathcal{A}}(u) = \lambda_{0,n} + \lambda(u)$ and $\lambda_{\mathcal{A}}(u)|Du| \lesssim |\mathcal{A}(u)Du|$, we prove (5.3.15) of the theorem. Next, by integrating (5.3.27), we then have for all $t \in (0, T_0)$ that

$$\begin{aligned}\int_t^{T_0}\int_\Omega \lambda_{\mathcal{A}}(u)|u_t|^2\,dx\,ds &+ \int_{\Omega\times\{T_0\}} |\mathcal{A}(u)Du|^2\,dx\\ &\le \int_{\Omega\times\{t\}} |\mathcal{A}(u)Du|^2\,dx + C(C_1) \le C(C_0, C_1).\end{aligned}$$

Letting $t \to 0$ and using the assumption (5.3.13) on the initial data, we obtain

$$\int_0^{T_0} \int_\Omega \lambda_{\mathcal{A}}(u)|u_t|^2 \, dx \, ds \le C(C_0, C_1) ,$$

and prove (5.3.16).

Finally, let $q_2 = \min\{q_1, q\}$ with q being the exponent in (5.3.22). By (5.0.5),

$$\int_\Omega |f(u)|^{q_2} \, dx \le C \int_\Omega (|u|^{q_2} + |u|^{q_2} \lambda^{q_2}(u)) \, dx , \tag{5.3.28}$$

so that the estimate (5.3.17) for $f(u)$ comes from the assumption (5.3.21) and the bound (5.3.22) of Lemma 5.3.3 and then (5.3.15). The proof is complete. □

Remark 5.3.4. We can include a drift term $B(u, Du)$ satisfying $|B(u, Du)| \lesssim \lambda_{\mathcal{A}}^{\frac{1}{2}}(u)|Du|$ into the above calculation. Indeed, we will have an extra integral of $|B(u, Du)||P_u(u)u_t|$ on the right-hand side of (5.3.20) in the proof of Lemma 5.3.2 and it can be estimated via Young's inequality as follows:

$$\varepsilon \int_\Omega \lambda_{\mathcal{A}}(u)|u_t|^2 \, dx + C(\varepsilon) \int_\Omega |\lambda_{\mathcal{A}}(u) Du|^2 \, dx .$$

Choosing sufficiently small ε, we obtain (5.3.27) again and see that Proposition 5.3.1 still holds.

We now turn to the proof of Theorem 5.2.2 on the existence of a weak solution to the degenerate system (we assume that $B \equiv 0$).

Proof of Theorem 5.2.2. Consider the sequence of strong solutions $\{u_n\}$ obtained from Theorem 5.2.1, with initial data $u_{0,n}$. This sequence exists because we are assuming that the (SBMO) condition a.1), for each n, and the condition a.2) is also satisfied because the exponents q_0, q_1 in (5.2.10) are greater than 1.

Denote $U_n := P(u_n)$. For any $q \in (1, 2)$, because

$$|(U_n)_t|^q \lesssim \lambda_n(u_n)^q |(u_n)_t|^q = \lambda_n(u_n)^{\frac{q}{2}} \lambda_n(u_n)^{\frac{q}{2}} |(u_n)_t|^q ,$$

we apply Hölder's inequality to obtain for $Q = \Omega \times [0, T_0]$ that

$$\iint_Q |(U_n)_t|^q \, dz \le \left(\iint_Q \lambda_n(u_n)^{\frac{q}{2-q}} \, dz \right)^{1-\frac{q}{2}} \left(\iint_Q \lambda_n(u_n)|(u_n)_t|^2 \, dz \right)^{\frac{q}{2}} .$$

As we assume that $\|\lambda(u_n)\|_{L^{q_0}(\Omega)}$ is uniformly bounded for some $q_0 > 1$, there is $q > 1$ such that $\frac{q}{2-q} \in (1, q_0)$ and therefore the first integral on the right-hand side is bounded uniformly by a constant C. By (5.3.16) of Proposition 5.3.1, the second integral is also bounded. By Hölder's inequality,

$$\int_0^{T_0} \left(\int_\Omega |(U_n)_t| \, dx \right)^q dt \le |\Omega|^{q-1} \iint_Q |(U_n)_t|^q \, dz .$$

Thus, the above shows that there is $q > 1$ such that $\{(U_n)_t\}$ is bounded in $L^q([0, T_0], X_1)$, where $X_1 = L^1(\Omega)$.

Let $X_0 := W^{1,2}(\Omega)$. Using $|DP(u)| \le C\Lambda(u)|Du|$ and (5.3.15), we get

$$\int_\Omega |DP(u_n)|^2\,dx \le C\int_\Omega |\Lambda(u_n)Du_n|^2\,dx \le \int_\Omega |\Lambda_n(u_n)Du_n|^2\,dx \le C(C_0, C_1)\,.$$

We see that $\{U_n\}$ is bounded in $L^\infty([0, T_0], X_0)$.

Hence, $\{U_n\}$ is a bounded set in

$$\{U \in L^\infty([0, T_0], X_0) : U_t \in L^q(0, T_0), X_1\}\,.$$

Since X_0 is compactly embedded in $X = L^p(\Omega)$ for any $p \in (1, 2_*)$ and X is continuously embedded in X_1, we can use the Aubin–Lions–Simon lemma to see that $\{U_n\}$ is compactly embedded in $C([0, T_0], X)$.

Thus, for any given $p \in (1, 2_*)$ we can find a subsequence of $\{U_n\}$ such that, after relabeling,

$$U_n \to U \quad \text{in } C([0, T_0], L^p(\Omega))\,. \tag{5.3.29}$$

Since P^{-1} is Hölder continuous, we have from the condition Ph) that

$$\int_\Omega |u_n - u_m|^q\,dx = \int_\Omega |P^{-1}(U_n) - P^{-1}(U_m)|^q\,dx \le [P^{-1}]_\alpha \int_\Omega |U_n - U_m|^{\alpha_P q}\,dx\,.$$

For $q > 1$ such that $\alpha_P q < 2_*$ the above and (5.3.29) imply $u_n \to u$ in $L^q(Q)$ for some u. Of course, because $\lambda_{0,n} \to 0$, $\lambda_{0,n}u_n + U_n \to P(u)$ in $L^q(Q)$. This implies $D(\lambda_{0,n}u_n + U_n)$ converges to $DP(u)$ in the sense of distribution. In fact, (5.3.15) shows that $D(\lambda_{0,n}u_n + U_n)$ converges weakly to $DP(u)$ in $L^2(Q)$.

On the other hand, by (5.3.17), $f(u_n)$ is uniformly bounded in $L^{q_2}(\Omega)$ for some $q_2 > 1$, and it converges weakly to $f(u)$ in $L^{q_2'}(\Omega)$.

For any $\eta \in C^1(\bar{Q})$, $\eta = 0$ on $\partial\Omega \times (0, T)$ and $\Omega \times \{T\}$, we multiply η to the equation of the strong solution u_n and derive

$$\begin{aligned}&\iint_Q (-\langle u_n, \eta_t\rangle + \langle D(\lambda_{0,n}u_n + U_n), D\eta\rangle)\,dz\\ &= \int_\Omega u_{0,n}\eta(x, 0)\,dx + \iint_Q \langle f(u_n), \eta\rangle\,dz\,.\end{aligned}$$

Let $n \to \infty$. By the convergences established above and the condition on the initial data in IC) we obtain

$$\iint_Q (-\langle u, \eta_t\rangle + \langle DP(u), D\eta\rangle)\,dz = \int_\Omega u_0\eta(x, 0)\,dx + \iint_Q \langle f(u), \eta\rangle\,dz\,.$$

Thus u is a weak solution. The proof is complete. □

The key points of the proof are the existence of strong solutions to the approximation systems (5.3.12) and the use of the Aubin–Lions–Simon lemma, which requires the uniform estimates for $\|(U_n)_t\|_{L^q(\Omega)}$ and $\|DU_n\|_{L^2(\Omega)}$. These bounds are provided by

Proposition 5.3.1, which holds for any N under a mild integrability assumption (5.3.14) that $\sup_{t\in(0,T_0)} \|\lambda(u_n)\|_{L^{q_0}(\Omega)}$ and $\sup_{t\in(0,T_0)} \|u_n\|_{L^{q_1}(\Omega)}$ are a priori uniformly bounded for some $q_0 > N/2$ and $q_1 > 2N/(N+2)$. Hence, the crucial condition (SBMO) in a.1) of Theorem 5.2.1 must be established in order to establish the existence of a sequence of strong solutions to the approximation systems. This condition is not easy to validate in general. However, when $N = 2$, Proposition 5.3.1 provides a bound for $\sup_{t\in(0,T_0)} \|Du_n\|_{L^2(\Omega)}$ and allows us to give the proof of the existence of weak solutions to (5.3.10) and their VMO property.

Proof of Theorem 5.2.3. For any given $\lambda_{0,n} > 0$ we assume in the theorem that there are constants $q_0 > 1$ and C_1 such that strong solutions of

$$\begin{cases} u_t - \Delta(\lambda_{0,n} u + P(u)) = \sigma f(u) , & x \in \Omega , \\ \text{(BIC)} \end{cases} \tag{5.3.30}$$

a priori satisfy

$$\|\lambda(u)\|_{L^{q_0}(\Omega)},\ \|u\|_{L^{q_0}(\Omega)} \le C_1 . \tag{5.3.31}$$

First of all, we show that the condition a.1) of Theorem 5.2.1 holds so that strong solutions of (5.3.30) exist for $\sigma = 1$. Consider a strong solution u of the family (5.3.30), $\sigma \in (0,1]$. Under the condition (5.3.31) the proof of Lemma 5.3.2 and Lemma 5.3.3 applies here with $q_0 = q_1 > 1$ (because $N = 2$) and $f(u)$ being $\sigma^2 f(u)$. We obtain from (5.3.15)

$$\sup_{t\in[0,T_0]} \int_\Omega (\lambda_{0,n}^2 + \lambda^2(u))|Du|^2\,dx \le \sigma^2 C(C_1)$$

so that

$$\sup_{t\in[0,T_0]} \int_\Omega |D(\sigma^{-1}u)|^2\,dx \le \lambda_{0,n}^{-2} C(C_1) .$$

As $N = 2$, a simple use of Poincaré's inequality, the continuity of integral and the last estimate show that $\sigma^{-1}u$ satisfies the (SBMO) condition (uniformly in $\sigma \in (0,1]$). Thus, a.1) is verified. The condition a.2) of Theorem 5.2.1 is just (5.3.31) here. We obtain a sequence of strong solutions $\{u_n\}$ to (5.3.30) for $\sigma = 1$. From Theorem 5.2.2, letting $\lambda_{0,n} \to 0$ we then obtain a weak solution u. To finish the proof we will show that u is VMO.

Because $|D(P(u_n))| \lesssim \lambda(u_n)|Du_n|$, by (5.3.15), the strong solutions satisfy

$$\sup_{(0,T_0)} \|D(P(u_n))\|_{L^2(\Omega)} \le C(C_1) .$$

Let $U_n = P(u_n)$. For any $q > 1$ it is well known that there is $c(q)$ such that

$$\int_{\Omega_R} |u_n - (u_n)_R|^q\,dx \le c(q) \int_{\Omega_R} |u_n - P^{-1}(U_n)_R|^q\,dx .$$

We use the Hölder property of P^{-1} in Ph) to estimate the last integral.

$$\int_{\Omega_R} |P^{-1}(U_n) - P^{-1}(U_n)_R|^q\,dx \le [P^{-1}]_{\alpha_P}^q \int_{\Omega_R} |U_n - (U_n)_R|^{q\alpha_P}\,dx .$$

By the Poincaré–Sobolev inequality and the uniform continuity of the integrals, for any $\mu_0 > 0$ there is $R > 0$ that depends only on μ_0 such that

$$R^{-2}\int_{\Omega_R} |U_n - (U_n)_R|^2 \, dx \le \int_{\Omega_R} |DU_n|^2 \, dx \le \mu_0 \,, \quad \forall n \,.$$

Take $q = 2/\alpha_P$. The above estimates yield

$$R^{-2}\int_{\Omega_R} |u_n - (u_n)_R|^q \, dx \le C(\alpha_P, [P^{-1}]_{\alpha_P})\mu_0 \,, \quad \forall n \,.$$

From the proof of Theorem 5.2.2, we see that $u_n \to u$ in $L^q(\Omega)$ for any $q > 1$ because $q_* = \infty$. Letting $n \to \infty$ in the above estimate, we see that u satisfies the same estimate. By the equivalence of BMO norm definitions, we have $[u]_{BMO(\Omega_R)} \le C(\alpha_P, [P^{-1}]_{\alpha_P})\mu_0$. As μ_0 can be arbitrarily small, u is VMO. The proof is complete. □

We consider the presence of the drift term.

Corollary 5.3.5. *The conclusion of Theorem 5.2.2 still holds if $B(u, Du) \neq 0$ and:*

Bw.1) *Let $K = \{u \in \mathbb{R}^m : \lambda(u) = 0\}$ be the degenerate set of $A(u) = P_u(u)$. We assume that K is bounded and there is a bounded neighborhood N_K of K and $c_0 > 0$ such that $\lambda(u) \ge c_0$ if $u \in \mathbb{R}^m \setminus N_K$; and*

$$|B(u,\zeta)| \le C\lambda(u)|\zeta| \text{ on } N_K \,, \quad |B(u,\zeta)| \le C\lambda^{\frac{1}{2}}(u)|\zeta| \text{ on } \mathbb{R}^m \setminus N_K \,. \tag{5.3.32}$$

Bw.2) *The map $B_A : \mathbb{R}^m \times \mathbb{R}^{mN} \to \mathbb{R}^m$ defined by*

$$B_A(u,\zeta) := \begin{cases} B(u, A^{-1}(u)\zeta) & \text{if } u \notin K \,, \\ 0 & \text{if } u \in K \,, \end{cases} \tag{5.3.33}$$

satisfies that condition that if $\{u_n\}$ and $\{\zeta_n\}$ are sequences of vector-valued functions on Ω such that $u_n \to u$ a.e. in Ω and $\zeta_n \to \zeta$ weakly in $L^2(\Omega)$ then $B_A(u_n, \zeta_n) \to B_A(u, \zeta)$ weakly in $L^1(\Omega)$.

Proof. First of all, if $u \in \mathbb{R}^m \setminus N_K$ then $\lambda(u) \ge c_0$ so that, by (5.3.32), $|B(u,\zeta)| \le Cc_0^{-\frac{1}{2}}\lambda(u)|\zeta|$. We then see that the condition Bw.1) implies $|B(u,\zeta)| \lesssim \lambda(u)|\zeta|$ for all $u \in \mathbb{R}^m$ so that B) is satisfied.

Next, because λ is continuous we can find a constant $c_1 > 0$ such that $\lambda(u) \le c_1$ if $u \in N_K$. Then, by (5.3.32), $|B(u,\zeta)| \le C\lambda(u)|\zeta| \le Cc_1^{\frac{1}{2}}\lambda^{\frac{1}{2}}(u)|\zeta|$. This and the second inequality in (5.3.32) then imply $|B(u,\zeta)| \lesssim \lambda^{\frac{1}{2}}(u)|\zeta|$ for all $u \in \mathbb{R}^m$. By Remark 5.3.4, the estimates of Proposition 5.3.1 hold so that Theorem 5.2.2 provides a sequence of strong solutions $\{u_n\}$ to (5.3.11), which is now

$$u_t - \Delta(\lambda_{0,n}u + P(u)) = B(u, Du) + f(u) \,, \quad (x,t) \in \Omega \times (0, T_0) \,. \tag{5.3.34}$$

For $q \ge 2$ such that $\alpha_P q < 2_*$, we showed in Theorem 5.2.2 that $u_n \to u$ in $L^q(Q)$ and $DP(u_n)$ converges weakly to $DP(u)$ in $L^2(Q)$. We can assume further that $u_n \to u$ a.e. in Q.

For any $\eta \in C^1(\bar{Q})$ we have

$$\iint_Q (-\langle u_n, \eta_t\rangle + \langle \lambda_{0,n} Du_n + DP(u_n), D\eta\rangle)\, dz$$
$$= \int_\Omega u_{0,n}\eta(x,0)\, dx + \iint_Q \langle B(u_n, Du_n), \eta\rangle\, dz + \iint_Q \langle f(u_n), \eta\rangle\, dz\,.$$

To show that u is a weak solution and conclude the proof we need only prove that

$$\iint_Q \langle B(u_n, Du_n), \eta\rangle\, dz \to \iint_Q \langle B(u, A^{-1}(u)DP(u)), \eta\rangle\, dz\,. \tag{5.3.35}$$

From the growth condition (5.3.32) in Bw.1) we have $B(u_n, Du_n) = 0$ if $\lambda(u_n) = 0$, i.e., $u(x,t) \in K$. Thus, if $u(x,t) \notin K$ then $Du_n = A^{-1}(u_n)D(P(u_n))$ so that, by the definition (5.3.33) of B_A, it is clear that

$$\iint_Q \langle B(u_n, Du_n), \eta\rangle\, dz = \iint_{u(z)\notin K} \langle B(u_n, Du_n, \eta\rangle\, dz = \iint_Q \langle B_A(u_n, D(P(u_n))), \eta\rangle\, dz\,.$$

Because $u_n \to u$ a.e. in Q and $DP(u_n)$ converges weakly to $DP(u)$ in $L^2(Q)$, the assumption Bw.2) then implies that $B_A(u_n, DP(u_n))$ converges weakly in $L^1(Q)$ to $B_A(u, DP(u))$. According to the definition W), u is a weak solution. □

Remark 5.3.6. We present a typical example of $B(u, D(u))$ in applications. For $u \in \mathbb{R}^m$ we consider a $m \times m$ matrix $\mathbf{b}(u) := [b_{ij}(u)]$, where $b_{ij}: \mathbb{R}^m \to \mathbb{R}^N$ are continuous. Then, for $u \in \mathbb{R}^m$ and $\zeta = [\zeta_1, \ldots, \zeta_m]^T$ with $\zeta_i \in \mathbb{R}^N$ we define the vector

$$B(u, \zeta) = \mathbf{b}(u)\zeta = \left[\sum_j \langle b_{ij}(u), \zeta_j\rangle\right]_{i=1}^m\,.$$

It is clear that (5.3.32) of Bw.1) holds if there is $C > 0$ such that $|b_{ij}(u)| \le C\lambda(u)$ on N_K and $|b_{ij}(u)| \le C\lambda^{\frac{1}{2}}(u)$ on $\mathbb{R}^m \setminus N_K$.

On the other hand, $B_A(u, \zeta) = B(u, A^{-1}(u)\zeta) = \mathbf{b}(u)A^{-1}(u)\zeta$. The entries of this vector are just $\langle g_i(u), \zeta_j\rangle$ where $g_i(u)$'s are the sum of products of entries of $\mathbf{b}(u)$ and $A^{-1}(u)$. It is clear that the $g_i(u)$'s are bounded if $b_{ij}(u)$'s satisfy the growth condition discussed above and $|A^{-1}(u)| \lesssim \lambda^{-1}(u)$. Therefore, if $\{u_n\}$ and $\{\zeta_n\}$ are sequences of vector-valued functions on Ω such that $u_n \to u$ a.e. in Ω and $\zeta_n \to \zeta$ weakly in $L^2(\Omega)$ then $g_i(u_n) \to g_i(u)$ a.e. in Ω. By the dominated convergence theorem, $g_i(u_n)$ converges to $g_i(u)$ in $L^2(Q)$ so that $\langle g_i(u_n), \zeta_{j,n}\rangle$ converges weakly to $\langle g_i(u), \zeta_j\rangle$ in $L^1(\Omega)$. The condition Bw.2) is then satisfied.

5.4 Uniqueness of limiting solutions

We discuss the uniqueness of weak solutions obtained by the approximation process described in the previous section. We will show that any sequence of these strong solutions in fact converges to a unique weak solution.

Theorem 5.4.1. *The weak solution obtained by the approximation process in Theorem 5.2.2 is unique if $f(u) = Ku + g(u)$ for any constant $m \times m$ matrix K and g satisfies $|g_u(u)| \lesssim \lambda(u)$ for all $u \in \mathbb{R}^m$.*

In the proof of this theorem, for any $u, v \in \mathbb{R}^m$ we write

$$\mathcal{A}(u, v) := \int_0^1 \frac{\partial}{\partial u} P(su + (1 - s)v)\, ds\,, \tag{5.4.1}$$

$$\mathcal{G}(u, v) := \int_0^1 \frac{\partial}{\partial u} g(su + (1 - s)v)\, ds\,. \tag{5.4.2}$$

Proof. Let $\{\lambda_{1,n}\}$ and $\{\lambda_{2,n}\}$ be two subsequences $(0, 1)$, which converge to 0. Accordingly, let $\{u_{1,n}\}$ and $\{u_{2,n}\}$ be the sequences of strong solutions that converge to the two weak solutions u_1, u_2 respectively. We will show that $u_1 \equiv u_2$.

For any integers m, n, subtracting the equations of $u_{1,n}$ and $u_{2,m}$, we get for $w := u_{1,n} - u_{2,m}$

$$w_t = \Delta(\lambda_{1,n}u_{1,n} - \lambda_{2,m}u_{2,m} + P(u_{1,n}) - P_{2,n}(u_{2,m})) + f(u_{1,n}) - f(u_{2,m})\,.$$

Using the definitions (5.4.1) and (5.4.2), we define

$$\mathbf{A}_{m,n} := \lambda_{1,n}\mathbf{I} + \mathcal{A}(u_{1,n}, u_{2,m}), \quad \mathbf{G}_{m,n} := \mathcal{G}(u_{1,n}, u_{2,m})\,. \tag{5.4.3}$$

Clearly, we can write

$$\lambda_{1,n}u_{1,n} - \lambda_{2,m}u_{2,m} + P(u_{1,n}) - P_{2,n}(u_{2,m}) = \mathbf{A}_{m,n}w + (\lambda_{1,n} - \lambda_{2,m})u_{2,m}\,,$$
$$f(u_{1,n}) - f(u_{2,m}) = Kw + \mathbf{G}_{m,n}w\,.$$

Hence, for any $T \in (0, T_0)$ and $\Psi \in L^2(\Omega \times (0, T))$

$$\iint_{Q(s)} \langle w_t, \Psi\rangle\, dz = \iint_{Q(s)} \langle \Delta(\mathbf{A}_{m,n}w) + \mathbf{G}_{m,n}w, \Psi\rangle\, dz + (\lambda_{1,n} - \lambda_{2,m})\iint_{Q(s)} \langle \Delta u_{2,m}, \Psi\rangle\, dz + \iint_{Q(s)} \langle Kw, \Psi\rangle\, dz\,,$$

where we denoted $Q(s) = \Omega \times (0, s)$ for any $s \in (0, T)$.

Assume sufficient smoothness and that Ψ and w satisfy the same boundary condition on $\partial\Omega \times (0, T)$. Using integration by parts twice in x and rearranging, we get

$$\iint_{Q(s)} \langle w_t, \Psi\rangle\, dz = \iint_{Q(s)} (\langle \mathbf{A}_{m,n}^T \Delta\Psi, w\rangle + \langle \mathbf{G}_{m,n}^T \Psi, w\rangle)\, dz + (\lambda_{1,n} - \lambda_{2,m})\iint_{Q(s)} \langle u_{2,m}, \Delta\Psi\rangle\, dz + \iint_{Q(s)} \langle Kw, \Psi\rangle\, dz\,.$$

This yields

$$\iint_{Q(s)} \langle w, \Psi\rangle_t\, dz = \iint_{Q(s)} \langle \Psi_t + \mathbf{A}_{m,n}^T \Delta\Psi + \mathbf{G}_{m,n}^T \Psi, w\rangle\, dz + (\lambda_{1,n} - \lambda_{2,m})\iint_{Q(s)} \langle u_{2,m}, \Delta\Psi\rangle\, dz + \iint_{Q(s)} \langle Kw, \Psi\rangle\, dz\,. \tag{5.4.4}$$

Concerning the first integral on the right-hand side, Lemma 5.4.3 following this proof shows that for any given $\psi \in W^{1,2}(\Omega)$ there is a sequence of function $\Psi_{m,n}$ satisfying the same BC of $u_{i,n}$ and the systems

$$\begin{cases} \Psi_t + \mathbf{A}^T_{m,n}\Delta\Psi + \mathbf{G}^T_{m,n}\Psi = 0 & \text{on } \Omega \times (0, T)\,, \\ \Psi(x, T) = \psi(x)\,. \end{cases} \tag{5.4.5}$$

Moreover, there is a constant $C(\|D\psi\|_{L^2(\Omega)})$ such that for all $s \in [0, T]$

$$\int_{\Omega\times\{s\}} |D\Psi_{m,n}|^2 \, dx \le C(\|D\psi\|_{L^2(\Omega)})\,. \tag{5.4.6}$$

In addition, for any $p \in (1, 2_*)$ there are a subsequence of $\{\Psi_{m,n}\}$ and a function Ψ_* such that

$$\Psi_{m,n} \to \Psi_* \quad \text{in } C([0, T], L^p(\Omega))\,. \tag{5.4.7}$$

Combining (5.4.4) and (5.4.11) we have

$$\int_{\Omega\times\{s\}} w\Psi_{m,n} \, dx = I_{m,n}(s) + \iint_{Q(s)} \langle Kw, \Psi_{m,n}\rangle \, dz\,, \tag{5.4.8}$$

where

$$I_{m,n}(s) := (\lambda_{1,n} - \lambda_{2,m})\iint_{Q(s)} \langle u_{2,m}, \Delta\Psi_{m.n}\rangle \, dz\,.$$

We will show that the quantity $I_{m,n}(s)$ tends to 0 as $m, n \to \infty$. Indeed, integrating by parts, we see that

$$I_{m,n}(s) = -(\lambda_{1,n} - \lambda_{2,m})\iint_{Q(s)} \langle Du_{2,m}, D\Psi_{m.n}\rangle \, dz\,.$$

We can assume $\lambda_{1,n} \le \lambda_{2,m}$ so that we can use Young's inequality to get

$$\begin{aligned} |I_{m,n}(s)| &\le 2\lambda_{2,m}^{\frac{1}{2}} \iint_{Q(s)} \lambda_{2,m}^{\frac{1}{2}} |Du_{2,m}||D\Psi_{m.n}| \, dz \\ &\le 2\lambda_{2,m}^{\frac{1}{2}} \left(\iint_Q \lambda_{2,m}|Du_{2,m}|^2 \, dz\right)^{\frac{1}{2}} \left(\iint_Q |D\Psi_{m,n}|^2 \, dz\right)^{\frac{1}{2}}\,. \end{aligned}$$

Testing the system for $u_{2,m}$ with $u_{2,m}$, we easily obtain a uniform bound (see also (5.3.15) of Proposition 5.3.1)

$$\iint_Q \lambda_{2,m}|Du_{2,m}|^2 \, dz \le \iint_Q (\lambda_{2,m} + \lambda(u_{2,m}))|Du_{2,m}|^2 \, dz \le C\,.$$

Together with (5.4.6), we get $|I_{m,n}| \le C^{\frac{1}{2}} C^{\frac{1}{2}}(\|D\psi\|_{L^2(\Omega)})\lambda_{2,m}^{\frac{1}{2}}$. As $\lambda_{2,m} \to 0$, we see that $I_{m,n} \to 0$ as $m, n \to \infty$.

Denote $w_* = u_1 - u_2$. Letting $m, n \to \infty$, we obtain from (5.4.8) that

$$\int_{\Omega\times\{s\}} \langle w_*, \Psi_*\rangle \, dx = \int_0^s \int_{\Omega\times\{t\}} \langle w_*, K^T\Psi_*\rangle \, dx\, dt\,. \tag{5.4.9}$$

For any constant matrix L, since $L\Psi_{m.n}$ satisfies the same *linear* equation in (5.4.11), the initial data does not matter, and we repeat the above argument to get

$$\int_{\Omega\times\{s\}} \langle w_*, L\Psi_*\rangle\, dx = \int_0^s \int_{\Omega\times\{t\}} \langle w_*, K^T(L\Psi_*)\rangle\, dx\, dt\,.$$

Let $L = K^T$. Inserting the above into the right-hand side of (5.4.9), we deduce

$$\int_{\Omega\times\{s\}} \langle w_*, \Psi_*\rangle\, dx = \int_0^s \int_0^t \int_{\Omega\times\{\tau\}} \langle w_*, (K^T)^2\Psi_*\rangle\, dx\, d\tau dt\,.$$

Keep arguing this way with L being $(K^T)^2, \dots, (K^T)^{n-1}$ and use induction to obtain for any integer $n \ge 1$ that

$$\int_{\Omega\times\{s\}} \langle w_*, \Psi_*\rangle\, dx = \int_{S_n} \int_{\Omega\times\{t_1\}} \langle w_*, (K^T)^n\Psi_*\rangle\, dx\, dt_1 \cdots dt_n\,,$$

where $S_n = \{(t_1, \cdots, t_n) :\ 0 \le t_1 \le \cdots \le t_n \le s\}$. We then have

$$\left|\int_{\Omega\times\{s\}} \langle w_*, \Psi_*\rangle\, dx\right| \le C_1 |K|^n \int_{S_n} 1\, dt_1 \cdots dt_n\,,$$

where (noting that w_*, Ψ_* are in $C([0, T_0], L^p(\Omega))$ for some $p > 2$)

$$C_1 = \sup_{t_1\in(0,T)} \int_{\Omega\times\{t_1\}} |w_*||\Psi_*|\, dx\,.$$

As

$$\int_{S_n} 1\, dt_1 \cdots dt_n = \int_0^s \int_0^{t_n} \cdots \int_0^{t_1} 1\, dt_1 \cdots dt_n = \frac{s^n}{n!}\,,$$

we then have for all $s \in (0, T)$

$$\left|\int_{\Omega\times\{s\}} \langle w_*, \Psi_*\rangle\, dx\right| \le C_1 |K|^n \frac{s^n}{n!}\,, \qquad \forall n \ge 1\,.$$

Letting $n \to \infty$ we see that

$$\int_{\Omega\times\{s\}} \langle w_*, \Psi_*\rangle\, dx = 0\,, \qquad \forall s \in [0, T]\,. \tag{5.4.10}$$

In particular, as $\Psi_*(x, T) = \psi(x)$,

$$\int_{\Omega\times\{T\}} \langle w_*, \psi\rangle\, dx = 0\,.$$

Because $\psi \in W^{1,2}(\Omega, \mathbb{R}^m)$ is arbitrary, we conclude that $w_* \equiv 0$, or $u_1 \equiv u_2$, on $\Omega\times\{T\}$. Since T is arbitrary in $(0, T_0)$, the uniqueness then follows. □

Remark 5.4.2. We should note that the key equation (5.4.10) follows immediately from (5.4.9) if $K = kI$ for some $k \in \mathbb{R}$. This is not the case even if K is a diagonal matrix, as in the Lotka–Volterra reactions in the SKT models. For a general matrix K, we have to use a somewhat trickier argument as in the proof to obtain (5.4.10).

Lemma 5.4.3. *For any m, n and any given $\psi \in W^{1,2}(\Omega, \mathbb{R}^m)$ there is a function $\Psi = \Psi_{m,n}$ such that*

$$\begin{cases} \Psi_t + \mathbf{A}^T_{m,n}\Delta\Psi + \mathbf{G}^T_{m,n}\Psi = 0 & \text{on } \Omega \times (0, T)\,, \\ \Psi(x, T) = \psi(x)\,. \end{cases} \tag{5.4.11}$$

Moreover, for any given $p \in (1, 2_*)$ there are a constant $C(\|D\psi\|_{L^2(\Omega)})$, a subsequence of $\{\Psi_{m,n}\}$ and a function Ψ_* such that

$$\int_{\Omega\times\{s\}} |D\Psi_{m,n}|^2\, dx \le C\|D\psi\|_{L^2(\Omega)} \qquad \text{for all } s \in [0, T]\,, \tag{5.4.12}$$

$$\Psi_{m,n} \to \Psi_* \text{in } C([0, T], L^p(\Omega))\,. \tag{5.4.13}$$

Proof. Using a change of variables $t \to T - t$ the system (5.4.11) is equivalent to the linear parabolic system for $\hat{\Psi}(x, t) = \Psi(x, T - t)$.

$$\begin{cases} \hat{\Psi}_t = \mathbf{A}^T_{m,n}\Delta\hat{\Psi} + \mathbf{G}^T_{m,n}\hat{\Psi} = 0 & \text{on } \Omega \times (0, T)\,, \\ \hat{\Psi}(x, 0) = \psi(x)\,. \end{cases} \tag{5.4.14}$$

Because $u_{1,n}$ and $u_{2,m}$ are strong solutions, from P) and the definition (5.4.1) and $\lambda_{1,n} > 0$, the above system is a regular linear parabolic system so that it has a strong solution $\hat{\Psi}$. Thus, $\Psi_{m,n} = \Psi$ exists.

Dropping the subscripts m, n and multiplying (5.4.11) with $\Delta\Psi$ and integrating by parts ($\Psi_t = 0$ on the boundary), we get for $Q^{(s)} = \Omega \times (s, T)$

$$-\iint_{Q^{(s)}} \frac{d}{dt}|D\Psi|^2\, dz + \iint_{Q^{(s)}} \langle \mathbf{A}^T\Delta\Psi, \Delta\Psi\rangle\, dz = -\iint_{Q^{(s)}} \langle \mathbf{G}^T\Psi, \Delta\Psi\rangle\, dz\,.$$

By P), for any vector ζ we have

$$\langle \mathcal{A}(u, v)\zeta, \zeta\rangle \ge \int_0^1 \lambda(su + (1 - s)v)\, ds|\zeta|^2\,.$$

From the growth assumption on $\mathcal{G}$, we then find a nonnegative function λ_* such that

$$\langle \mathbf{A}^T\Delta\Psi, \Delta\Psi\rangle \ge \lambda_*|\Delta\Psi|^2, \ \langle \mathbf{G}^T\Psi, \Delta\Psi\rangle \le \varepsilon\lambda_*|\Delta\Psi|^2 + C(\varepsilon)\lambda_*|\Psi|^2\,. \tag{5.4.15}$$

For small $\varepsilon > 0$ we deduce that

$$\int_{\Omega\times\{s\}} |D\Psi|^2\, dx \le \int_{\Omega\times\{T\}} |D\Psi|^2\, dx + C\iint_{Q^{(s)}} \lambda_*|\Psi|^2\, dz\,.$$

Choosing q such that $N/2 < q < q_0$, we easily see that $2q' < 2_*$ so that we can estimate the last integrand by Hölder and Sobolev's inequalities

$$\left(\int_{\Omega\times\{\tau\}} \lambda_*^q\, dx\right)^{\frac{1}{q}} \left(\int_{\Omega\times\{\tau\}} |\Psi|^{2q'}\, dx\right)^{\frac{1}{q'}} \le C\int_{\Omega\times\{\tau\}} |D\Psi|^2\, dx\,. \tag{5.4.16}$$

Here, we used the fact that λ_* satisfies the same uniform bound of $\lambda(u_{i,n})$, i.e., $\|\lambda(u_{i,n})\|_{L^{q_0}(\Omega)}$ is assumed to be bounded in Theorem 5.2.2, so that

$$\|\lambda_*\|_{L^{q_0}(\Omega)} \leq C\,. \tag{5.4.17}$$

As $\|D\Psi(\cdot,T)\|_{L^2(\Omega)} = \|D\psi\|_{L^2(\Omega)}$ we have

$$\int_{\Omega\times\{s\}} |D\Psi|^2\,dx \leq \|D\psi\|^2_{L^2(\Omega)} + C\int_0^s \int_{\Omega\times\{s\}} |D\Psi|^2\,dx\,dt\,.$$

By Gronwall's inequality we can find a constant C_0 depending on $\|D\psi\|_{L^2(\Omega)}$ such that

$$\int_{\Omega\times\{s\}} |D\Psi|^2\,dx \leq C_0\,.$$

This is (5.4.12).

Next, by P) and in comparison with (5.4.15), we see that there is a constant C such that for any $\zeta \in \mathbb{R}^m$

$$\langle(\mathbf{A}^T)^{-1}\zeta,\zeta\rangle \geq \lambda_*^{-1}|\zeta|^2,\quad |(\mathbf{A}^T)^{-1}\mathbf{G}^T| \leq C\,. \tag{5.4.18}$$

The system for Ψ can be rewritten as

$$(\mathbf{A}^T)^{-1}\Psi_t + \Delta\Psi + (\mathbf{A}^T)^{-1}\mathbf{G}^T\Psi = 0\,.$$

Testing this system with Ψ_t and using (5.4.18), we get

$$\iint_{Q^{(s)}} \lambda_*^{-1}|\Psi_t|^2\,dz \leq \iint_{Q^{(s)}} \frac{d}{dt}|D\Psi|^2\,dz + C\iint_{Q^{(s)}} |\Psi||\Psi_t|\,dz\,.$$

Applying Young's inequality to the last integral, we have

$$\iint_{Q^{(s)}} |\Psi||\Psi_t|\,dz \leq \varepsilon\iint_{Q^{(s)}} \lambda_*^{-1}|\Psi_t|^2\,dz + C(\varepsilon)\iint_{Q^{(s)}} \lambda_*|\Psi|^2\,dz\,.$$

For small ε we then derive

$$\iint_{Q^{(s)}} \lambda_*^{-1}|\Psi_t|^2\,dz \leq \iint_{Q^{(s)}} \frac{d}{dt}|D\Psi|^2\,dz + C\iint_{Q^{(s)}} \lambda_*|\Psi|^2\,dz\,.$$

Using the bounds (5.4.12) and (5.4.16), we can estimate the integrals on the right-hand side by

$$\iint_{Q^{(s)}} \frac{d}{dt}|D\Psi|^2\,dz = \int_\Omega (|D\Psi(x,T)|^2 - |D\Psi(x,s)|^2)\,dx \leq C(C_0)\,,$$

$$\iint_{Q^{(s)}} \lambda_*|\Psi|^2\,dz \leq C(C_0)\,.$$

We then find a constant $C(C_0)$ such that

$$\iint_{Q^{(s)}} \lambda_*^{-1}|\Psi_t|^2\,dz \leq C(C_0)\,,\quad \forall s \in [0,T]\,.$$

For some $q \in (1, 2)$ such that $\frac{q}{2-q} \le q_0$ we obtain from Hölder's inequality and the above that

$$\iint_Q |\Psi_t|^q \, dz \le \left(\iint_Q \lambda_*^{\frac{q}{2-q}} \, dz \right)^{\frac{2-q}{2}} \left(\iint_Q \lambda_*^{-1} |\Psi_t|^2 \, dz \right)^{\frac{q}{2}} \le C(C_0) \, .$$

We just show that there are a sequence $\{\Psi_{m,n}\}$ and a constant $C(C_0)$ such that for all $s \in [0, T]$

$$\int_{\Omega \times \{s\}} |D\Psi_{m,n}|^2 \, dx, \quad \iint_Q |(\Psi_{m,n})_t|^q \, dz \le C(C_0) \, . \tag{5.4.19}$$

Again, by the Aubin–Lions–Simon lemma (see (5.3.29) in the proof of Theorem 5.2.2), for any given $p \in (1, 2_*)$ there is a subsequence of $\{\Psi_{m,n}\}$ such that

$$\Psi_{m,n} \to \Psi_* \quad \text{in} \quad C([0, T], L^p(\Omega)) \, . \tag{5.4.20}$$

This completes the proof of the lemma. □

We end this section by a short discussion of weak solutions in the sense of definition W). We have proved that the weak solution obtained by the limiting process in the previous section is unique. To the best of our knowledge, the existence of weak solutions to the degenerate/singular systems has always been established this way in the literature. It is desirable to establish the uniqueness of general weak solutions. This is done for scalar equations if boundedness of weak solutions is assumed. As boundedness of solutions to systems generally is an open problem, the methods for scalar equations are not applicable here and the uniqueness question is wide open. However, we would like to point out below how the uniqueness result for limiting solutions in Theorem 5.4.1 can be modified to obtain such uniqueness results. Let u_1, u_2 be two weak solutions to the system

$$u_t = \Delta(P(u)) + f(u) \, .$$

As in (5.4.4) of the proof, for any $T \in (0, T_0)$ and $\Psi \in L^2(\Omega \times (0, T))$ we have

$$\iint_{Q(s)} \langle w, \Psi \rangle_t \, dz = \iint_{Q(s)} \langle \Psi_t + \mathbf{A}^T \Delta \Psi + K\Psi + \mathbf{G}^T \Psi, w \rangle \, dz \, , \tag{5.4.21}$$

where we denoted $Q(s) = \Omega \times (0, s)$ for any $s \in (0, T)$, $w = u_1 - u_2$, $\mathbf{A} = \mathcal{A}_u(u_1, u_2)$ and $\mathbf{G} = \mathcal{G}(u_1, u_2)$.

For any given $\psi \in W^{1,2}(\Omega)$ if we can find Ψ solving the systems

$$\begin{cases} \Psi_t + \mathbf{A}^T \Delta \Psi + \mathbf{G}^T \Psi = 0 & \text{on } \Omega \times (0, T) \, , \\ \Psi(x, T) = \psi(x) \, , \end{cases} \tag{5.4.22}$$

then

$$\int_{\Omega \times \{s\}} \langle w, \Psi \rangle \, dx = \int_0^s \int_{\Omega \times \{t\}} \langle w, K^T \Psi \rangle \, dx \, dt \, . \tag{5.4.23}$$

The argument following (5.4.9) in the proof then shows that the above implies $w \equiv 0$ or $u_1 \equiv u_2$. We then see that the desired uniqueness result will follow if such Ψ exists. Hence, the key question here is the existence of a solution to (5.4.22) for any given two weak solutions u_1, u_2 in the sense of the definition W). In general, this is a widely open question because the linear system (5.4.22) is not regular parabolic as u_1, u_2 can be zero or unbounded.

5.5 Existence and regularity of weak steady-state solutions

We now turn to the steady-state problem of (5.2.6) and consider the elliptic system

$$\begin{cases} -\Delta(P(u)) = B(u, Du) + f(u) \text{ in } \Omega\,, \\ \text{homogeneous Dirichlet or Neumann conditions on } \partial\Omega\,. \end{cases} \tag{5.5.1}$$

The absence of the temporal derivative allows us to consider much weaker structural and integrability conditions, and yet obtain stronger assertions than those of the parabolic case.

We discuss first the existence of strong solutions to the *regular* (5.5.1). Again, we embed (5.5.1) in the following family:

$$\begin{cases} -\Delta(P(u)) = \sigma B(u, Du) + \sigma f(u) \text{ in } \Omega \text{ and } \sigma \in [0, 1]\,, \\ \text{homogeneous Dirichlet or Neumann conditions on } \partial\Omega\,. \end{cases} \tag{5.5.2}$$

The following theorem establishes the existence of strong solutions to (5.5.1) *for any dimension N* and thus extends our result in [30, Corollary 3.10], which only treated the case $N \le 4$ and $B(u, Du) = 0$.

Theorem 5.5.1. *Assume that the system* (5.5.1) *satisfies PR) for some* $\lambda_0 > 0$ *and B) and f). We suppose that there exist constants* $\beta_0 \in (0, 1]$ *and* C_0 *independent of* σ *such that any strong solution u of the family* (5.5.2) *satisfies*

$$\||P(u)|^{\beta_0}\|_{L^1(\Omega)}^{\frac{1}{\beta_0}}, \quad \||f(u)|^{\beta_0}\|_{L^1(\Omega)}^{\frac{1}{\beta_0}} \le C_0\,. \tag{5.5.3}$$

Then there exists a strong solution to the system (5.5.1). *Moreover, for any* $q \ge 2$ *there exists a constant* $C(C_0, q)$ *such that*

$$\|D(P(u))\|_{L^{q_*}(\Omega)} \le C(C_0, \lambda_0^{-1}, q)\,. \tag{5.5.4}$$

We now have the following result on the existence of a Hölder continuous weak solution to the *degenerate* system (5.5.1). Let $\{\lambda_{0,n}\}$ be a sequence converging to 0 in $(0, 1)$. We consider the approximation family of systems

$$\begin{cases} -\Delta(\lambda_{0,n}u + P(u)) = \sigma B(u, Du) + \sigma f_n(u) \text{ in } \Omega,\ \sigma \in [0, 1]\,, \\ \text{homogeneous Dirichlet or Neumann conditions on } \partial\Omega\,. \end{cases} \tag{5.5.5}$$

We impose the following hypotheses on (5.5.5).

Pfn) The map P satisfies P). The maps f_n satisfy f) with the ellipticity functions $\lambda_n(u) = \lambda_{0,n} + \lambda(u)$. That is, $|(f_n)_u(u)| \le C(1 + \lambda_n(u))$. In addition,

$$f_n \to f \text{ in } L^\infty_{\text{loc}}(\mathbb{R}^m)\,. \tag{5.5.6}$$

Bw) Assume B). Moreover, the following map is continuous on $\mathbb{R}^m \times \mathbb{R}^{Nm}$:

$$B_A(u,\zeta) := \begin{cases} B(u, A^{-1}(u)\zeta) & \text{if } u \notin K\,, \\ 0 & \text{if } u \in K\,, \end{cases} \tag{5.5.7}$$

where $K := \{u \in \mathbb{R}^m : \lambda(u) = 0\}$.

Theorem 5.5.2. *Assume Ph), Pfn), and Bw). Suppose that there is $\beta_0 \in (0,1]$ and a constant C_0 such that any strong solution u of (5.5.5) satisfies*

$$\||u|^{\beta_0}\|_{L^1(\Omega)},\ \||P(u)|^{\beta_0}\|_{L^1(\Omega)},\ \||f(u)|^{\beta_0}\|_{L^1(\Omega)} \le C_0\,. \tag{5.5.8}$$

Importantly, there is $r_0 > N/2$ such that any strong solution u of (5.5.5) with $\sigma = 1$ satisfies ($\lambda_n(u) = \lambda_{0,n} + \lambda(u)$)

$$\|\lambda_n^{-1}(u)(f_n)_u(u)\|_{L^{r_0}(\Omega)} \le C_0\,. \tag{5.5.9}$$

Then there is a Hölder continuous weak solution to (5.5.1). Actually, u is a strong solution away from the degenerate set $\{x : u(x) \in K\}$.

5.6 The proofs

We now present the proof of our theorems concerning the elliptic system

$$\begin{cases} -\Delta(P(u)) = B(u, Du) + f(u) \ \text{ in } \Omega\,, \\ \text{homogeneous Dirichlet or Neumann conditions on } \partial\Omega\,. \end{cases} \tag{5.6.1}$$

We start with the following technical lemma, which provides the key estimates for strong solutions to the general system (5.6.1) and the proof of our main theorems in this section. We note that no growth conditions on f and λ are assumed in this lemma. For some $\sigma \in (0,1]$ consider a strong solution u to

$$\begin{cases} -\Delta(P(u)) = \sigma B(u, Du) + \sigma f(u) \ \text{ in } \Omega\,, \\ \text{homogeneous Dirichlet or Neumann conditions on } \partial\Omega\,. \end{cases} \tag{5.6.2}$$

Lemma 5.6.1. *Assume that P) holds and f is a C^1 map. Also, B satisfies the growth condition*

$$|B(u,\zeta)| \le C\lambda(u)|\zeta| \quad \forall u \in \mathbb{R}^m\,,\ \zeta \in \mathbb{R}^{Nm}\,. \tag{5.6.3}$$

For any $\sigma \in (0,1]$ and any strong solution u to (5.6.2) and $\beta_0 \in (0,1)$ and $r_0 > N/2$ we denote

$$C_0 := \||f(u)|^{\beta_0}\|_{L^1(\Omega)}^{\frac{1}{\beta_0}},\ C_1 := \|\lambda^{-1}(u)f_u(u)\|_{L^{r_0}(\Omega)}\,. \tag{5.6.4}$$

Then for any $q \ge 2$ and $\beta \in (0,1)$ there are constants C, $C(C_1,\beta_0)$, which also depend on q, such that

$$\|P(u)\|_{W^{2,q}(\Omega)} \le C(C_1,\beta_0)\sigma\|P(u)\|_{L^\beta(\Omega)}^{\frac{1}{\beta}} + C\sigma C_0\,. \tag{5.6.5}$$

Proof. We will denote $U = P(u) = [P_i(u)]_{i=1}^m$ in the proof. Because u is a strong solution, u and Du are bounded. From the equation of u, we see that $\Delta P(u)$ is bounded. Hence, for any $q \geq 2$ we can multiply the i-th equation in (5.6.2) by $-|\Delta U|^{q-2}\Delta P_i(u)$, $i = 1, \ldots, m$, integrate over Ω and sum the results to get

$$\int_\Omega |\Delta U|^q \, dx = -\sum_i \int_\Omega \sigma(B_i(u, Du) + f_i(u))|\Delta U|^{q-2}\Delta P_i(u) \, dx \, . \tag{5.6.6}$$

Using the growth condition (5.6.3) ($|B(u, Du)| \lesssim \lambda(u)|Du| \leq |DU|$) and applying Young's inequality to the integrands on the right-hand side of the above, for any $\varepsilon > 0$ we find a constant $C(\varepsilon)$ such that

$$\sigma|B_i(u, Du)|\Delta U|^{q-2}\Delta P_i(u)| \leq C\lambda(u)|\Delta U|^{q-1}\sigma|DU| \leq \varepsilon|\Delta U|^q + C(\varepsilon)\sigma^q|DU|^q \, ,$$
$$\sigma|f_i(u)|\Delta U|^{q-2}\Delta P_i(u)| \leq \sigma|f(u)||\Delta U|^{q-1} \leq \varepsilon|\Delta U|^q + C(\varepsilon)\sigma^q|f(u)|^q \, .$$

Choosing sufficiently small ε, we obtain from the above inequalities and (5.6.6) that

$$\int_\Omega |\Delta U|^q \, dx \leq C\sigma^q \int_\Omega |DU|^q \, dx + C\sigma^q \int_\Omega |f(u)|^q \, dx \, . \tag{5.6.7}$$

Using Schauder's inequality $\|D^2 U\|_{L^q(\Omega)} \leq C\|\Delta U\|_{L^q(\Omega)}$ [15, Corollary 9.10], we obtain from the above that

$$\|D^2 U\|^q_{L^q(\Omega)} \leq C\sigma^q \|DU\|^q_{L^q(\Omega)} + C\sigma^q \|f(u)\|^q_{L^q(\Omega)} \, . \tag{5.6.8}$$

We consider the last norm in the above estimate. For any $p > \frac{qN}{N+q}$ we have $q < p_*$ so that we can use the inequality (5.3.2) with $W = f(u)$ to get

$$\int_\Omega |f(u)|^q \, dx \leq C\left(\int_\Omega |Df(u)|^p \, dx\right)^{\frac{q}{p}} + C\left(\int_\Omega |f(u)|^{\beta_0} \, dx\right)^{\frac{q}{\beta_0}} \, . \tag{5.6.9}$$

Using the facts that $|Df(u)| \leq |f_u(u)||Du|$ and $\lambda(u)|Du| \leq |DU|$ and then Hölder's inequality, for $r > 1$ we estimate the first integral on the right-hand side by

$$\int_\Omega |f_u(u)|^p |Du|^p \, dx \leq \left(\int_\Omega |f_u(u)\lambda^{-1}(u)|^{pr} \, dx\right)^{\frac{1}{r}} \left(\int_\Omega |DU|^{pr'} \, dx\right)^{\frac{1}{r'}} \, . \tag{5.6.10}$$

For $p \in [qN/(N+q), q_*)$ we note that the fraction

$$\gamma(p) := 1/\left(\frac{1}{p} - \frac{1}{q} + \frac{1}{N}\right) = pq_*/(q_* - p)$$

is increasing from $\frac{N}{2}$ to infinity. Hence, for any given $r_0 > N/2$, we can choose and fix a $p > \frac{qN}{N+q}$ in the sequel such that $\gamma(p) < r_0$.

We now choose r such that $pr \in (\gamma(p), r_0]$. We see that $r > \gamma(p)/p = q_*/(q_* - p) > 1$. For such r the first factor on the right-hand side of (5.6.10) is bounded by $\|\lambda^{-1}(u)f_u(u)\|^p_{L^{r_0}(\Omega)}$ because $pr \leq r_0$. By the assumption (5.6.4), we get

$$\int_\Omega |f_u(u)|^p |Du|^p \, dx \leq C_1^p \left(\int_\Omega |DU|^{pr'} \, dx\right)^{\frac{1}{r'}} \, .$$

Using this and the definition (5.6.4) of C_0 in (5.6.9), we have

$$\int_\Omega |f(u)|^q\,dx \le C[C_1^q\|DU\|^q_{L^{pr'}(\Omega)} + C_0^q]\,.$$

We now obtain from (5.6.8) and the above estimate that

$$\|D^2U\|^q_{L^q(\Omega)} \le C\sigma^q[\|DU\|^q_{L^q(\Omega)} + C_1^q\|DU\|^q_{L^{pr'}(\Omega)} + C_0^q]\,.$$

That is,

$$\|D^2U\|_{L^q(\Omega)} \le C\sigma[\|DU\|_{L^q(\Omega)} + C_1\|DU\|_{L^{pr'}(\Omega)} + C_0]\,. \tag{5.6.11}$$

Because $q, pr' < q_*$ (it is easy to see that $pr > \gamma(p)$ implies $pr' < q_*$) we can apply (5.3.2) with $W = DU$, $\beta = 1$ and the exponent p is q or pr' in (5.6.11) to estimate the norms $\|DU\|_{L^q(\Omega)}$, $\|DU\|_{L^{pr'}(\Omega)}$ and see that

$$\|D^2U\|_{L^q(\Omega)} \le C\varepsilon\sigma\|D^2U\|_{L^q(\Omega)} + C(\varepsilon)\sigma C_1\|DU\|_{L^1(\Omega)} + C\sigma C_0\,.$$

The above implies $\|D^2U\|_{L^q(\Omega)} \le C(C_1)\sigma\|DU\|_{L^1(\Omega)} + C\sigma C_0$ if we choose $\varepsilon = C/2$. Using the interpolation inequality

$$\|DU\|_{L^1(\Omega)} \le \varepsilon\|D^2U\|_{L^q(\Omega)} + C(\varepsilon)\|U\|_{L^1(\Omega)}$$

for small ε, we deduce $\|D^2U\|_{L^q(\Omega)} \le C(C_1)\sigma\|U\|_{L^1(\Omega)} + C\sigma C_0$. We now have

$$\|D^2U\|_{L^q(\Omega)} + \|DU\|_{L^1(\Omega)} \le C(C_1)\sigma\|U\|_{L^1(\Omega)} + C\sigma C_0\,. \tag{5.6.12}$$

Finally, we apply (5.3.2) to $W = |U|$, $p = 1$ and $\beta \in (0, 1]$ to get

$$\|U\|_{L^1(\Omega)} \le \varepsilon\|DU\|_{L^1(\Omega)} + C(\varepsilon)\||U|^\beta\|^{\frac{1}{\beta}}_{L^1(\Omega)}\,.$$

Using this in (5.6.12), for small ε we obtain (5.6.5). The proof is complete. □

Remark 5.6.2. Assume that we can write $f(u) = f_1(u) + f_2(u)$ and the quantities

$$C_0 := \||f_1(u)|^{\beta_0}\|^{\frac{1}{\beta_0}}_{L^1(\Omega)},\quad C_1 := \|\lambda^{-1}(u)(f_1)_u(u)\|_{L^{r_0}(\Omega)}\,, \tag{5.6.13}$$

and

$$C_2 := \||f_2(u)|\|_{L^{r_0}(\Omega)} \tag{5.6.14}$$

are finite for some $\beta_0 \in (0, 1)$ and $r_0 > N/2$. Then for any $q \ge 2$ and $\beta \in (0, 1)$ there are constants C, $C(C_1, \beta_0)$, which also depend on q, such that

$$\|P(u)\|_{W^{2,q}(\Omega)} \le C(C_1, \beta_0)\sigma\|P(u)\|^{\frac{1}{\beta}}_{L^\beta(\Omega)} + C\sigma(C_0 + C_2)\,. \tag{5.6.15}$$

Indeed, starting with (5.6.8), one needs only apply the argument to $f_1(u)$ to obtain (5.6.15).

We use Lemma 5.6.1 and Theorem 4.3.1 to show the existence of strong solutions to regularized systems ($\lambda(u) \geq \lambda_0 > 0$).

Proof of Theorem 5.5.1. Following Theorem 4.3.1, we embed the system (5.6.1) in the family of systems (5.6.2), equipped with the same boundary conditions

$$-\operatorname{div}(A(\sigma u)Du) = B(\sigma u, \sigma Du) + f(\sigma u) \text{ in } \Omega . \tag{5.6.16}$$

Let u be any strong solution to (5.6.16) for some $\sigma \in (0, 1]$. Multiply the above by σ, we see that $w = \sigma u$ is a strong solution to

$$\begin{cases} -\Delta(P(w)) = \sigma B(w, Dw) + \sigma f(w) \text{ in } \Omega \text{ and } \sigma \in (0,1] , \\ \text{homogeneous Dirichlet or Neumann conditions on } \partial\Omega , \end{cases} \tag{5.6.17}$$

which is the family (5.5.2) considered in the theorem. The assumption (5.5.3) of the theorem on any strong solution of (5.5.2), which is now (5.6.17), reads

$$\||P(w)|^{\beta_0}\|_{L^1(\Omega)}^{\frac{1}{\beta_0}}, \ \||f(w)|^{\beta_0}\|_{L^1(\Omega)}^{\frac{1}{\beta_0}} \leq C_0 . \tag{5.6.18}$$

As $w = \sigma u$ is a strong solution to (5.6.17) so that (5.6.18) holds for w. Furthermore, because of f), PR) and $\lambda(w) \geq \lambda_0$, for any $r_0 \geq 1$

$$\|\lambda^{-1}(w) f_w(w)\|_{L^{r_0}(\Omega)} \leq C\|\lambda^{-1}(w) + 1\|_{L^{r_0}(\Omega)} \leq C(\lambda_0^{-1}) . \tag{5.6.19}$$

We see that the proof of Lemma 5.6.1 applies to the solution w of the system (5.6.17). For any $q \geq 2$ the bounds in (5.6.18), (5.6.19) and the estimate (5.6.5) of the lemma give a constant $C(C_0, \lambda_0, q)$ such that $\|P(w)\|_{W^{2,q}(\Omega)} \leq \sigma C(C_0, \lambda_0^{-1}, q)$. By Sobolev's inequality, as $w = \sigma u$, we have

$$\|D(P(\sigma u))\|_{L^{q_*}(\Omega)} \leq \sigma C(C_0, \lambda_0^{-1}, q) . \tag{5.6.20}$$

Because $\lambda(\sigma u)|D(\sigma u)| \sim |D(P(\sigma u))|$ and $\lambda(\sigma u) \geq \lambda_0 > 0$, (5.6.20) implies

$$\|\lambda_0 D(\sigma u)\|_{L^{q_*}(\Omega)} \leq \sigma C(C_0, \lambda_0^{-1}, q) ,$$

which yields

$$\|Du\|_{L^{q_*}(\Omega)} \leq \lambda_0^{-1} C(C_0, \lambda_0^{-1}, q) . \tag{5.6.21}$$

We choose $q > \max\{2, N/2\}$ so that $q_* > N$. The above estimate shows that the strong solutions of the family (5.6.16) are uniformly Hölder continuous for $\sigma \in [0, 1]$ (the case $\sigma = 0$ holds due to the regularity theory of linear systems). Hence they satisfy (SBMO) uniformly in $\sigma \in [0, 1]$. By Theorem 4.3.1 (its condition M.0') is obvious due to (5.6.19) with $r_0 = \infty$) we obtain the existence of a strong solution u to (5.6.16) for $\sigma = 1$, which is (5.6.1). Of course, the estimate (5.5.4) comes from (5.6.20) with $\sigma = 1$. We prove the theorem. □

Remark 5.6.3. We note that the theory in Chapter 4 deals with general matrix $A(u)$ and the spectral gap condition SG') required in the proof of Theorem 4.3.1 is used only in order to obtain the energy estimate for Du in Lemma 4.3.2 so that the key a priori bound for $\|Du\|_{L^p(\Omega)}$ then follows. Here, thanks to the special $A(u) = P_u(u)$ we directly derive such a bound from Lemma 5.6.1, which does not use such an energy estimate and thus the condition SG') is not needed here.

We now give the proof of Theorem 5.5.2 on the existence of a regular weak solution. We assumed in Theorem 5.5.2 that the approximation systems (5.5.5), writing $P_n(u) := \lambda_{0,n}u + P(u)$,

$$\begin{cases} -\Delta(P_n(u)) = \sigma B(u, Du) + \sigma f_n(u) \text{ in } \Omega \quad \text{and} \quad \sigma \in [0, 1]\,, \\ \text{homogeneous Dirichlet or Neumann conditions on } \partial\Omega\,, \end{cases} \tag{5.6.22}$$

satisfy the hypotheses of Theorem 5.5.1 uniformly. That is, there exist constants $\beta_0 \in (0, 1]$, C_0 independent of $\sigma \in [0, 1]$ *and n* such that any strong solution u of the above family a priori satisfies

$$\||P_n(u)|^{\beta_0}\|_{L^1(\Omega)},\ \||f_n(u)|^{\beta_0}\|_{L^1(\Omega)} \le C_0\,; \tag{5.6.23}$$

and, when $\sigma = 1$, for some $r_0 > N/2$

$$\|\lambda_n^{-1}(u)(f_n)_u(u)\|_{L^{r_0}(\Omega)} \le C_0\,. \tag{5.6.24}$$

Proof of Theorem 5.5.2. As P_n satisfies PR), the assumption (5.6.23) allows us to apply Theorem 5.5.1 and obtain a sequence of strong solutions u_n to the systems

$$-\Delta(P_n(u)) = B(u, Du) + f_n(u) \text{ in } \Omega\,. \tag{5.6.25}$$

Define $U_n = P_n(u_n)$ and $W_n = P(u_n)$. Fix a $q > N$. By the two assumptions (5.6.23) and (5.6.24), (5.6.5) of Lemma 5.6.1 implies that there is a constant $C(C_0, q)$ such that $\|U_n\|_{W^{2,q}(\Omega)} \le C(C_0, q)$.

Observe that

$$|D(P(u_n))| \sim \lambda(u_n)|Du_n| \le (\lambda_{0,n} + \lambda(u_n))|Du_n| \sim |D(P_n(u_n))|\,.$$

Hence, $\|DW_n\|_{L^q(\Omega)} \le C\|DU_n\|_{L^{q_*}(\Omega)} \le C(C_0, q)$.

We then see that $\{U_n\}$ is bounded in $W^{2,q}(\Omega)$ and $\{W_n\}$ is bounded in $W^{1,q}(\Omega)$. As $q > N$, by compactness and relabeling, we can assume that $\{DU_n\}$ and $\{W_n\}$ converge in $C^{0,\gamma}(\Omega)$ for some $\gamma > 0$.

Since P^{-1} is Hölder continuous, from Ph) we have

$$|u_n - u_m| = |P^{-1}(W_n) - P^{-1}(W_m)| \le [P^{-1}]_{\alpha_P}|W_n - W_m|^{\alpha_P}\,.$$

So u_n converges to some u in $C(\Omega)$. In fact, for any x, y in Ω

$$\begin{aligned} |u_n(x) - u_n(y)| &\le [P^{-1}]_{\alpha_P}|W_n(x) - W_n(y)|^{\alpha_P} \\ &\le C[P^{-1}]_{\alpha_P}|x - y|^{\gamma\alpha_P}\,, \end{aligned}$$

because $\{W_n\}$ is bounded in $C^{0,\gamma}(\Omega)$. Letting $n \to \infty$ we see that $u \in C^{\gamma\alpha_P}(\Omega)$.

For any $\phi \in C^1(\Omega)$, we obtain from the system (5.6.25) for u_n that

$$\int_\Omega \langle DP_n(u_n), D\phi\rangle\, dx = \int_\Omega \langle B_A(u_n, DP(u_n)) + f(u_n), \phi\rangle\, dx\,. \tag{5.6.26}$$

Here, we used the fact that $B(u_n, Du_n) = B_A(u_n, DP(u_n))$ because $B(u_n, Du_n) = 0$ if $u_n(x) \in K$, the singular set of $A(u_n)$.

As $u_n \to u$ in $C(\Omega)$, we have that $U_n = P_n(u_n)$ converges to $P(u)$ in $C(\Omega)$ so that the limit of DU_n in $C(\Omega)$ is $DP(u)$. Thus, $DP(u) \in C(\Omega)$. Also, $f_n(u_n) \to f(u)$ in $C(\Omega)$ because of Pfn).

Since $q \ge 2$ and $\{DW_n\}$ is bounded in $L^q(\Omega)$, we can assume that $DP(u_n)$ converges weakly in $L^2(\Omega)$. The weak limit is of course $DP(u)$. The assumption Bw) then implies that $B_A(u_n, DP(u_n))$ converges in $C(\Omega)$ to $B_A(u, DP(u))$.

Letting $n \to \infty$ in (5.6.26) and using the established convergences, we see that u is a weak solution, in accordance with the definition W). We proved before that u is Hölder continuous. The proof is complete. □

Remark 5.6.4. In general, we cannot show that Du is bounded in Ω and $D^2u \in L^2(\Omega)$ so that u is not a strong solution in the usual sense. However, if $u(x) \notin K$ then $Du(x) = A^{-1}(u(x))DP(u(x))$ is bounded. Thus, the obtained weak solution u is a strong solution away from the degenerate set $\{x :\ u(x) \in K\}$.

5.7 The general SKT systems

We apply the results of Section 5.2 to the following system:

$$-\Delta(u_i[d_i + \lambda_i(u)]) = b_i(u, Du) + k_i u_i + g_i(u)\,, \quad i = 1, \dots, m\,. \tag{5.7.1}$$

Obviously, this system is a special case of those studied in Section 5.2 with

$$P(u) = [u_i(d_i + \lambda_i(u))]_{i=1}^m,\ f(u) = [k_i u_i + g_i(u)]_{i=1}^m\,.$$

Being inspired by the porous media type system discussed in the Introduction and the generalized SKT in Section 5.1, we consider the following hypotheses:

Pskt) The map $P(u) := [P_i(u)]_{i=1}^m$ with $P_i(u) = u_i\lambda_i(u)$ satisfies P). For some $k > 0$ the elliptic function $\lambda(u)$ in P) has a polynomial growth $\lambda(u) \sim |u|^k$ for some $k > 0$.

Bskt) The $B(u, Du) := [b_i(u, Du)]_{i=1}^m$ satisfies B). That is, $|B(u, Du)| \le C\lambda(u)|Du|$.

fskt) There is a constant C such that $|g_i(u)| \le C|u|\lambda(u)$ and $|(g_i)_u(u)| \le C\lambda(u)$.

Concerning the strong solutions to the regular (SKTgen) system, since the data P, f have polynomial growths we need only assume that $\|u\|_{L^1(\Omega)}$ is bounded. The following result greatly improves on those in [30, Section 3.2] where it was assumed that the dimension $N \le 4$ and the drift term $B \equiv 0$. Here, we allow any N.

Corollary 5.7.1. *Assume $d_i > 0$ and that there is a constant C such that any strong solution u to the family (parametrized by $\sigma \in [0, 1]$)*

$$\begin{cases} -\Delta(u_i[d_i + \lambda_i(u)]) = \sigma[b_i(u, Du) + k_i u_i + g_i(u)]\,, & \text{in } \Omega, i = 1, \dots, m\,,\\ \text{Homogeneous Dirichlet or Neumann conditions on } \partial\Omega \end{cases} \tag{5.7.2}$$

satisfies $\|u\|_{L^1(\Omega)} \le C$. Then there is a strong solution to (5.7.1).

Proof of Corollary 5.7.1. The existence of strong solutions of regular system (5.7.1) is just a consequence of Theorem 5.5.1. Indeed, the structural condition PR) holds with $\lambda_0 = \min\{d_i\} > 0$. B) and f) are verified by Bskt) and fskt) respectively. Finally, the condition (5.6.23), that $\|P^{\beta_0}(u)\|_{L^1(\Omega)}$ and $\|f^{\beta_0}(u)\|_{L^1(\Omega)}$ are uniformly bounded, holds here. Indeed, because P, f have polynomial growths in u and $\|u\|_{L^1(\Omega)}$ is assumed to be bounded uniformly, we need only choose β_0 sufficiently small such that $|P^{\beta_0}(u)|$ and $|f^{\beta_0}(u)|$ are bounded in terms of $1 + |u|$. □

Remark 5.7.2. We present an example where one can obtain a uniform bound for L^p norm of solution. Inspired by the competitive Lotka–Volterra type reaction, we consider its generalized version by assuming that

$$\sum_i \langle g_i(u), u_i\rangle \le -c_1|u|^{k+2} \text{ for some } k, c_1 > 0\,. \tag{5.7.3}$$

For $A(u) = \partial_u P(u)$ and some sufficiently small and positive α, depending on the ellipticity of the matrix $A(u)$ we observe that there is $c_0 > 0$ such that

$$\langle A(u)Du, D(|u|^\alpha u)\rangle \ge c_0\lambda(u)|u|^\alpha|Du|^2\,.$$

We then test the system $-\Delta P(u) = \sigma B(u, Du) + \sigma f(u)$ with $|u|^\alpha u$ to obtain

$$c_0\int_\Omega \lambda(u)|u|^\alpha|Du|^2\,dx \le \sigma\int_\Omega |u|^\alpha[\langle B(u, Du), u\rangle + C|u|^2 + \sum_i\langle g_i(u), u_i\rangle]\,dx\,.$$

We also assume that $|B(u, Du)| \le \varepsilon_0\lambda(u)|Du|$ and use Young's inequality to obtain from the above that

$$\int_\Omega \lambda(u)|u|^\alpha|Du|^2\,dx \le \sigma\int_\Omega |u|^\alpha[C\varepsilon_0^2\lambda(u)|u|^2 + C|u|^2 + \sum_i\langle g_i(u), u_i\rangle]\,dx\,.$$

By (5.7.3), we derive

$$0 \le \sigma\int_\Omega [C\varepsilon_0^2|u|^{k+2+\alpha} + C|u|^{2+\alpha} - c_1|u|^{k+2+\alpha}]\,dx\,.$$

Hence, if ε_0 is sufficiently small, we rearrange and get

$$\int_\Omega |u|^{k+2+\alpha}\,dx \le C\int_\Omega |u|^{2+\alpha}\,dx\,.$$

By Hölder's inequality, we arrive at

$$\left(\int_\Omega |u|^{2+\alpha}\,dx\right)^{\frac{k+2+\alpha}{2+\alpha}} \le C\int_\Omega |u|^{2+\alpha}\,dx\,.$$

Because $k > 0$, the above clearly implies a uniform bound for $\|u\|_{L^{2+\alpha}(\Omega)}$.

We should note that the argument applies to the general elliptic systems as well and they are not necessarily the SKT ones considered in this chapter.

Remark 5.7.3. The example in Remark 5.7.2 can be extended to the weak cooperative case. That is, instead of (5.7.3), we only assume that

$$\sum_i \langle g_i(u), u_i\rangle \le c_1|u|^{k+2} \text{ for some } k, c_1 > 0\,. \tag{5.7.4}$$

Assume also that there is a constant C_1 such that strong solutions of the system satisfy

$$\|u\|_{L^1(\Omega)} \le C_1\,. \tag{5.7.5}$$

As before, for sufficiently small positive α we have

$$\int_\Omega \lambda(u)|u|^\alpha|Du|^2\,dx \le \sigma\int_\Omega |u|^\alpha[C\varepsilon_0^2\lambda(u)|u|^2 + C|u|^2 + \sum_i\langle g_i(u), u_i\rangle]\,dx\,.$$

If we assume that $\lambda(u) \ge C|u|^k$ then the above and (5.7.4) imply

$$\int_\Omega |u|^{k+\alpha}|Du|^2\,dx \le C\sigma\int_\Omega |u|^2\,dx + C\int_\Omega |u|^{k+\alpha+2}\,dx\,. \tag{5.7.6}$$

Concerning the last integral, we apply the interpolation inequality to $w = |u|^{\frac{k+\alpha}{2}+1}$, then $|Dw|^2 \sim |u|^{k+\alpha}|Du|^2$, to see that for any positive ε, β we can find a constant $C(\varepsilon, \beta)$ such that

$$\int_\Omega |u|^{k+\alpha+2}\,dx \le \varepsilon\int_\Omega |u|^{k+\alpha}|Du|^2\,dx + C(\varepsilon,\beta)\left(\int_\Omega |w|^\beta\,dx\right)^{\frac{2}{\beta}}\,.$$

If β is small then $|w| \sim |u|$ so that the last integral is bounded by the constant C_1 in the assumption (5.7.5). Hence, by choosing sufficiently small ε (and then $\varepsilon = 1$ to estimate the left-hand side of (5.7.6)), we easily get

$$\int_\Omega |u|^{k+\alpha+2}\,dx \le C\int_\Omega |u|^2\,dx + C(C_1)\,.$$

With this, as before, we obtain a bound for $\|u\|_{L^{2+\alpha}(\Omega)}$.

Next, assuming no drift term, we establish the existence of Hölder continuous weak solutions. We apply Theorem 5.5.2 here by considering the following approximation systems:

$$-\Delta(u_i[\lambda_{0,n}^k + \lambda_i(u)]) = \sigma[k_i(u_i + \lambda_{0,n}) + g_i(u)]\,,\quad \sigma \in [0,1] \text{ and } i = 1,\dots,m\,, \tag{5.7.7}$$

where $\{\lambda_{0,n}\}$ is a sequence converging to 0 in $(0, 1)$. The limit system is

$$-\Delta(u_i\lambda_i(u)) = k_iu_i + g_i(u)\,, \quad i = 1, \dots, m\,, \tag{5.7.8}$$

which is singular at $u = 0$ and degenerate when $|u| \to \infty$.

Corollary 5.7.4. *Assume Pskt) and fskt). The following assertions hold:*

i.1) *For each n, if the L^1 norms of strong solutions to the family (5.7.7) are bounded uniformly in $\sigma \in [0, 1]$ then there is a sequence $\{u_n\}$ of strong solutions to (5.7.7) for $\sigma = 1$.*

i.2) *Assume that there exists a subsequence of $\{u_n\}$ consisting of nonnegative strong solutions to (5.7.7) for $\sigma = 1$ and their L^1 norms are bounded uniformly in n. Furthermore, assume that P^{-1} is Hölder continuous. Then there is a Hölder continuous weak solution to (5.7.8).*

Proof of Corollary 5.7.4. The assertion i.1) is just a restatement of Corollary 5.7.1 with $d_i = \lambda_{0,n}$ and $B \equiv 0$. We need only prove i.2) by using Theorem 5.5.2.

Let $\{u_n\}$ be a sequence of nonnegative strong solutions to (5.7.7) for $\sigma = 1$. The assumption that $\|u_n\|_{L^1(\Omega)}$ is uniformly bounded implies the bounds in (5.5.8) of Theorem 5.5.2 by choosing small β_0.

Thus, in order to complete the proof we need to establish the crucial uniform bound (5.5.9). That is, there is a constant C such that $\|\lambda_n^{-1}(u_n)(f_n)_u(u_n)\|_{L^{r_0}(\Omega)} \le C$ for some $r_0 > N/2$ and any n. Here, $\lambda_n(u) = \lambda_{0,n}^k + \lambda(u)$ and $f_n(u) = [k_i(u_i + \lambda_{0,n}) + g_i(u)]_{i=1}^m$.

From the condition fskt) we have $|(f_n)_u(u)|\lambda_n^{-1}(u)| \lesssim \lambda_n^{-1}(u) + 1$. Hence, we need only show that there is a constant $C(r_0)$ independent of $\lambda_{0,n}$ such that

$$\|\lambda_n^{-1}(u)\|_{L^{r_0}(\Omega)} \le C(r_0)\,, \quad \text{for some } r_0 > N/2\,. \tag{5.7.9}$$

In the sequel, we fix an n and denote $\lambda_0 = \lambda_{0,n}$, $f_{\lambda_0,i}(u) = k_i(u_i + \lambda_0) + g_i(u)$, and

$$P_{\lambda_0,i}(u) := u_i(\lambda_0^k + \lambda_i(u)),\ \mathcal{P}_{\lambda_0}(u) = \lambda_0^{k+1} + \sum_{i=1}^m P_{\lambda_0,i}(u)\,.$$

For any number μ and any strong solution u to (5.7.7), $\sigma = 1$, we can test the system with $\mathcal{P}^\mu_{\lambda_0}(u)$ to obtain

$$\mu\int_\Omega \langle DP_{\lambda_0,i}(u), D\mathcal{P}_{\lambda_0}\rangle \mathcal{P}^{\mu-1}_{\lambda_0}(u)\,dx = \int_{\partial\Omega} D_\nu P_{\lambda_0,i}(u)\mathcal{P}^\mu_{\lambda_0}(u)d\sigma + \int_\Omega f_{\lambda_0,i}(u)\mathcal{P}^\mu_{\lambda_0}(u)\,dx\,,$$

where $D_\nu P_{\lambda_0,i}(u) = \langle DP_{\lambda_0,i}(u), \nu\rangle$ is the normal derivative of $P_{\lambda_0,i}(u)$ on $\partial\Omega$. Summing the above and noting that $\sum_i DP_{\lambda_0,i} = D\mathcal{P}_{\lambda_0}$, we rearrange and get

$$\int_\Omega \mathcal{P}^\mu_{\lambda_0}(u)\sum_i f_{\lambda_0,i}(u)\,dx + \int_{\partial\Omega} D_\nu\mathcal{P}_{\lambda_0}(u)\mathcal{P}^\mu_{\lambda_0}(u)d\sigma = \mu\int_\Omega |D\mathcal{P}_{\lambda_0}(u)|^2\mathcal{P}^{\mu-1}_{\lambda_0}(u)\,dx\,.$$

For Neumann boundary conditions, the boundary integral is zero. For Dirichlet boundary conditions, because $P_{\lambda_0,i}(u) = 0$ on $\partial\Omega$ and $P_{\lambda_0,i}(u) \ge 0$ (as u is nonnegative in Ω) so the normal derivative $D_\nu \mathcal{P}_{\lambda_0}(u) = \sum_i D_\nu P_{\lambda_0,i}(u) \geq 0$. Hence, the boundary integral on the left-hand side of the above is nonnegative. Meanwhile, as the right-hand side is nonpositive if $\mu \leq 0$, we arrive at

$$\int_\Omega \mathcal{P}^\mu_{\lambda_0}(u) \sum_i f_{\lambda_0,i}(u)\,dx \leq 0 \quad \text{if } \mu \leq 0\,. \tag{5.7.10}$$

Because $f_{\lambda_0,i}(u) = k_i(u_i + \lambda_0) + g_i(u)$, (5.7.10) then yields

$$\int_\Omega \mathcal{P}^\mu_{\lambda_0}(u) \sum_i k_i(u_i + \lambda_0)\,dx \leq \left| \int_\Omega \mathcal{P}^\mu_{\lambda_0}(u) \sum_i g_i(u)\,dx \right| . \tag{5.7.11}$$

We consider two cases, either $|u| \geq \lambda_0$ or $|u| < \lambda_0$, and see easily that

$$\mathcal{P}_{\lambda_0}(u) \sim (\lambda_0 + |u|)^{k+1}\,. \tag{5.7.12}$$

From the assumption on g_i's in fskt), $\lambda_i(u) \sim \lambda(u) \sim |u|^k$, and (5.7.12), we have

$$\left|\sum_i g_i(u)\right| \leq C(\lambda_0 + |u|)^{k+1} \lesssim \mathcal{P}_{\lambda_0}(u), \; \sum_i k_i(u_i + \lambda_0) \sim (\lambda_0 + |u|) \sim \mathcal{P}^{1/(k+1)}_{\lambda_0}(u)\,.$$

We then find a constant C independent of λ_0 and use (5.7.11) to get

$$\int_\Omega \mathcal{P}^{\mu+1-\frac{k}{k+1}}_{\lambda_0}(u)\,dx \leq C \int_\Omega \mathcal{P}^{\mu+1}_{\lambda_0}(u)\,dx\,. \tag{5.7.13}$$

For integers $l = 0, 1, \ldots$ we let $\mu = -1 - \frac{lk}{k+1}$, which is negative, so that (5.7.13) holds. Iterating and using induction, for any integer l we can find a constant $C(l)$, independent of λ_0, such that

$$\int_\Omega \mathcal{P}^{-\frac{(l+1)k}{k+1}}_{\lambda_0}(u)\,dx \leq C(l)\,.$$

Using the fact that $\lambda_n(u) \sim (\lambda_{0,n} + |u|)^k \lesssim \mathcal{P}^{\frac{k}{k+1}}_{\lambda_0}(u)$, with $\lambda_0 = \lambda_{0,n}$, we conclude that $\|\lambda_n(u)\|_{L^q(\Omega)} \leq C(q)$ for any $q < 0$. The constant $C(q)$ is independent of $\lambda_{0,n}$. We obtain the desired estimate (5.7.9) and complete the proof. □

Remark 5.7.5. The assumption on the existence of nonnegative solutions in i.2) can be verified in many applications by using some invariant principles. For instance, we refer the readers to [30, Lemma 4.7]. In fact, we can see that this assumption was used only to get nonnegativity of the normal derivative $D_\nu \mathcal{P}_{\lambda_0}$, if Dirichlet boundary conditions are considered, so that the key estimate (5.7.10) follows.

In applications, especially the general SKT systems on lower-dimensional domains, it is possible to control the L^r norms of the strong solutions to the approximation system for some sufficiently large r so that the following result can be established via a much simpler argument.

Corollary 5.7.6. *The conclusions in Corollary 5.7.4 hold if $\|u\|_{L^{r_0}(\Omega)}$ is uniformly bounded for some $r_0 > N/2$. That is, there is a constant c_0 such that any strong solution u to*

$$-\Delta(u_i[\lambda_{0,n} + \lambda_i(u)]) = \sigma[k_i u_i + g_i(u)] , \tag{5.7.14}$$

$\sigma \in [0, 1]$ *and* $i = 1, \dots, m$ *satisfies*

$$\|u\|_{L^{r_0}(\Omega)} \le c_0 . \tag{5.7.15}$$

Then there exists a weak solution u to

$$-\Delta(u_i\lambda_i(u)) = k_i u_i + g_i(u), \quad i = 1, \dots, m . \tag{5.7.16}$$

Moreover, there are constants $\gamma > 0$ and $C(c_0)$ such that

$$\|u\|_{C^{0,\gamma}(\Omega)} \le C(c_0) . \tag{5.7.17}$$

Proof. We can write the reaction term $f(u)$ in the approximation system (5.7.14) as $f(u) = f_1(u) + f_2(u)$ with

$$f_1(u) = g_i(u), \quad f_2(u) = \sigma k_i u_i .$$

From the assumption on g_i in fskt) that $|(g_i)_u| \lesssim \lambda(u)$, the quantity

$$C_1 := \|\lambda^{-1}(u)(f_1)_u(u)\|_{L^{r_0}(\Omega)} \tag{5.7.18}$$

is finite for any $r_0 \ge 1$. Furthermore, as we assume that $\|u\|_{L^{r_0}(\Omega)}$ is bounded for some $r_0 > N/2$, the norms

$$C_2 := \||f_2(u)|\|_{L^{r_0}(\Omega)} \quad \text{and} \quad C_0 := \||f_1(u)|^{\beta_0}\|_{L^1(\Omega)}^{\frac{1}{\beta_0}} \tag{5.7.19}$$

are also finite. From Remark 5.6.2, we obtain a uniform bound for $\|P_n(u)\|_{W^{2,q}(\Omega)}$ for some $q > N/2$. By Sobolev's embedding theorem, we see that $\|P_n(u)\|_{W^{1,p}(\Omega)}$ is uniformly bounded for some $p > N$ and thus, for each n, there is a sequence $\{u_n\}$ of strong solutions to (5.7.7) for $\sigma = 1$. Importantly, $\{\|P(u_n)\|_{W^{1,p}(\Omega)}\}$ is bounded. The argument in the proof of Theorem 5.5.2 then provides a limiting weak solution u to (5.7.8). Furthermore, as P^{-1} is Hölder continuous, u is also Hölder continuous. The uniform bound for $\|P(u_n)\|_{W^{1,p}(\Omega)}$ in terms of c_0 then provides a constant $C(c_0)$ such that $\|u\|_{C^{0,\gamma}(\Omega)} \le C(c_0)$. This is (5.7.17) and the proof is complete. □

6 Nontrivial steady-state solutions

In this chapter, we will study the existence of *strong* solutions and *other nontrivial* solutions to the following nonlinear *strongly coupled* and *nonregular* but *uniform* elliptic system:

$$\begin{cases} -\operatorname{div}(A(u, Du)) = \hat{f}(u, Du) & \text{in } \Omega\,, \\ u \text{ satisfies Dirichlet or Neumann boundary conditions} & \text{on } \partial\Omega\,. \end{cases} \tag{6.0.1}$$

Throughout this chapter, we always assume the conditions A'), SG') and F) from Chapter 4 so that the existence theory there applies here.

We now see that Theorem 4.1.3 establishes the existence of a strong solution in a suitable Banach space **X** to (6.0.1) under a priori mild integrability conditions of strong solutions (or more importantly, the smallness condition on the BMO norm of Theorem 4.3.1). However, this result provides no interesting information if some 'trivial' or 'semitrivial' solutions, which are solutions to a subsystem of (6.0.1), may be obviously guaranteed by other means. Consequently, we will be interested in finding other nontrivial solutions to (6.0.1) and the uniform estimates in Section 4.2 still play a crucial role here. Although many results in this chapter, in particular the abstract results on fixed-point indices in Section 6.2, can apply to the general (6.0.1), we restrict ourselves to the system

$$\begin{cases} -\operatorname{div}(A(u)Du) = f(u) & \text{in } \Omega\,, \\ \text{homogeneous Dirichlet or Neumann boundary conditions} & \text{on } \partial\Omega\,. \end{cases} \tag{6.0.2}$$

This problem is the prototype of a general class of nonlinear elliptic systems, which arise in numerous applications, where u usually denotes a population/chemical density vector of species/agents. Therefore, we will also be interested in finding *positive* solutions of this system, i.e., those that are in the positive cone

$$\mathbf{P} := \{u \in \mathbf{X}\colon u = (u_1, \dots, u_m),\ u_i(x) \geq 0\ \forall x \in \Omega\}\,.$$

Under suitable assumptions on $\hat{f}$, we will show that the associated fixed-point map T, defined in Chapter 4, can be regarded as a map on a bounded set of **P** into **P**, i.e., T is a positive map.

If $\hat{f}(0) = 0$ then (6.0.2) has the *trivial solution* $u = 0$. A solution u is a *semitrivial* solution if some components of u are zero. A nontrivial positive solution to (6.0.2) will be found via a fixed point of the map T in the interior of **P**.

Our main tool in this chapter is the theory of fixed-point index 'in the **P** direction' or 'the expansion and the compression of a cone', which are excellently discussed in the survey [3] by Amann and with more details, among other things, in the book [47] by Zeidler. These concepts are closely related to the stability in the **P** direction of those trivial and semitrivial solutions.

https://doi.org/10.1515/9783110608762-006

The existence of nontrivial solutions then follows if the sum of the local fixed-point indices at trivial and semitrivial solutions does not add up to the fixed-point index of T in $\mathbf{P}$. Several results with specific structures of (6.0.1) are given to show that this will be the case in Section 6.3.

Next, in Section 6.4, if Neumann boundary conditions are considered then it could be that a nontrivial and *constant* solution of (6.0.2) exists and solves $\hat{f}(u) = 0$. In this case, the conclusion in the previous section does not provide useful information. We are then interested in finding nontrivial *nonconstant* solutions to (6.0.2). The results in this section greatly improve on those in [32, 35], which dealt only with systems of two equations, and establish the effect of cross-diffusions in 'pattern formation' problems in mathematical biology and chemistry. Besides the fact that our results here can be used for large systems, the analysis provides a systematic way to study pattern formation problems. We conclude the chapter by considering two extreme cases: either the self-diffusion is too large or it vanishes. We present a simple proof of the fact that nonconstant solutions do not exist if the diffusion is sufficiently large. If the self-diffusivities are all vanishing then a nonzero solution must exist.

6.1 On trivial and semitrivial solutions

We heuristically discuss the meaning of semitrivial solutions, their stability and our strategy in finding conditions for nontrivial solutions to exist.

Roughly speaking, we decompose $\mathbf{X} = \mathbf{X}_1 \oplus \mathbf{X}_2$, accordingly $\mathbf{P} = \mathbf{P}_1 \oplus \mathbf{P}_2$ with $\mathbf{P}_i$ being the positive cone of $\mathbf{X}_i$, and write an element of $\mathbf{X}$ as (u, v) with $u \in \mathbf{X}_1$, $v \in \mathbf{X}_2$. Then $w = (u, 0)$, with $u > 0$, is a semitrivial positive fixed point if w is a fixed point of T in $\mathbf{P}$ and u is a fixed point of $T|_{\mathbf{P}_1}$, the restriction of T to $\mathbf{P}_1$. We then show that the local indices of T at these semifixed points are solely determined by $\mathbf{P}_2$-*stable* fixed points of $T|_{\mathbf{P}_1}$.

We consider the system (6.0.2)

$$\begin{cases} -\operatorname{div}(A(u)Du) = \hat{f}(u) & \text{in } \Omega\,, \\ \text{homogenenous Dirichlet or Neumann boundary conditions} & \text{on } \partial\Omega\,. \end{cases} \tag{6.1.1}$$

As in the previous Chapter 4, we will regard solutions to this system as fixed points of nonlinear compact maps on an appropriate Banach space $\mathbf{X}$.

We fix some $\alpha_0 > 0$ and let $\mathbf{X}$ be $C^{1,\alpha_0}(\Omega, \mathbb{R}^m)$ (or $C^{1,\alpha_0}(\Omega) \cap C_0(\Omega)$ if Dirichlet boundary conditions are considered). We should note that this space is the subspace of the space $C^{1,\alpha_0}(\Omega) \cap W^{1,2}(\Omega)$ considered in Theorem 4.1.3, however, once we can establish that the Hölder and $W^{1,2}$ norm of all solutions u are uniformly bounded then their $C^{1,\alpha}(\Omega)$ norms are also bounded ([16, Chapter 8], in particular [16, Theorems 8.5 and 8.6]). Thus, under the same appropriate assumptions as in Chapter 4,

Theorem 4.1.3 gives the existence of a strong solution in $\mathbf{X}$ to (6.1.1) and a uniform estimate for $\|u\|_{\mathbf{X}}$. Therefore we can take $\mathbf{X}$ as defined here.

The obtained solutions by Theorem 4.1.3 may be trivial. For example, the *trivial solution* $u = 0$ is a solution to the system if $\hat{f}(0) = 0$.

Let us discuss the existence of *semitrivial* solutions. We write $\mathbb{R}^m = \mathbb{R}^{m_1} \oplus \mathbb{R}^{m_2}$ for some $m_1, m_2 \geq 0$ and denote $\mathbf{X}_i = C^{1,\alpha_0}(\Omega, \mathbb{R}^{m_i})$. By reordering the equations and variables, we write $\mathbf{X} = \mathbf{X}_1 \oplus \mathbf{X}_2$, an element of $\mathbf{X}$ as (u, v) with $u \in \mathbf{X}_1$, $v \in \mathbf{X}_2$, and

$$A(u,v) = \begin{bmatrix} P^{(u)}(u,v) & P^{(v)}(u,v) \\ Q^{(u)}(u,v) & Q^{(v)}(u,v) \end{bmatrix} \quad \text{and} \quad \hat{f}(u,v) = \begin{bmatrix} f^{(u)}(u,v) \\ f^{(v)}(u,v) \end{bmatrix} .$$

Here, $P^{(u)}(u,v)$ and $Q^{(v)}(u,v)$ are matrices of sizes $m_1 \times m_1$ and $m_2 \times m_2$ respectively.

Suppose that

$$Q^{(u)}(u,0) = 0 \ \text{ and } \ f^{(v)}(u,0) = 0 \quad \forall u \in \mathbf{X}_1 , \tag{6.1.2}$$

then $(u, 0)$, with $u \neq 0$, is a semitrivial solution if u solves the subsystem

$$-\operatorname{div}(P^{(u)}(u,0)Du) = f^{(u)}(u,0) .$$

For each $u \in \mathbf{X}$ and some constant matrix K we consider the following *linear* elliptic system for w:

$$\begin{cases} -\operatorname{div}(A(u)Dw) + Kw = \hat{f}(u) + Ku & \text{in } \Omega , \\ \text{homogenenous boundary conditions} & \text{on } \partial\Omega . \end{cases} \tag{6.1.3}$$

For a suitable choice of K, ((6.1.5) below), we can always assume that (6.1.3) has a unique weak solution $w \in \mathbf{X}$. This is equivalent to saying that the elliptic system

$$\begin{cases} -\operatorname{div}(A(u)Dw) + Kw = 0 & x \in \Omega , \\ \text{homogenenous boundary conditions} & \text{on } \partial\Omega \end{cases} \tag{6.1.4}$$

has $w = 0$ as the only solution. This is the case if we assume that there is $k > 0$ such that

$$\langle Ku, u \rangle \geq k|u|^2 \quad \forall u \in \mathbb{R}^m . \tag{6.1.5}$$

We then define $T(u) := w$ with w being the weak solution to (6.1.3). It is clear that the fixed-point solutions of $T(u) = u$ are solutions to (6.1.1), where $w = u$.

Since $A(u)$ is C^1 in u, $A(u(x))$ is Hölder continuous on Ω. The regularity theory of linear elliptic systems then shows that $w \in C^{1,\alpha}(\Omega, \mathbb{R}^m)$ for all $\alpha \in (0,1)$ so that T is compact in $\mathbf{X}$. Furthermore, T is a differentiable map.

If (6.1.1) satisfies the assumptions of Theorem 4.1.3 or its consequences in Chapter 4 then there is $M > 0$ such that

$$T(u) = u \Rightarrow \|u\|_{\mathbf{X}} < M . \tag{6.1.6}$$

In applications, we are also interested in finding solutions that are positive. We then consider the positive cone in **X**,

$$\mathbf{P} := \{u \in \mathbf{X} : u = (u_1, \ldots, u_m), u_i \geq 0 \; \forall i\},$$

which has nonempty interior

$$\dot{\mathbf{P}} := \{u \in \mathbf{X} : u = (u_1, \ldots, u_m), u_i > 0 \; \forall i\}.$$

Let M be the number provided by (6.1.6). We denote by $\mathbf{B} := B_{\mathbf{X}}(0, M)$ the ball in **X** centered at 0 with radius M. If T maps $\mathbf{B} \cap \mathbf{P}$ into **P** then, since **P** is closed in **X** and convex and it is a retract of **X** [7], we can define the cone index $\mathrm{ind}(T, U, \mathbf{P})$ for any open subset U of $\mathbf{B} \cap \mathbf{P}$ as long as T has no fixed point on ∂U, the boundary of U in **P** [3, Theorem 11.1].

The argument in the proof of Theorem 4.1.3 can apply here to give

$$\mathrm{ind}(T, \mathbf{B} \cap \mathbf{P}, \mathbf{P}) = 1.$$

This yields the existence of a fixed point of T, or a solution to (6.1.1), in **P**. From the previous discussion, this solution may be trivial or semitrivial. To establish the existence of a nontrivial positive solution u, i.e., $u \in \dot{\mathbf{P}}$, we will have to compute the local indices of T at its trivial and semitrivial fixed points. If these indices do not add up to $\mathrm{ind}(T, \mathbf{B} \cap \mathbf{P}, \mathbf{P}) = 1$ then the existence of nontrivial solutions follows from [3, Corollary 11.2].

The next section is then devoted to the computation of the local indices of semitrivial fixed points of general abstract maps defined on bounded subsets of positive cones of **X**.

6.2 Some general index results

We consider the case when (6.1.1) has trivial or semitrivial solutions, that is when $u = 0$ or some component of u is zero. We will compute the local indices of the map $T(u)$ at these trivial or semitrivial solutions. The abstract results in this section are in fact independent of (6.1.1) and thus can apply to (6.1.1) and other general situations as well.

We decompose **X** as $\mathbf{X} = \mathbf{X}_1 \oplus \mathbf{X}_2$ and denote by $\mathbf{P}_i$ and $\dot{\mathbf{P}}_i$, $i = 1, 2$, the positive cones and their nonempty interiors in $\mathbf{X}_i$'s. We assume ((6.1.6)) that there is $M > 0$ such that the map T is well defined as a map from the ball **B** centered at 0 with radius M into **P**. Accordingly, we denote $\mathbf{B}_i = \mathbf{B} \cap \mathbf{X}_i$.

For $(u, v) \in \mathbf{B}_1 \oplus \mathbf{B}_2$, we write

$$T(u, v) = (F_1(u, v), F_2(u, v)),$$

where F_i's are maps from **B** into $\mathbf{X}_i$. We also write $\partial_u F_i, \partial_v F_i$ for the partial Fréchet derivatives of these maps.

It is clear that for $\phi = (\phi_1, \phi_2) \in \mathbf{X}_1 \oplus \mathbf{X}_2$

$$T'(u, v)\phi = (\partial_u F_1(u, v)\phi_1 + \partial_v F_1(u, v)\phi_2, \partial_u F_2(u, v)\phi_1 + \partial_v F_2(u, v)\phi_2).$$

For any fixed $u \in \mathbf{B}_1$ and $v \in \mathbf{B}_2$, we will think of $F_1(\cdot, v)$ and $F_2(u, \cdot)$ as maps from $\mathbf{B}_1$ into $\mathbf{X}_1$ and from $\mathbf{B}_2$ into $\mathbf{X}_2$ respectively. With a slight abuse of the notation, we still write $\partial_u F_1$, $\partial_v F_2$ for the Fréchet derivatives of these maps.

We consider semitrivial fixed points in $\mathbf{X}_1$. Taking into account (6.1.2), we will therefore assume in the sequel that

$$F_2(u, 0) = 0 \quad \forall u \in \mathbf{P}_1 . \tag{6.2.1}$$

This implies for any real number t

$$F_2(u, tv) = t \int_0^1 \partial_v F_2(u, tsv)v \, ds , \tag{6.2.2}$$

where $\partial_v F_2(u, \cdot)$ is the derivative of $F_2(u, \cdot) \colon \mathbf{B}_2 \to \mathbf{X}_2$.

Let Z_1 be the set of fixed points of $F_1(\cdot, 0)$ in $\mathbf{P}_1$. We always assume that Z_1 is a nonempty compact subset of $\mathbf{P}_1$. Of course, $u \in Z_1$ iff $F_1(u, 0) = u$ and $F_2(u, 0) = 0$.

For each $u \in \mathbf{B}_1$ we consider the spectral radius $r_v(u)$ *in the complement direction* v of $\partial_v F_2(u, 0)$:

$$r_v(u) := \lim_{k \to \infty} \|\partial_v F_2(u, 0)\|_{L(\mathbf{X}_2)}^{1/k} .$$

We also consider the following subsets of Z_1:

$$Z_1^+ = \{u \in Z_1 \, : \, r_v(u) > 1\} , \quad Z_1^- = \{u \in Z_1 \, : \, r_v(u) < 1\} . \tag{6.2.3}$$

Roughly speaking, Z_1^+ (respectively Z_1^-) consists of unstable (respectively stable) fixed points of T in the $\mathbf{P}_2$-direction. Sometimes we simply say that an element in Z_1^+ (respectively Z_1^-) is *v-unstable* (respectively *v-stable*).

Let us fix an open neighborhood U of Z_1 in $\mathbf{P}_1$. We first need to show that the index $\operatorname{ind}(T, U \oplus V)$ is well defined for some appropriate neighborhood of V in $\mathbf{P}_2$, i.e., $U \oplus V$ is a neighborhood of Z_1 as a subset of $\mathbf{P}$ and T has no fixed point on its boundary. To this end, we will *always* assume that

Z) If $u \in Z_1$ then $\partial_v F_2(u, 0)$, the Frechet derivative of $F_2(u, \cdot) \colon \mathbf{B}_2 \to \mathbf{X}_2$, does not have a positive eigenvector to the eigenvalue 1.

The following main result of this section shows that $\operatorname{ind}(T, U \oplus V, \mathbf{P})$ is determined by the index of the restriction $T|_{\mathbf{X}_1}$, i.e., $F_1(\cdot, 0)$, at v-stable fixed points (in Z_1^-).

Theorem 6.2.1. *Assume Z). There is a neighborhood of V of 0 in $\mathbf{P}_2$ such that* $\operatorname{ind}(T, U \oplus V, \mathbf{P})$ *is well defined.*

Suppose also the following conditions:

i) *T is a positive map. That is, T maps $\mathbf{B} \cap \mathbf{P}$ into $\mathbf{P}$.*
ii) *$F_2(u, 0) = 0$ for all $u \in \mathbf{B}_1$.*
iii) *At each semitrivial fixed point $u \in Z_1$, $\partial_v F_2(u, 0)$ is a strongly positive map on $\mathbf{B}_2$ into $\mathbf{X}_2$.*

Then there exist two disjoint open sets U^+, U^- in U such that $Z_1^+ \subset U^+$ and $Z_1^- \subset U^-$, and

$$\operatorname{ind}(T, U \oplus V, \mathbf{P}) = \operatorname{ind}(T, U^- \oplus V, \mathbf{P}) = \operatorname{ind}(F_1(\cdot, 0), U^-, \mathbf{P}_1) .$$

In iii) and the sequel, we say that G is a *strongly* positive map on W into $\mathbf{X}_i$ if G maps $W \cap \dot{\mathbf{P}}_i$ into $\dot{\mathbf{P}}_i$).

Remark 6.2.2. For $u \in Z_1$, i) implies that $\partial_v F_2(u,0)$ is a positive endomorphism on $\mathbf{X}_2$. In fact, for any $u \in Z_1$, $x > 0$ and positive small t such that $tx \in V$ we have by our assumptions that $F_2(u,tx) \geq 0$ and $F_2(u,0) = 0$. Hence, $\partial_v F_2(u,0)x = \lim_{t\to 0^+} t^{-1}F_2(u,tx) \geq 0$. So that $\partial_v F_2(u,0)$ is positive. If a certain strong maximum principle for the linear elliptic system defining $\partial_v F_2(u,0)$ is available, [3, Theorem 4.2], then $\partial_v F_2(u,0)$ is *strongly* positive and iii) follows. This assumption can be relaxed if Z_1 is a singleton (Remark 6.2.8).

Remark 6.2.3. If $\partial_v F_2(u,0)$ is strongly positive then $r_v(u)$ is the only eigenvalue with positive eigenfunction. Therefore, the assumptions $r_v(u) < 1$ and $r_v(u) > 1$ are respectively equivalent to the following:
I'.1) $\partial_v F_2(u,0)$ does not have any positive eigenvector to any eigenvalue $\lambda > 1$.
I'.2) $\partial_v F_2(u,0)$ has a positive eigenvector to some eigenvalue $\lambda > 1$.

The proof of Theorem 6.2.1 will be divided into several lemmas, which can be of interest in themselves.

In what follows, if G is a map from an open subset W of $\mathbf{P}_i$ into $\mathbf{P}_i$ and if no ambiguity can arise then we will abbreviate $\mathrm{ind}(G, W, \mathbf{P}_i)$ by $\mathrm{ind}(G, W)$. We will always regard $F_1(u,0)$ as a map from $\mathbf{X}_1$ into $\mathbf{X}_1$ and, for a given $u \in \mathbf{X}_1$, $\partial_v F_2(u,0)$ as a map from $\mathbf{X}_2$ into $\mathbf{X}_2$.

Our first lemma shows that there exists a neighborhood V claimed in Theorem 6.2.1 such that $\mathrm{ind}(T, U \oplus V)$ is well defined.

Lemma 6.2.4. *Assume Z). There is $r > 0$ such that for $V = B(0,r) \cap \mathbf{P}_2$, the ball in $\mathbf{P}_2$ centered at 0 with radius $r > 0$, there is no fixed point of $T(u,v) = (u,v)$ with $v > 0$ in the closure of $U \oplus V$ in $\mathbf{P}$.*

Proof. By contradiction, we can suppose that there are sequences $\{r_n\}$ of positives $r_n \to 0$ and $\{u_n\} \subset U$, $\{v_n\} \subset \mathbf{P}_2$ with $\|v_n\| = r_n$ such that, using (6.2.2),

$$u_n = F_1(u_n, v_n)\,, \quad v_n = F_2(u_n, v_n) = \int_0^1 \partial_v F_2(u_n, s v_n) v_n \, ds\,.$$

Setting $w_n = v_n/\|v_n\|$ we have

$$w_n = \int_0^1 \partial_v F_2(u_n, s r_n w_n) w_n \, ds\,.$$

By compactness, via a subsequence of $\{u_n\}$ and continuity we can let $n \to \infty$ and obtain $u_n \to u$ for some $u \in Z_1$, $v_n \to 0$ and $w_n \to w$ in $\mathbf{X}_2$ such that $u = F(u,0)$ and $\|w\| = 1$. Hence, $w > 0$ and satisfies

$$w = \int_0^1 \partial_v F_2(u,0) w \, ds = \partial_v F_2(u,0) w\,.$$

Thus, w is a positive eigenvector of $\partial_v F_2(u, 0)$ to the eigenvalue 1. This is a contradiction to Z) and completes the proof. □

In the sequel, we will always denote by V the neighborhood of 0 in $\mathbf{P}_2$ as in the above lemma.

Our next lemma shows that the index of T can be computed by using its restriction and partial derivatives.

Lemma 6.2.5. *We have*

$$\operatorname{ind}(T, U \oplus V) = \operatorname{ind}(T_*, U \oplus V),$$

where $T_*(u, v) = (F_1(u, 0), \partial_v F_2(u, 0)v)$.

Proof. Consider the following homotopy:

$$H(t, u, v) = \left(F_1(u, tv), \int_0^1 \partial_v F_2(u, tsv)v\, ds\right) \quad \text{for } t \in [0, 1]. \tag{6.2.4}$$

We show that this homotopy is well defined on $U \oplus V$. Indeed, if $H(t, u, v)$ has a fixed point (u, v) on the boundary of $U \oplus V$ for some $t \in [0, 1]$ then

$$F_1(u, tv) = u, \quad \int_0^1 \partial_v F_2(u, tsv)v\, ds = v, \quad (u, v) \in \partial(U \oplus V).$$

Assume first that $t > 0$. If $v = 0$ then the first equation gives that $F_1(u, 0) = u$ so that $u \in Z_1$. But then $(u, 0) \notin \partial(U \oplus V)$. Thus, $v > 0$ and the second equation ((6.2.2)) yields $F_2(u, tv) = tv$. This means (u, tv) is a fixed point of T in the closure of $U \oplus V$ with $tv > 0$. But there is no such fixed point of $T(u, v) = (u, v)$ in the closure of $U \oplus V$ by Lemma 6.2.4. Hence, $H(t, u, v)$ cannot have a fixed point (u, v) on the boundary of $U \oplus V$ if $t > 0$. We then consider $H(0, u, v)$ whose fixed points $(u, v) \in \partial(U \oplus V)$ must satisfy $u = F_1(u, 0)$ so that $u \in Z_1$ and $\partial_v F_2(u, 0)v = v$ with $v > 0$. But this contradicts Z).

Thus the homotopy is well defined and we have that

$$\operatorname{ind}(T, U \oplus V) = \operatorname{ind}(H(1, \cdot), U \oplus V) = \operatorname{ind}(H(0, \cdot), U \oplus V).$$

By (6.2.4), $H(0, u, v) = (F_1(u, 0), \partial_v F_2(u, 0)v) = T_*(u, v)$. The proof is complete. □

We now compute $\operatorname{ind}(T_*, U \oplus V)$.

Lemma 6.2.6. *Assume that $\partial_v F_2(u, 0)$ is a strongly positive endomorphism on $\mathbf{B}_2$ into $\mathbf{X}_2$ for each $u \in Z_1$. The following holds:*

I.1) *If $r_v(u) < 1$ for any $u \in Z_1$ then* $\operatorname{ind}(T_*, U \oplus V) = \operatorname{ind}(F_1(\cdot, 0), U)$.

I.2) *If $r_v(u) > 1$ for any $u \in Z_1$ then* $\operatorname{ind}(T_*, U \oplus V) = 0$.

Proof. First of all, we see that $\partial_v F_2(u,0)$ is a compact map. In fact, we have $F_2(u,0) = 0$ so that $\partial_v F_2(u,0)x = \lim_{t\to 0^+} t^{-1}F_2(u,tx)$. Since F_2 is compact, so is $\partial_v F_2(u,0)$.

To prove I.1), we consider the following homotopy:

$$H(u,v,t) = (F_1(u,0), t\partial_v F_2(u,0)v)\,, \quad t \in [0,1]\,.$$

This homotopy is well defined on $U \oplus V$. Indeed, a fixed point of (u,v) of $H(\cdot,\cdot,t)$ on $\partial(U\oplus V)$ must satisfy $u \in Z_1$ and $tv > 0$. But this means $v > 0$ is a positive eigenfunction to the eigenvalue $t^{-1} \geq 1$. This is a contradiction to Z) and the Krein–Rutman theorem, (see [3, Theorem 3.2, ii)]) for strongly positive compact endomorphism on $\mathbf{X}_2$, because $\partial_v F_2(u,0)$ has no positive eigenvector different from $r_v(u)$, which is assumed to be less than 1 in this case. Thus, by the definition of H and the homotopy invariance,

$$\mathrm{ind}(T_*, U\oplus V) = \mathrm{ind}(H(\cdot,\cdot,1), U\oplus V) = \mathrm{ind}(H(\cdot,\cdot,0), U\oplus V)\,.$$

As $H(u,v,0) = (F_1(u,0),0)$, by index product theorem, $\mathrm{ind}(H(\cdot,\cdot,0), U\oplus V) = \mathrm{ind}(F_1(\cdot,0), U)$. Hence,

$$\mathrm{ind}(T_*, U\oplus V) = \mathrm{ind}(F_1(\cdot,0), U)\,,$$

and this proves I.1).

We now consider I.2). Let h be any element in $\dot{\mathbf{P}}_2$, the interior of $\mathbf{P}_2$. We first consider the following homotopy:

$$H(u,v,t) = (F_1(u,0), t\partial_v F_2(u,0)v + th)\,, \quad t \geq 1\,. \tag{6.2.5}$$

If $H(\cdot,t)$ has a fixed point (u,v) in $\partial(U\oplus V)$ then $u \in Z_1$ and $v > 0$. Thus, there is some $v_* > 0$ such that $v_* = t\partial_v F_2(u,0)v_* + th$. This means $t^{-1}v_* - \partial_v F_2(u,0)v_* = h$. Since $t^{-1} \leq 1 < r_v(u)$, this contradicts the following consequence of the Krein–Rutman theorem (see [3, Theorem 3.2, iv)]) for strongly positive compact operators:

$$\lambda x - \partial_v F_2(u,0)x = h \ \text{ has no positive solution if } \lambda \leq r_v(u)\,.$$

Thus the homotopy is well defined on $U\oplus V$. Because $\partial_v F_2(u,0)v_* \geq 0$, $t\partial_v F_2(u,0)v_* + th$ becomes unbounded as $t \to \infty$, it is clear that $H(u,v,t)$ has no fixed point in $U\oplus V$ for t large. We then have

$$\mathrm{ind}(H(\cdot,\cdot,1), U\oplus V) = 0\,. \tag{6.2.6}$$

We now consider the homotopy

$$G(u,v,t) = (F_1(u,0), \partial_v F_2(u,0)v + th) \quad t \in [0,1]\,.$$

We will see that this homotopy is well defined on $U\oplus V$ if $\|h\|_{\mathbf{X}_2}$ is sufficiently small. First of all, we define $f: Z_1 \times \mathbf{X}_2 \to \mathbf{X}_2$ by $f(x,y) = y - \partial_v F_2(x,0)y$ and it is not difficult to show that $f(Z_1 \times \partial V)$ is closed in $\mathbf{X}_2$; see Remark 6.2.7 after this proof. By Z), if $u \in Z_1$, then $0 \notin f(Z_1 \oplus \partial V)$ so that there is $\varepsilon > 0$ such that $B_\varepsilon(0) \cap f(Z_1 \oplus \partial V) = \emptyset$. This means

$$\|x - \partial_v F_2(u, 0)x\|_{\mathbf{X}_2} > \varepsilon \quad \forall x \in \partial V \text{ and } \forall u \in Z_1 \,. \tag{6.2.7}$$

We now take h such that $\|h\|_{\mathbf{X}_2} < \varepsilon/2$. Then $G(\cdot, \cdot, t)$ has no fixed point $(u, v) \in \partial(U \oplus V)$. Indeed, if $G(\cdot, \cdot, t)$ has a fixed point $(u, v) \in \partial(U \oplus V)$ then $u \in Z_1$, $v \in \partial V$ and $v - \partial_v F_2(u, 0)v = th$. This fact and (6.2.7) then yield a contradiction

$$\|v - \partial_v F_2(u, 0)v\|_{\mathbf{X}_2} > \varepsilon > \|th\|_{\mathbf{X}_2} \quad \forall t \in [0, 1] \,.$$

This means $v - \partial_v F_2(u, 0)v \neq th$ for all $u \in Z_1$, $v \in \partial V$. Hence, the homotopy defined by G is well defined on $U \oplus V$. We then have

$$\mathrm{ind}(T_*, U \oplus V) = \mathrm{ind}(G(\cdot, \cdot, 0), U \oplus V) = \mathrm{ind}(G(\cdot, \cdot, 1), U \oplus V) = 0 \,,$$

where (6.2.6) was used, noting that $G(\cdot, \cdot, 1) = H(\cdot, \cdot, 1)$. We establish I.2) and complete the proof. □

Remark 6.2.7. We will show that $C = f(Z_1 \times \partial V)$ is closed in $\mathbf{X}_2$, where f is defined by $f(x, y) = y - \partial_v F_2(x, 0)y$. Indeed, let $z_n = f(x_n, y_n)$ be a convergent sequence in C for some sequences $\{x_n\} \subset Z_1$ and $\{y_n\} \subset \partial V$. As Z_1 is compact, we can assume that x_n converges. Because $\partial_v F_2(u, 0)$ is a compact map, $\{\partial_v F_2(x_n, 0)y_m\}$ is compact for each fixed n and can be assumed to be convergent. A diagonal argument then shows that we can find sequences $\{x_n\} \subset Z_1$ and $\{y_n\} \subset \partial V$ so that $\{\partial_v F_2(x_n, 0)y_n\}$ converge. Thus, there are $x \in Z_1$ and $y \in \partial V$ such that the sequences $x_n \to x$ and $y_n = z_n + \partial_v F_2(x_n, 0)y_n \to y$ in Z_1 and ∂V respectively. This implies $\{z_n\}$ converges to $y - \partial_v F_2(x, 0)y$ in C and therefore C is closed.

Remark 6.2.8. If we drop the assumption that $\partial_v F_2(u, 0)$ is *strongly* positive then the conclusion of Lemma 6.2.6 continues to hold if I.1) is replaced by I'.1), which is essentially used in the argument. Furthermore, if $\partial_v F_2(u, 0)$ is only positive then this is also the case, if we assume I'.2) in place of I.2) and Z_1 is a singleton, $Z_1 = \{u\}$. In fact, let h be a positive eigenvector of $\partial_v F_2(u, 0)$ for some $\lambda_u > 1$. We consider the following homotopy:

$$H(u, v, t) = (F_1(u, 0), \partial_v F_2(u, 0)v + th) \,, \quad t \geq 0 \,.$$

We will show that the homotopy is well defined. The case $t = 0$ is easy. Indeed, if $H(\cdot, \cdot, 0)$ has a fixed point (u, v) in $\partial(U \oplus V)$ then $u \in Z_1$ and $\partial_v F_2(u, 0)v = v$. But this gives $v = 0$, by Z), and u is not in ∂U.

We consider the case $t > 0$. If $H(\cdot, t)$ has a fixed point (u, v) in $\partial(U \oplus V)$ then $u \in Z_1$ and $v > 0$. Thus, there is some $v_* > 0$ such that $v_* = \partial_v F_2(u, 0)v_* + th$. Let τ_0 be the maximal number such that $v_* > \tau_0 h$. Because $\partial_v F_2(u, 0)$ is positive, we then have $\partial_v F_2(u, 0)v_* \geq \partial_v F_2(u, 0)\tau_0 h$ so that (as $\lambda > 1$)

$$v_* = \partial_v F_2(u, 0)v_* + th \geq \partial_v F_2(u, 0)\tau_0 h + th = (\lambda\tau_0 + t)h > (\tau_0 + t)h \,.$$

Since $t > 0$, the above contradicts the maximality of τ_0. Thus the homotopy is well defined. Again, when t is sufficiently large $H(u, v, t)$ has no solution in $U \oplus V$. Therefore, $\mathrm{ind}(H(\cdot, \cdot, 1), U \oplus V) = 0$.

Proof of Theorem 6.2.1. The assumption i) and the regularity results in the previous section show that T is a compact map on $\mathbf{X}$ so that $\mathrm{ind}(T, O, \mathbf{P})$ is well defined whenever T has no fixed point on the boundary of an open set O in $\mathbf{P}$. The assumptions ii) and iii) allow us to make use of the lemmas in this section.

We first prove that $r_v(u)$ is continuous in $u \in Z_1$ (Remark 6.2.9). Let $\{u_n\} \subset Z_1$ be a sequence converging to some $u_* \in Z_1$. Accordingly, let h_n be the normalized eigenfunction (i.e., $\|h_n\| = 1$ and $h_n > 0$) to the eigenvalue $\lambda_n = r_v(u_n)$. Because $\|\partial_v F_2(u_n, 0)\|_{L(\mathbf{X}_2)}$ is bounded for all n, we see that $\{\lambda_n\}$ is bounded from the definition of the spectral radius. Let $\{\lambda_{n_k}\}$ be a convergent subsequence of $\{\lambda_n\}$ that converges to some λ. The regularity of elliptic systems yields that the corresponding positive eigenfunction sequence $\{h_{n_k}\}$ is compact and has a convergent subsequence that converges to a solution $h > 0$ of the eigenvalue problem $\partial_v F_2(u_*, 0)h = \lambda h$. By uniqueness of the positive eigenfunction [3, Theorem 3.2, ii)], $\lambda = r_v(u_*)$. We now see that all convergent subsequences of $\{\lambda_n\}$ converge to $r_v(u_*)$. Thus, $\limsup \lambda_n = \liminf \lambda_n$ and $\lambda_n = r_v(u_n) \to r_v(u_*)$ as $n \to \infty$. Hence, $r_v(u)$ is continuous in $u \in Z_1$.

Therefore, Z_1^+, Z_1^- are disjoint open sets in Z_1. By Z), their union is the compact set Z_1 so that they are also closed in Z_1 and compact in $\mathbf{X}_1$. Hence, there are disjoint open sets U^+, U^- in $\mathbf{X}_1$ such that $Z_1^+ \subset U^+$, $Z_1^- \subset U^-$. We then have

$$\mathrm{ind}(T_*, U \oplus V) = \mathrm{ind}(T_*, (U^+ \cup U^-) \oplus V) = \mathrm{ind}(T_*, U^+ \oplus V) + \mathrm{ind}(T_*, U^- \oplus V)\,.$$

Applying I.2) of Lemma 6.2.6 with $U = U^+$, we see that $\mathrm{ind}(T_*, U^+ \oplus V)$ is zero. It follows that

$$\mathrm{ind}(T_*, U \oplus V) = \mathrm{ind}(T_*, U^- \oplus V) = \mathrm{ind}(F_1(\cdot, 0), U^-)\,.$$

Here, we used Lemma 6.2.5, $\mathrm{ind}(T, U \oplus V) = \mathrm{ind}(T_*, U \oplus V)$, in the first equality and I.1) of Lemma 6.2.6 with $U = U^-$ in the second equality. The theorem is proved. □

Remark 6.2.9. The strong positiveness of $\partial_v F_2$ is essential in several places of our proof. Under this assumption, we provided a simple proof of the continuity of $r_v(u)$ on Z_1. In general, as $\partial_v F_2$ is always compact, the continuity of $r_v(u)$ follows from [18, Theorem 2.1], where it was proved that the spectral radius is continuous on the subspace of compact operators.

We end this section by the following well-known result, which is a special case of Theorem 6.2.1.

Corollary 6.2.10. *Let X be a Banach space with positive cone P and f be a positive compact map on P. Suppose that $f(0) = 0$ and the directional Fréchet derivative $f'_+(0)$ exists (i.e., $f'_+(0)x = \lim_{t \to 0^+} t^{-1} f(tx)$). Assume also that $f'_+(0)$ does not have any positive* *eigenvector to the eigenvalue* 1 *and that the following holds:*

$$f'_+(0) \text{ does not have any positive eigenvector to any eigenvalue } \lambda > 1\,. \tag{6.2.8}$$

Then we can find a neighborhood V of 0 *in P such that* 0 *is the only fixed point of f in V and*

$$\operatorname{ind}(f, V) = \begin{cases} 1 & \text{if (6.2.8) holds}\,, \\ 0 & \text{otherwise}\,. \end{cases} \tag{6.2.9}$$

To see this, we let $\mathbf{X} = X \oplus X$, i.e., $\mathbf{X}_1 = X$ and $\mathbf{X}_2 = X$, and $T(u, v) = (0, f(v)$. Obviously, with $F_1 = 0$ being the constant map and $F_2 = f$, a similar argument as in the proof of Lemma 6.2.4 (Remark 6.2.8) provides a neighborhood $U = V$ of 0 in P such that the fixed-point set Z_1 in V is the singleton $\{0\}$. Theorem 6.2.1 then provides a neighborhood U^- of Z_1^- in $\mathbf{X}_1$ such that $\operatorname{ind}(T, U \oplus V) = \operatorname{ind}(0, U^-)$. Clearly, as F_1 is a constant map, if (6.2.8) holds then 0 is v-stable so that $U^- = U$ and $\operatorname{ind}(F_1(\cdot, 0), U^-) = 1$; otherwise $U^- = \emptyset$ and $\operatorname{ind}(F_1(\cdot, 0), U^-) = 0$. By the product theorem of indices, $\operatorname{ind}(T, U \oplus V) = \operatorname{ind}(f, V)$ and (6.2.9) then follows.

Of course, the above result can be generalized as follows. We first recall the concept of a *stable* fixed point. Let X be a Banach space with positive cone P. Consider a map $f : U \to P$, a positive compact map on a subset $U \subset P$. Suppose that $f(x) = x$ for some $x \in U$ and the (P-directional) derivative $f_P(x)$ exists. Here, for $h \in P$,

$$f_P(x)h = \lim_{t \to 0} \frac{f(x + th) - f(x)}{t} = \lim_{t \to 0} \frac{f(x + th) - x}{t}\,.$$

We say that x is P-stable if

$$\text{the spectral radius } r_P \text{ of } f_P(x) \text{ is less than } 1\,. \tag{6.2.10}$$

Of course, x is P-unstable if

$$\text{the spectral radius } r_P \text{ of } f_P(x) \text{ is greater than } 1\,. \tag{6.2.11}$$

Suppose that the fixed-point set Z of f in U can be decomposed into the union of two disjoint stable and unstable sets Z_- and Z_+. That is, $Z = Z_- \cup Z_+$ with

$$Z_- := \{x \in Z : x \text{ is } P \text{ stable}\}, \; Z\_+ = \{x \in Z : x \text{ is } P \text{ unstable}\}\,.$$

Then there are disjoint open sets U^-, U^+ such that $Z_- \subset U^-$ and $Z_+ \subset U^+$ and $\operatorname{ind}(f, U) = \operatorname{ind}(f, U^-)$.

6.3 Applications of index results to the existence of nontrivial steady-state solutions

In this section, we will show that the abstract results on the local indices of T at trivial and semitrivial solutions in Theorem 6.2.1 can apply to the map T defined by (6.1.1).

The content of this section is a somewhat straightforward translation of the abstract results in the previous section to the general elliptic systems (6.1.1) in consideration. We will present several sets of assumptions under which (6.1.1) possesses a nontrivial solution. A more applicable setting for SKT-like systems will be discussed in the next section.

Going back to the definition of T, for each $(u, v) \in \mathbf{X}$ and some suitable constant matrix K we consider the following *linear* elliptic system for $w = T(u, v)$:

$$\begin{cases} -\operatorname{div}(A(u, v)Dw) + Kw = \hat{f}(u, v) + K(u, v) & \text{in } \Omega\,, \\ \text{homogenenous boundary conditions for } w & \text{on } \partial\Omega\,. \end{cases} \tag{6.3.1}$$

Remark 6.3.1. We observe that the choice of the matrix K is not important here as long as the map T is well defined (as a positive map). In fact, let K_1, K_2 be two different matrices and T_1, T_2 be the corresponding maps defined by (6.3.1). It is clear that these maps have the same set of fixed points consisting of solutions to (6.1.1). Hence, for any given open set U, via a simple homotopy $tT_1 + (1 - t)T_2$ for $t \in [0, 1]$, the indices $\operatorname{ind}(T_i, U)$ are equal whenever one of their indices is defined (i.e., (6.1.1) does not have any solution on ∂U).

Trivial solution: It is clear that 0 is a solution if $\hat{f}(0) = 0$. In this case, we can apply Corollary 6.2.10 with $F = T$. The eigenvalue problem of $T'(0)h = \lambda h$ is now

$$-\operatorname{div}(A(0)Dh) + Kh = \lambda^{-1}(\hat{f}_u(0) + K)h\,. \tag{6.3.2}$$

We then have the following result from Corollary 6.2.10.

Lemma 6.3.2. *There is a neighborhood V_0 of 0 in $\mathbf{P}$ such that if* (6.3.2) *has a positive solution h to some eigenvalue $\lambda > 1$ then* $\operatorname{ind}(T, V_0) = 0$. *Otherwise,* $\operatorname{ind}(T, V_0) = 1$. *In the first case, there is a nontrivial/semitrivial solution to the system* (6.1.1).

Semitrivial solution: By reordering the equations and variables, we will write an element of $\mathbf{X} = \mathbf{X}_1 \oplus \mathbf{X}_2$ as (u, v) and

$$A(u, v) = \begin{bmatrix} P^{(u)}(u, v) & P^{(v)}(u, v) \\ Q^{(u)}(u, v) & Q^{(v)}(u, v) \end{bmatrix} \quad \text{and} \quad \hat{f}(u, v) = \begin{bmatrix} f^{(u)}(u, v) \\ f^{(v)}(u, v) \end{bmatrix}\,.$$

The existence of semitrivial solutions $(u, 0)$ usually comes from the assumption that

$$Q^{(u)}(u, 0) = 0 \text{ and } f^{(v)}(u, 0) = 0 \quad \forall u \in \mathbf{X}_1\,. \tag{6.3.3}$$

If (6.3.3) holds then it is clear that $(u, 0)$ is a solution of (6.1.1) if and only if u solves the following subsystem:

$$-\operatorname{div}(P^{(u)}(u, 0)Du) = f^{(u)}(u, 0)\,. \tag{6.3.4}$$

Let us then assume that the set Z_1 of positive solutions to (6.3.4) is nonempty.

To compute the local index of T at a semitrivial solution we consider the following constant matrix:

$$K = \begin{bmatrix} K_1^{(u)} & K_1^{(v)} \\ 0 & K_2^{(v)} \end{bmatrix}, \tag{6.3.5}$$

where the matrices $K_1^{(u)}, K_1^{(v)}$ and $K_2^{(v)}$ are of sizes $m_1 \times m_1$, $m_1 \times m_2$ and $m_2 \times m_2$ respectively. The system in (6.3.1) for $w = (w_1, w_2) = T(u, v)$ now reads

$$-\operatorname{div}(P^{(u)}(u,v)Dw_1 + P^{(v)}(u,v)Dw_2) + K_1^{(u)}w_1 + K_1^{(v)}w_2 = f^{(u)}(u,v) + K_1^{(u)}u + K_1^{(v)}v, \tag{6.3.6}$$

and

$$-\operatorname{div}(Q^{(u)}(u,v)Dw_1 + Q^{(v)}(u,v)Dw_2) + K_2^{(v)}w_2 = f^{(v)}(u,v) + K_2^{(v)}v. \tag{6.3.7}$$

We will consider the following assumptions on the above subsystems.

The first condition simply implies that K is a positive matrix ((6.1.5)) so that T is well defined.

K.0) Assume that there are $k_1, k_2 > 0$ such that

$$\langle K_1^{(u)}x_1, x_1\rangle \ge k_1|x_1|^2, \quad \langle K_2^{(v)}x_2, x_2\rangle \ge k_2|x_2|^2 \quad \forall x_i \in \mathbb{R}^{m_i}, i = 1, 2. \tag{6.3.8}$$

The next assumptions require that a maximum principle holds for certain *linear* elliptic systems so that the corresponding fixed-point map is positive. We refer the reader to Lemma 6.3.8 below where we discuss a general maximum principle for elliptic systems.

K.1) For all $(u, v) \in \mathbf{B} \cap \mathbf{P}$

$$\hat{f}(u,v) + K(u,v) \ge 0,$$

and the following maximum principle holds: if $(u, v) \in \mathbf{B} \cap \mathbf{P}$ and w solves

$$\begin{cases} -\operatorname{div}(A(u,v)Dw) + Kw \ge 0 & \text{in } \Omega, \\ \text{homogenenous boundary conditions} & \text{on } \partial\Omega \end{cases}$$

then $w \ge 0$.

According to the notations of the previous section, the system (6.3.1) defines $w = T(u, v) = (F_1(u, v), F_2(u, v))$ with $w_1 = F_1(u, v)$ and $w_2 = F_2(u, v)$ respectively solving (6.3.6) and (6.3.7).

For $u \in Z_1$, $\phi \in \dot{\mathbf{P}}_2$ we will prove later on that $\mathbf{U} := \partial_v F_2(u, 0)\phi$ solves the system

$$-\operatorname{div}(Q^{(v)}(u,0)D\mathbf{U} + Q_v^{(u)}(u,0)Du\phi) + K_2^{(v)}\mathbf{U} = f_v^{(v)}(u,0)\phi + K_2^{(v)}\phi. \tag{6.3.9}$$

Regarding the term $Q_v^{(u)}(u, 0)Du\phi$ in K.2) we have used the following notation: if $B(u, v) = (b_{ij}(u, v))$, with $i = 1, \dots, m_2$ and $j = 1, \dots, m_1$, and $\phi = (\phi^{(1)}, \dots, \phi^{(m_2)})$ then

$$B_v(u,v)Du\phi = \left(\partial_{v^{(k)}} b_{ij}(u,v)\phi^{(k)}Du_j\right)_i = [(\partial_{v^{(k)}} b_{ij}(u,v)Du_j)_{k,i}]\phi. \tag{6.3.10}$$

Hence, concerning the strong positivity of $\partial_v F_2$ in the previous section, we assume

K.2) For any $u \in Z_1$ and $\phi \in \dot{\mathbf{P}}_2$ a strong maximum principle holds for (6.3.9). That is, if $f_v^{(v)}(u,0)\phi + K_2^{(v)}\phi \in \dot{\mathbf{P}}_2$ then $\mathbf{U} \in \dot{\mathbf{P}}_2$.

Turning to the crucial condition Z) on the eigenvector of $\partial_v F_2(u,0)$ at a semitrivial steady state in the previous section, we also assume that

K.3) For any $u \in Z_1$ the linear system

$$-\operatorname{div}(Q^{(v)}(u,0)Dh + Q_v^{(u)}(u,0)Duh) = f_v^{(v)}(u,0)h$$

has no positive solution h in $\mathbf{P}_2$.

For $u \in Z_1$ we will also consider the following eigenvalue problem:

$$-\operatorname{div}(\lambda Q^{(v)}(u,0)Dh + Q_v^{(u)}(u,0)Duh) + \lambda K_2^{(v)}h = f_v^{(v)}(u,0)h + K_2^{(v)}h\,, \tag{6.3.11}$$

and consider the set of v-stable semitrivial steady states

$$Z_1^- := \{u \in Z_1 \,:\, \text{(6.3.11) has a positive solution } h \text{ to an eigenvalue } \lambda < 1\}\,.$$

The main theorem of this subsection follows.

Theorem 6.3.3. *Assume K.0)–K.3) with k_1 in K.0) being sufficiently large. Then the map T described in (6.3.1) is well defined on $\mathbf{B} \cap \mathbf{P}$ and maps $\mathbf{B} \cap \mathbf{X}$ into $\mathbf{P}$. There are neighborhoods U, U^- respectively of Z_1, Z_1^- in $\mathbf{P}_1$ and a neighborhood V on 0 in $\mathbf{P}_2$ such that*

$$\operatorname{ind}(T, U \oplus V) = \operatorname{ind}(F_1(\cdot,0), U^-)\,.$$

Here, $F_1(\cdot,0)$ maps $\mathbf{B} \cap \mathbf{P}_1$ into $\mathbf{P}_1$ and $w_1 = F_1(u,0)$, $u \in \mathbf{B} \cap \mathbf{P}_1$, is the unique solution to

$$-\operatorname{div}(P^{(u)}(u,0)Dw_1) + K_1^{(u)}w_1 = f^{(u)}(u,0) + K_1^{(u)}u\,. \tag{6.3.12}$$

The above theorem is just a consequence of Theorem 6.2.1 applied to the system (6.3.1). We need only to verify the assumption of Theorem 6.2.1. For this purpose and later use in the section we will divide its proof into lemmas, which also contain additional and useful facts.

We first have the following lemma, which shows that the assumption (6.2.1) in the previous section, that $F_2(u,0) = 0$ for all $u \in \mathbf{X}_1$, is satisfied.

Lemma 6.3.4. *Let T be defined by (6.3.1). If K.0) holds for some sufficiently large k_1 then T is well defined by (6.3.1) for any given matrix $K_1^{(v)}$. The components F_1, F_2 of T satisfy:*

i) $F_2(u,0) = 0$ *for all* $u \in \mathbf{X}_1$.

ii) $w_1 = F_1(u,0)$ *solves (6.3.12). In addition,* $\operatorname{ind}(F_1(\cdot,0), \mathbf{B} \cap \mathbf{P}_1) = 1$.

Proof. We write $w = T(u, v) = (F_1(u, v), F_2(u, v))$ in (6.3.1) by (w_1, w_2), with $w_i \in \mathbf{X}_i$. Because

$$\langle Kw, w\rangle = \langle K_1^{(u)} w_1, w_1\rangle + \langle K_1^{(v)} w_2, w_1\rangle + \langle K_2^{(v)} w_2, w_2\rangle ,$$

a simple use of Young's inequality and (6.3.8) show that if k_1 is sufficiently large then $\langle Kx, x\rangle \ge |x|^2$ for any given $K_1^{(v)}$. Hence, T is well defined by (6.3.1).

At $(u, 0)$, since $\hat{f}^{(v)}(u, 0) = 0$ and $Q^{(u)}(u, 0) = 0$, the subsystem (6.3.7) defining $w_2 = F_2(u, 0)$ is now

$$-\operatorname{div}(Q^{(v)}(u, 0)Dw_2) + K_2^{(v)} w_2 = 0 .$$

This system has $w_2 = 0$ as the only solution because of the assumption (6.3.8) on $K_2^{(v)}$ and the ellipticity of $Q^{(v)}(u, 0)$. This gives i).

Next, as w_2 and Dw_2 are zero, (6.3.6) gives that $w_1 = F_1(u, 0)$ solves

$$-\operatorname{div}(P^{(u)}(u, 0)Dw_1) + K_1^{(u)} w_1 = f^{(u)}(u, 0) + K_1^{(u)} u .$$

Again, for a given $u \in \mathbf{X}_1$ this subsystem has a unique solution w_1 if $\langle K_1^{(u)} x, x\rangle \ge k_1|x|^2$ for some $k_1 > 0$. Moreover, the fixed point of $u = F_1(u, 0)$ solves

$$-\operatorname{div}(P^{(u)}(u, 0)Du) = f^{(u)}(u, 0) .$$

This system satisfies the same set of structural conditions for the full system (6.1.1) so that Theorem 4.1.3 can apply here to give ii). □

From the proof of Theorem 6.2.1 we need study the Fréchet (directional) derivative of T defined by (6.3.1). For this purpose and later use, we consider a general linear system defining $w = T(u)$,

$$\begin{cases} -\operatorname{div}(A(u)Dw + B(u, Du)w) + C(u, Du)w = \hat{f}(u, Du) & x \in \Omega , \\ w = 0 \text{ on } \partial\Omega , \end{cases} \tag{6.3.13}$$

for some matrix-valued C^1 functions $A, B, C, \hat{f}$.

We then recall the following elementary result on the linearization of the above system at u.

Lemma 6.3.5. *Let u, ϕ be in $\mathbf{X}$. If $w = T(u)$ is defined by (6.3.13) then $W = T'(u)\phi$ solves the following system:*

$$-\operatorname{div}(A(u)DW + B(u, Du)W + \mathcal{B}(u, w, \phi)) + \mathcal{C}(u, W, w, \phi) = \mathcal{F}(u, \phi) ,$$

where

$$\begin{aligned} \mathcal{B}(u, w, \phi) &= A_u(u)\phi Dw + B_u(u, Du)\phi w + B_\zeta(u, Du)D\phi w , \\ \mathcal{C}(u, W, w, \phi) &= C(u, Du)W + C_u(u, Du)\phi w + C_\zeta(u, Du)D\phi w , \\ \mathcal{F}(u, \phi) &= \hat{f}_u(u, Du)\phi + \hat{f}_\zeta(u, Du)D\phi . \end{aligned}$$

The proof of this lemma is standard. Because $A, \hat{f}$ are C^1 in u, it is easy to see that T is differentiable. In fact, for any $u, \phi \in \mathbf{X}$ we can compute $T'(u)\phi = \lim_{h\to 0} \delta_{h,\phi} T(u)$, where $\delta_{h,\phi}$ is the difference quotient operator

$$\delta_{h,\phi}T(u) = h^{-1}(T(u+h\phi) - T(u)) .$$

Subtracting (6.1.3) with u being $u + h\phi$ and u and dividing the result by h, we get

$$-\operatorname{div}(\delta_{h,\phi}[A(u)DT(u)+B(u,Du)T(u)])+\delta_{h,\phi}[C(u,Du)T(u)] = \delta_{h,\phi}\hat{f}(u,Du). \quad (6.3.14)$$

It is elementary to see that if g is a C^1 function in the variables u, $\zeta = Du$, $w = T(u)$, $\xi = Dw$ then

$$\lim_{h\to 0} \delta_{h,\phi}g(u,Du,w,Dw) = g_u\phi + g_\zeta D\phi + g_w T'(u)\phi + g_\xi D(T'(u)\phi) .$$

Using the above in (6.3.14) and rearranging the terms, we obtain the lemma.

Applying Lemma 6.3.5 to the system (6.3.7), we have the following lemma concerning the map $\partial_v F_2(u,0)$ at $u \in Z_1$.

Lemma 6.3.6. *Let $u \in Z_1$. An eigenvector function h of $\partial_v F_2(u,0)h = \lambda h$ satisfies the system*

$$-\operatorname{div}(\lambda Q^{(v)}(u,0)Dh + Q_v^{(u)}(u,0)Duh) + \lambda K_2^{(v)}h = f_v^{(v)}(u,0)h + K_2^{(v)}h . \quad (6.3.15)$$

In addition, if K.2) holds then $\partial_v F_2(u,0)$ is strongly positive.

Proof. Let $\phi = (0, \phi_2)$ and $u \in Z_1$. We have

$$w := T(u,0), \quad W := T'(u,0)\phi = (\partial_v F_1(u,0)\phi_2, \partial_v F_2(u,0)\phi_2)$$

satisfying, using Lemma 6.3.5 with $B(u,Du) = 0$ and $C(u,Du) = K$,

$$-\operatorname{div}(A(u,0)DW + \partial_{u,v}A(u,0)\phi Dw) + KW = \hat{f}_{u,v}(u,0)\phi + K\phi .$$

At $(u,0)$, $T(u,0) = (u,0)$ so that v, Dv are zero and $Dw = D(T(u,0)) = (Du,0)$ we have

$$A(u,0) = \begin{bmatrix} P^{(u)}(u,0) & P^{(v)}(u,0) \\ 0 & Q^{(v)}(u,0) \end{bmatrix},$$

$$\partial_{u,v}A(u,0)\phi = \begin{bmatrix} P_v^{(u)}(u,0)\phi_2 & P_v^{(v)}(u,0)\phi_2 \\ Q_v^{(u)}(u,0)\phi_2 & Q_v^{(v)}(u,0)\phi_2 \end{bmatrix},$$

$$\partial_{u,v}A(u,0)\phi Dw = \begin{bmatrix} P_v^{(u)}(u,0)Du\phi_2 \\ Q_v^{(u)}(u,0)Du\phi_2 \end{bmatrix}.$$

Thus, $\mathbf{U}_1 := \partial_v F_1(u,0)\phi_2$ and $\mathbf{U}_2 := \partial_v F_2(u,0)\phi_2$, the components of $T'(u,0)\phi$, satisfy

$$-\operatorname{div}(P^{(u)}(u,0)D\mathbf{U}_1 + P^{(v)}(u,0)D\mathbf{U}_2 + P_v^{(u)}(u,0)Du\phi_2) + K_1^{(u)}\mathbf{U}_1 + K_1^{(v)}\mathbf{U}_2 = f_v^{(u)}(u,0)\phi_2 + K_1^{(v)}\phi_2 ,$$

and

$$-\operatorname{div}(Q^{(v)}(u,0)D\mathbf{U}_2 + Q_v^{(u)}(u,0)Du\phi_2) + K_2^{(v)}\mathbf{U}_2 = f_v^{(v)}(u,0)\phi_2 + K_2^{(v)}\phi_2 . \quad (6.3.16)$$

The above two systems are in fact decoupled. One can solve $\mathbf{U}_2$ from (6.3.16) and then use it in the first to obtain $\mathbf{U}_1$.

We consider the eigenvalue problem $\partial_v F_2(u,0)h = \lambda h$. Set $\phi_2 = h$ then $\mathbf{U}_2 = \lambda h$ and it is clear from (6.3.16) that h is the solution to (6.3.15).

Finally, the system (6.3.16) defining $\mathbf{U}_2 := \partial_v F_2(u,0)\phi_2)$ is exactly (6.3.9) in K.2). Thus, the strong maximum principle, assumed in K.2), for (6.3.9) yields that $\partial_v F_2(u,0)$ is strongly positive. □

Proof of Theorem 6.3.3. Lemma B.1.1 shows that T is well defined and maps $\mathbf{B}\cap\mathbf{P}$ into $\mathbf{P}$ if K.1) is assumed. Lemma 6.3.6 and K.2) then gives the strong positivity of $\partial_v F_2(u,0)$ for any $u \in Z_1$. In addition, the equation in the condition K.3) is (6.3.15) of Lemma 6.3.6 when $\lambda = 1$ so that K.3) means that the condition Z) of the previous section holds here. Thus, our theorem is just a consequence of Theorem 6.2.1. □

We now turn to semitrivial fixed points of T in $\mathbf{X}_2$. These fixed points are determined by the following system, setting $w_1 = u = 0$, $w_2 = v$ in (6.3.6) and (6.3.7):

$$\begin{cases} -\operatorname{div}(P^{(v)}(0,v)Dv) = f^{(u)}(0,v)\,, \\ -\operatorname{div}(Q^{(v)}(0,v)Dv) = f^{(v)}(0,v)\,. \end{cases} \tag{6.3.17}$$

We will assume that this system has no positive solution v. In fact, if $P^{(v)}(0,v) \neq 0$ the above system is *overdetermined* so that the existence of a nonzero solution v of the second subsystem satisfying the first subsystem is very unlikely. In addition, assuming $f^{(v)}(0,0) = 0$, it could happen that the second subsystem already has $v = 0$ as the only solution.

At $(u,v) = (0,0)$, the eigenvalue problem $\partial_v F_2(0,0)h_2 = \lambda h_2$ for $h_2 \in \mathbf{X}_2$ is

$$-\operatorname{div}(Q^{(v)}(0,0)Dh_2) + K_2^{(v)}h_2 = \lambda^{-1}(f_v^{(v)}(0,0) + K_2^{(v)})h_2\,. \tag{6.3.18}$$

From ii) of Lemma B.1.1, the eigenvalue problem $\partial_u F_1(0,0)h_1 = \lambda h_1$ for $h_1 \in \mathbf{X}_1$ is

$$-\operatorname{div}(P^{(u)}(0,0)Dh_1) + K_1^{(u)}h_1 = \lambda^{-1}(f_u^{(u)}(0,0) + K_1^{(u)})h_1\,. \tag{6.3.19}$$

Again, we will say that $0 \in \mathbf{X}_1$ is u-stable if the above has no positive eigenvector h_1 to any eigenvalue $\lambda > 1$. Otherwise, we say that 0 is u-unstable.

Our first application of Theorem 6.2.1 gives sufficient conditions such that semitrivial and nontrivial solutions exist.

Theorem 6.3.7. *Suppose that* (6.3.17) *has no positive solution. Assume K.0)–K.4) and that the system*

$$-\operatorname{div}(P^{(u)}(0,0)Dh_1) = f_u^{(u)}(0,0)h_1 \quad \textit{has no solution } h_1 \in \dot{\mathbf{P}}_1\,. \tag{6.3.20}$$

If either one of the following holds then there is a nontrivial positive solution to (6.1.1)*:*

i.1) *0 is u-stable and $Z_1^+ = \{0\}$;*

i.2) *0 is u-unstable and $Z_1^- = \{0\}$.*

Proof. We denote $Z_p = \{u \in Z_1 : u > 0\}$. Thus Z_p is the set of semitrivial solutions and $Z_1 = \{0\} \cup Z_p$. Accordingly, we denote by Z_p^+ (respectively Z_p^-) the v-unstable (respectively v-stable) subset of Z_p. The assumption (6.3.20) means $\partial_u F_1(0,0)$ does not have a positive eigenfunction to the eigenvalue 1 in $\mathbf{P}_1$. Applying Corollary 6.2.10 with $X = \mathbf{X}_1$ and $F(\cdot) = F_1(\cdot, 0)$, we can find a neighborhood U_0 in $\mathbf{X}_1$ of 0 such that 0 is the only fixed point of $F_1(\cdot, 0)$ in U_0 and (6.2.9) gives

$$\operatorname{ind}(F_1(\cdot, 0), U_0) = \begin{cases} 1 & \text{if } 0 \text{ is } u\text{-stable}\,, \\ 0 & \text{if } 0 \text{ is } u\text{-unstable}\,. \end{cases} \tag{6.3.21}$$

Since Z_1 is compact and $Z_1 = \{0\} \cup Z_p$, the previous argument shows that Z_p is compact. From the proof of Theorem 6.2.1, there are disjoint open neighborhoods U_p^- and U_p^+ in $\mathbf{X}_1$ of Z_p^- and Z_p^+ respectively. Of course, we can assume that U_0, U_p^- and U_p^+ are disjoint so that for $U = U_0 \cup U_p^+ \cup U_p^-$

$$\operatorname{ind}(F_1(\cdot, 0), U) = \operatorname{ind}(F_1(\cdot, 0), U_0) + \operatorname{ind}(F_1(\cdot, 0), U_p^+) + \operatorname{ind}(F_1(\cdot, 0), U_p^-)\,. \tag{6.3.22}$$

By Lemma B.1.1, $\operatorname{ind}(F_1(\cdot, 0), \mathbf{B} \cap \mathbf{P}_1) = 1$. This implies $\operatorname{ind}(F_1(\cdot, 0), U) = 1$. If i.1) holds then $Z_p^+ = \emptyset$ and we can take $U_p^+ = \emptyset$ and $U^- = U_p^-$ in Theorem 6.2.1. From (6.3.21), $\operatorname{ind}(F_1(\cdot, 0), U_0) = 1$ so that (6.3.22) implies $\operatorname{ind}(F_1(\cdot, 0), U_p^-) = 0$. This yields $\operatorname{ind}(F_1(\cdot, 0), U^-) = 0$. Similarly, if i.2) holds then $Z_1^- = \{0\}$ and $Z_p^- = \emptyset$ and we can take $U_p^- = \emptyset$ and $U^- = U_0$ in Theorem 6.2.1. From (6.3.22), $\operatorname{ind}(F_1(\cdot, 0), U^-) = \operatorname{ind}(F_1(\cdot, 0), U_0) = 0$.

Hence, $\operatorname{ind}(F_1(\cdot, 0), U^-) = 0$ in both cases. By Theorem 6.3.3, we find a neighborhood V of 0 in $\mathbf{P}_2$ such that $\operatorname{ind}(T, U \oplus V) = \operatorname{ind}(F_1(\cdot, 0), U^-) = 0$. Since $\operatorname{ind}(T, \mathbf{B} \cap \mathbf{X}) = 1$, we see that T has a fixed point in $\mathbf{B} \setminus \overline{U \oplus V}$. This fixed point is nontrivial because we are assuming that T has no semitrivial fixed point in $\mathbf{P}_2$. □

6.3.1 Notes on a more special case and a different way to define *T*

The positivity properties assumed in K.1)–K.3) of the previous section are crucial. Since the involved systems defining the map T and its derivatives in these conditions are all strongly coupled, such maximum principles are very hard to verify. We present here an alternative definition of the map T such that the systems are weakly coupled and such needed principles are available.

In many applications, it is reasonable to assume that the cross-diffusion effects by other components should be proportional to the density of a given component. This is to say that if $A(u) = (a_{ij}(u))$ then there are smooth functions b_{ij} such that

$$a_{ij}(u) = u_i b_{ij}(u) \quad \text{if } j \neq i\,. \tag{6.3.23}$$

In this case, instead of using (6.3.1), we can define the i-th component w_i of $T(u)$ by

$$L_i(u) w_i + \sum_j k_{ij} w_j = f_i(u) + \sum_j k_{ij} u_j\,, \tag{6.3.24}$$

where the operator $L_i(u)$ is defined as follows:

$$L_i(u)w = -\operatorname{div}\left(a_{ii}(u)Dw + w\left(\sum_{j\neq i} b_{ij}(u)Du_j\right)\right). \tag{6.3.25}$$

We now set $\mathbf{X} = C^{1,\gamma}(\Omega)$ (or $C^{1,\gamma}(\Omega) \cap C^0(\Omega)$ if Dirichlet boundaries conditions are considered). As $u \in \mathbf{B} \cap \mathbf{P}$, (6.3.24) is a *weakly coupled* system with Hölder continuous coefficients. We will see that the condition K.1) on the positivity of solutions in the previous section is verified. To this end, we recall the maximum principles for cooperative linear systems in [9, 10] and give here an alternative and simple proof to [9, Theorem 1.1]. In fact, we consider a more general setting that covers both Dirichlet and Neumann boundary conditions.

We should remark on the definition of $\mathbf{X}$ in this section. In Chapter 4, we set $W^{1,2}(\Omega) \cap C^{0,\gamma}(\Omega)$. The key issue is to prove that T is a compact map on $\mathbf{X}$. In the present case, $\mathbf{X} \subset C^{1,\gamma}(\Omega)$, if $u \in \mathbf{X}$ then (6.3.24) is a linear elliptic equation with Hölder continuous coefficients and it is well known from the theory of linear elliptic equations that $T(u) = w$ is in $C^{2,\gamma}(\Omega)$. Thus, the compactness of T is obvious. The key point here is, of course, the uniform bound in $\mathbf{X}$ for fixed points of T and we have to deal with strongly coupled systems again. This is the main object we addressed in our theory of Chapter 4 once we established their Hölder continuity under the very weak VMO assumption.

Let us define

$$L_i w = -\operatorname{div}(\alpha_i(x)Dw + \beta_i(x)w), \tag{6.3.26}$$

where $\alpha_i \in L^\infty(\Omega)$, $\beta_i \in L^\infty(\Omega, \mathbb{R}^n)$. Denote $\mathcal{F} = (\mathcal{F}_1, \cdots, \mathcal{F}_m)$. We then have the following weak minimum principle for weakly coupled parabolic systems.

Lemma 6.3.8. *Let w be a weak solution to the system*

$$\begin{cases} L_i w_i + Kw = \mathcal{F}_i, \quad i = 1, \dots, m, \text{ in } \Omega, \\ \text{homogeneous Dirichlet or Neumann boundary conditions on } \partial\Omega. \end{cases}$$

Assume that $\alpha_i(x) \geq \lambda_i$ for some $\lambda_i > 0$ and $\mathcal{F}_i \geq 0$ for all i. If $k_{ij} \leq 0$ for $i \neq j$ and k_{ii} are sufficiently large, in terms of $\sup_\Omega \beta_i(u(x))$, then $w \geq 0$.

Proof. Let ϕ^+, ϕ^- denote the positive and negative parts of a scalar function ϕ, i.e., $\phi = \phi^+ - \phi^-$. We note that $\langle Dw_i, Dw_i^- \rangle = -|Dw_i^-|^2$ and $w_i Dw_i^- = -w_i^- Dw_i^-$. Integrating by parts, we have

$$\int_\Omega L_i w_i w_i^- \, dx = \int_\Omega (-\alpha_i |Dw_i^-|^2 - \langle \beta_i, Dw_i^- \rangle w_i^-) \, dx.$$

Hence, multiplying the i-th equation of the system by $-w_i^-$, we obtain

$$\sum_i \int_\Omega (\alpha_i |Dw_i^-|^2 + \langle \beta_i, Dw_i^- \rangle w_i^-) \, dx - \sum_{i,j} \int_\Omega k_{ij} w_i w_j^- \, dx = -\sum_i \int_\Omega \langle \mathcal{F}_i, w_i^- \rangle \, dx.$$

Since $\mathcal{F}_i, w_i^- \geq 0$, we get

$$\sum_i \int_\Omega (\alpha_i |Dw_i^-|^2 + \langle \beta_i, Dw_i^- \rangle w_i^-)\, dx - \sum_{i,j} \int_\Omega k_{ij} w_i w_j^-\, dx \leq 0\,.$$

Since $w_i w_i^- = -w_i^- w_i^-$ and $w_i w_j^- = w_i^+ w_j^- - w_i^- w_j^- \geq w_i^- w_j^-$, the above inequality yields, using the assumption that $k_{ij} \leq 0$ for $i \neq j$,

$$\sum_i \int_\Omega (\alpha_i |Dw_i^-|^2 + \langle \beta_i, Dw_i^- \rangle w_i^- + k_{ii}|w_i^-|^2)\, dx - \sum_{i \neq j} \int_\Omega k_{ij} w_i^- w_j^-\, dx \leq 0\,. \tag{6.3.27}$$

By Young's inequality, for any $\varepsilon > 0$ we can find a constant $C(\varepsilon, \beta_i)$, depending on $\sup_\Omega \beta_i(u(x))$, such that

$$\left| \int_\Omega \langle \beta_i, Dw_i^- \rangle w_i^-\, dx \right| \leq \varepsilon \int_\Omega |Dw_i^-|^2\, dx + C(\varepsilon, \beta_i) \int_\Omega |w_i^-|^2\, dx\,.$$

Thus, (6.3.27) implies

$$\sum_i \int_\Omega (\alpha_i - \varepsilon)|Dw_i^-|^2\, dx + \sum_i (k_{ii} - C(\varepsilon, \beta_i)) \int_\Omega |w_i^-|^2\, dx - \sum_{i \neq j} \int_\Omega k_{ij} w_i^- w_j^-\, dx \leq 0\,.$$

Combining the ellipticity assumption and Poincaré's inequality, we have

$$c_0(\lambda_i - \varepsilon) \int_\Omega |w_i^-|^2\, dx \leq \int_\Omega (\alpha_i - \varepsilon)|Dw_i^-|^2\, dx$$

for some $c_0 > 0$. Therefore,

$$c_0(\lambda_i - \varepsilon) \sum_i \int_\Omega |w_i^-|^2\, dx + \sum_i (k_{ii} - C(\varepsilon, \beta_i)) \int_\Omega |w_i^-|^2\, dx - \sum_{i \neq j} \int_\Omega k_{ij} w_i^- w_j^-\, dx \leq 0\,.$$

This implies

$$\int_\Omega \sum_{i,j} \gamma_{ij} w_i^- w_j^-\, dx \leq 0\,, \tag{6.3.28}$$

where

$$\gamma_{ij} = \begin{cases} c_0(\lambda_i - \varepsilon) + k_{ii} - C(\varepsilon, \beta_i) & i = j\,, \\ -k_{ij} & i \neq j\,. \end{cases}$$

It is clear that if k_{ii} is sufficiently large then the matrix $\gamma = (\gamma_{ij})$ is positive definite, i.e., $\langle \gamma x, x \rangle \geq c|x|^2$ for some positive c. Thus, (6.3.28) forces $w_i^- = 0$ a.e. and therefore $w \geq 0$. □

Thanks to this lemma, we now see how to construct a matrix K such that T maps $\mathbf{B} \cap \mathbf{P}$ into $\mathbf{P}$. To this end, we note that $f^{(u)}(0, 0) = 0$ and $f^{(v)}(u, 0) = 0$ so that we can write

$$f^{(u)}(u, v) + K_1^{(u)} u + K_1^{(v)} v = \int_0^1 (f_u^{(u)}(tu, tv) + K_1^{(u)})\, dt\, u + \int_0^1 (f_v^{(u)}(tu, tv) + K_1^{(v)})\, dt\, v\,,$$

$$f^{(v)}(u, v) + K_2^{(v)} v = \int_0^1 (f_v^{(v)}(u, tv) + K_2^{(v)})\, dt\, v\,.$$

Since $\|\hat{f}_u(u)\|_{\mathbf{X}}$ is bounded for $u \in \mathbf{B} = B(0, M)$ (the bound M is independent of K), it is not difficult to see that if the reaction is 'cooperative', i.e., $\partial_{u_i}\hat{f}_j \geq 0$ for $i \neq j$, then we can always find K with $k_{ij} = 0$ for $i \neq j$ and $k_{ii} > 0$ sufficiently large such that the matrix integrands in the above equations are all positive. Therefore, $\mathcal{F} := \hat{f}(u, v) + K(u, v) \geq 0$ for $(u, v) \in \mathbf{B} \cap \mathbf{P}$ and the lemma can apply here.

Finally, for future use in the next section we now explicitly describe the map $T'(u)$ in this case. For $\Phi = (\phi_1, \ldots, \phi_m)$, by Lemma 6.3.5, the components $w_i = T_i(u)$ of $T(u)$ and $W_i = T_i'(u)\Phi$ of $T'(u)\Phi$, solves

$$-\operatorname{div}(\mathcal{A}_i(u, W_i) + \mathcal{B}_i(u, w_i, \Phi)) + \sum_j k_{ij} W_j = \sum_j (\partial_{u_j} f_i(u) + k_{ij})\phi_j \,, \tag{6.3.29}$$

where

$$\mathcal{A}_i(u, W_i) := a_{ii}(u) DW_i + \left(\sum_{j \neq i} b_{ij}(u) Du_j \right) W_i \,, \tag{6.3.30}$$

$$\mathcal{B}_i(u, w_i, \Phi) := \sum_j \partial_{u_j} a_{ii}(u)\phi_j Dw_i + w_i \sum_{j \neq i} [\partial_{u_k} b_{ij}(u)\phi_k Du_j + b_{ij}(u) D\phi_j] \,. \tag{6.3.31}$$

Consider a semitrivial solution $u \in Z_1$, i.e., for some integer $m_1 \geq 0$ and $m_1 < m$

$$T(u, 0) = (u, 0) = (u_1, \ldots, u_{m_1}, 0, \ldots, 0) \,.$$

Let $\Phi = (0, \ldots, 0, \phi_{m_1+1}, \ldots, \phi_m)$. For $i > m_1$ we have that $w_i = T_i(u, 0)$ and $Dw_i = D(T_i(u, 0))$ are zero so that $\mathcal{B}_i(u, w_i, \Phi) \equiv 0$ and W_i solves

$$-\operatorname{div}(a_{ii}(u, 0) DW_i + W_i \sum_{j \leq m_1} b_{ij}(u, 0) Du_j) + \sum_j k_{ij} W_j = \sum_j (\partial_{u_j} f_i(u, 0) + k_{ij})\phi_j \,. \tag{6.3.32}$$

The above is a weakly coupled linear elliptic system for W_i and it can be written in the form (6.3.24) so that Lemma 6.3.8 can be applied here to give the positivity of the map $T'(u)$ under suitable assumptions on f and K.

6.4 Nonconstant and nontrivial solutions

We devote this section to the study of (6.1.1) with Neumann boundary conditions. Theorem 6.3.7 gives the existence of a positive nontrivial solution but this solution may be a constant solution. This is the case when there is a constant vector $u^* = (u_1^*, \ldots, u_m^*)$ such that $f(u^*) = 0$. Obviously $u = u_*$ is a nontrivial solution to (6.1.1) and Theorem 6.3.7 then yields no useful information. In applications, we are interested in finding a nonconstant solution other than this obvious solution. This question has drawn much attention lately in mathematical biology and often referred to as pattern formation problems.

We will assume throughout this section that the semitrivial solutions are all constant and show that cross-diffusion will play an important role for the existence of nonconstant and nontrivial solutions.

Inspired by the SKT systems, we assume that the diffusion is given by (6.3.23) as in Section 6.3.1 and the reaction term in the i-th equation is also proportional to the density u_i. This means

$$f_i(u) = u_i g_i(u) \tag{6.4.1}$$

for some C^1 functions g_i's. A constant solution u^* exists if it is a solution to the equations $g_i(u^*) = 0$ for all i.

Throughout this section, we denote by ψ_i's the eigenfunctions of $-\Delta$, satisfying Neumann boundary conditions, to the eigenvalues $\hat{\lambda}_i$'s such that $\{\psi_i\}$ is a basis for $W^{1,2}(\Omega)$. That is,

$$\begin{cases} -\Delta\psi_i = \hat{\lambda}_i\psi_i \text{ in } \Omega, \\ \psi_i \text{ satisfies homogeneous Neumann boundary conditions}. \end{cases}$$

6.4.1 Semitrivial constant solutions

We consider a semitrivial solution $(u, 0)$ with $u = (u_1, \dots, u_{m_1})$ for some integer $m_1 = 1, \dots, m$. Following the analysis of Section 6.2, we need to consider the eigenvalue problem $\partial_v F_2(u, 0)\Phi_2 = \mu\Phi_2$ with $\Phi_2 = (\phi_{m_1+1}, \dots, \phi_m)$ and $\Phi = (0, \Phi_2)$, a positive vector in the complement direction of $(u, 0)$. Then the equation (6.3.32), with $W_i = \mu\phi_i$ for $i = m_1 + 1, \dots, m$, gives

$$-\operatorname{div}(a_{ii}(u, 0)D\phi_i + \sum_{j\le m_1} b_{ij}(u, 0)Du_j\phi_i) + \sum_{j>m_1} k_{ij}\phi_j = \mu^{-1}\sum_{j>m_1}(\partial_{u_j} f_i(u, 0) + k_{ij})\phi_j .$$

If u is a constant vector then $Du_j = 0$ and the above reduces to

$$-\operatorname{div}(a_{ii}(u, 0)D\phi_i) + \sum_{j>m_1} k_{ij}\phi_j = \mu^{-1}\sum_{j>m_1}(\partial_{u_j} f_i(u, 0) + k_{ij})\phi_j , \tag{6.4.2}$$

which is an elliptic system with constant coefficients. We then need the following lemma.

Lemma 6.4.1. *Let A, B be constant matrices. Then the solution space of the problem*

$$\begin{cases} -\operatorname{div}(AD\Phi) = B\Phi , \\ \textit{Neumann boundary conditions} \end{cases}$$

has a basis $\{\mathbf{c}_{i,j}\psi_i\}$ where $\mathbf{c}_{i,j}$'s are the basis vectors of $\operatorname{Ker}(\hat{\lambda}_i A - B)$.

The proof of this lemma is elementary. If Φ solves its equation of the lemma then we can write $\Phi = \sum k_i\psi_i$, in $W^{1,2}(\Omega)$, with $k_i \in \mathbb{R}^m$. We then have $\sum_i \hat{\lambda}_i A k_i \psi_i = \sum_i B k_i \psi_i$. Since $\{\psi_i\}$ is a basis of $W^{1,2}(\Omega)$, this equation implies $\hat{\lambda}_i A k_i = B k_i$ for all i. Thus, k_i is a linear combination of $\mathbf{c}_{i,j}$'s. It is easy to see that $\{\mathbf{c}_{i,j}\psi_j\}$ is linearly independent if $\{\mathbf{c}_{i,j}\}$, $\{\psi_j\}$ are. The lemma then follows.

Lemma 6.4.2. *Let m_1 be a positive integer less than m and $u = (u_1, \dots, u_{m_1})$ be a constant function such that $T(u, 0) = (u, 0)$. Then*

$$\partial_v f^{(v)}(u, 0)\mathbf{c} = \lambda K_2^{(v)}\mathbf{c}$$

has a positive eigenvector $\mathbf{c}$ to a positive (respectively negative) eigenvalue λ if and only if the eigenvalue problem $\partial_v F_2(u, 0)\Phi_2 = \mu\Phi_2$ has a positive solution for some $\mu > 1$ (respectively $\mu < 1$).

Proof. By (6.4.2), the eigenvalue problem $\partial_v F_2(u, 0)\Phi_2 = \mu\Phi_2$ (or $W_i = \mu\phi_i$) is determined by the following system:

$$-\mu \operatorname{div}(A^{(m_1)}(u, 0)D\Phi_2) + \mu K_2^{(v)}\Phi_2 = [\partial_v f^{(v)}(u, 0) + K_2^{(v)}]\Phi_2 , \tag{6.4.3}$$

where $A^{(m_1)}(u, 0) = \operatorname{diag}[a_{ii}(u, 0)]_{i>m_1}$. The coefficients of the above system are constant and Lemma 6.4.1 yields that the solutions to the above system are $\sum \mathbf{c}_i\psi_i$ with $\mathbf{c}_i$'s solving

$$\mu[\hat{\lambda}_i A^{(m_1)}(u, 0) + K_2^{(v)}]\mathbf{c} = [\partial_v f^{(v)}(u, 0) + K_2^{(v)}]\mathbf{c} .$$

Note that the only positive eigenfunction of $-\Delta$ to the eigenvalue $\hat{\lambda}_0 = 0$ is $\psi_0 = 1$. Therefore, from the above system with $i = 0$ we see that if the system

$$\mu K_2^{(v)}\mathbf{c} = [\partial_v f^{(v)}(u, 0) + K_2^{(v)}]\mathbf{c} \Leftrightarrow \partial_v f^{(v)}(u, 0)\mathbf{c} = (\mu - 1)K_2^{(v)}\mathbf{c}$$

has a positive solution $\mathbf{c}$ then the constant function $\mathbf{c}$ is a positive eigenfunction for $\partial_v F_2(u, 0)$. Conversely, if $\partial_v F_2(u, 0)\Phi_2 = \mu\Phi_2$ has a positive solution Φ_2 then we integrate (6.4.3) over Ω, using the Neumann boundary conditions, to see that $\mathbf{c} = \int_\Omega \Phi_2\,dx$ is a positive solution to the above system. The lemma then follows. □

Remark 6.4.3. By the Krein–Rutman theorem, if $\partial_v F_2(u, 0)$ is strongly positive then $\mu = r_v(u)$ is the only eigenvalue with positive eigenvector. The eigenvalue problem $\partial_v F_2(u, 0)\Phi_2 = \mu\Phi_2$ has a positive solution for $\mu = 1$ if and only if the matrix $\partial_v f^{(v)}(u, 0)$ has a positive eigenvector to the zero eigenvalue.

We now discuss the special case $f_i(u) = u_i g_i(u)$.

Lemma 6.4.4. *Assume that $f_i(u) = u_i g_i(u)$. Let m_1 be a nonnegative integer less than m and $u = (u_1, \dots, u_{m_1})$ be a constant vector such that $T(u, 0) = (u, 0)$. Then the eigenvalue problem $\partial_v F_2(u, 0)\Phi_2 = \mu\Phi_2$:*

i.1) *has no nonzero solution for $\mu = 1$ if and only if $g_i(u, 0) \neq 0$ for any $i > m_1$;*
i.2) *has a positive solution for some $\mu > 1$ if and only if $g_i(u, 0) > 0$ for some $i > m_1$;*
i.3) *has no positive solution for $\mu > 1$ if and only $g_i(u, 0) < 0$ for any $i > m_1$.*

Proof. We now let $K = kI$. By Lemma 6.4.2 the existence of positive eigenvectors of $\partial_v F_2(u, 0)\Phi_2 = \mu\Phi_2$ is equivalent to that of

$$\partial_v f^{(v)}(u, 0)\mathbf{c} = k\lambda\mathbf{c} \quad \text{with } \mathbf{c} = (c_{m_1+1}, \dots, c_m) > 0 \text{ and } \lambda = \mu - 1 . \tag{6.4.4}$$

Since $f_i(u) = u_i g_i(u)$ and $u_i = 0$ for $i > m_1$, we have $\partial_{u_k} f_i(u, 0) = \delta_{ik} g_i(u, 0)$, where δ_{ik} is the Kronecker symbol, for $i, k > m_1$. Thus, $\partial_v f^{(v)}(u, 0)$ is a diagonal matrix and (6.4.4) is simply

$$g_i(u, 0)c_i = k\lambda c_i \quad \forall i > m_1 .$$

Clearly i.1) holds because then the above system has nonzero eigenvector to $\lambda = 0$. For i.2) we can take $\lambda = g_i(u, 0)/k > 0$ and $c_i = 1$, where other components of $\mathbf{c}$ can be zero. i.3) is obvious. The proof is complete. □

We then have the following theorem for systems of two equations.

Theorem 6.4.5. *Assume that $f_i(u) = u_i g_i(u)$ for $i \in \{1, 2\}$. Suppose that the trivial and semitrivial solutions are only the constant ones $(0, 0)$, $u_{1,*}$ and $u_{2,*}$. This means $g_i(u_{i,*}) = 0$. Then there is a nontrivial solution $(u_1, u_2) > 0$ in the following situations:*
a) *$g_i(0) > 0$, $i = 1, 2$, and $g_1(u_{2,*})$ and $g_2(u_{1,*})$ are positive.*
b) *$g_i(0) > 0$, $i = 1, 2$, and $g_1(u_{2,*})$ and $g_2(u_{1,*})$ are negative.*
c) *$g_1(0) > 0$, $g_2(0) < 0$, and $g_2(u_{1,*}) > 0$.*

Proof. We just need to compute the local indices of T at the trivial and semitrivial solutions and show that the sum of these indices is not 1.

First of all, by i.1) of Lemma 6.4.4, it is clear that the condition Z) at these solutions is satisfied in the above situations. We also see that the stability of $u_{i,*}$ is determined by the sign of $g_i(u_{j,*})$, $j \neq i$. We know that $u_{i,*}$ is stable (respectively unstable) in its complement direction if $g_i(u_{j,*}) < 0$ (respectively $g_i(u_{j,*}) > 0$).

The conditions in case a) and i.2) of Lemma 6.4.4 imply that 0 and the semitrivial solutions are unstable in their complement directions. Theorem 6.3.3, with $Z_1^- = \emptyset$, gives that the local indices at these solutions are all zero. Similarly, in case b), the local index at 0 is 0 and the local indices at the semitrivial solutions, which are stable in their complement directions, are 1. In these cases, the sum of the indices is either 0 (in case a)) or 2 (in case b)).

In case c), because $g_2(0) < 0$ we see that 0 is u_2-stable so that $T_2 := T|_{\mathbf{X}_2}$, the restriction of the map T to $\mathbf{X}_2$, has its local index at 0 equal 1 and therefore its local index at $u_{2,*}$ is zero (see also the proof of Theorem 6.3.7). The assumption $g_1(0) > 0$ also yields a neighborhood V_1 in $\mathbf{X}_1$ of 0 such that $\mathrm{ind}(T, V_1 \oplus \mathbf{X}_2) = 0$ (the stability of $u_{2,*}$ in the u_1 direction does not matter). On the other hand, because $g_1(0) > 0$ we see that 0 is u_1-unstable so that $T_1 := T|_{\mathbf{X}_1}$, the restriction of the map T to $\mathbf{X}_1$, has its local index at 0 equal 0 and therefore its local index at $u_{1,*}$ is 1. But $u_{1,*}$ is u_2-unstable, because $g_2(u_{1,*}) > 0$, so that there is a neighborhood V_2 in $\mathbf{X}_2$ of 0 such that $\mathrm{ind}(T, \mathbf{X}_1 \oplus V_2) = 0$.

In three cases, we have shown that the sum of the local indices at the trivial and semitrivial solutions is not 1. Hence, there is a positive nontrivial fixed point (u_1, u_2). □

Remark 6.4.6. If the system $g_i(u) = 0$, $i = 1, 2$, has no positive constant solution then the above theorem gives conditions for the existence of nonconstant and nontrivial solutions. This means pattern formations occur.

One of the most famous examples of reaction terms in mathematical biology is of course the Lotka–Volterra ones for systems of two equations. We consider

$$\begin{cases} f_1(u_1, u_2) = u_1(a_1 - b_1u_1 - c_1u_2)\,, \\ f_2(u_1, u_2) = u_2(a_2 - b_2u_1 - c_2u_2)\,, \end{cases} \tag{6.4.5}$$

where a_i, b_i, c_i are positive constants. Thus, for $f_i(u) = u_ig_i(u)$ with $g_1(u_1, u_2) = a_1 - b_1u_1 - c_1u_2$ and $g_2(u_1, u_2) = a_2 - b_2u_1 - c_2u_2$. The semitrivial (constant) solutions in Theorem 6.4.5 are $u_{1,*} = (\frac{a_1}{b_1}, 0)$ and $u_{2,*} = (0, \frac{a_2}{c_2})$.

That the conditions $g_1(u_{2,*})$ and $g_2(u_{1,*})$ are positive in a) of the theorem is then equivalent to $\frac{b_1}{b_2} > \frac{a_1}{a_2} > \frac{c_1}{c_2}$, which was usually referred to as the *weak competitive* case in mathematical biology literature. Similarly, that the conditions $g_1(u_{2,*})$ and $g_2(u_{1,*})$ are negative in b) is then equivalent to $\frac{b_1}{b_2} < \frac{a_1}{a_2} < \frac{c_1}{c_2}$, the *strong competitive* case. Since $g_i(0) = a_i > 0$, the theorem applies here for both strong and weak cases to establish the existence of a nontrivial solution.

However, a *positive constant* nontrivial solution $u_* = (\frac{a_1c_2 - a_2c_1}{b_1c_2 - b_2c_1}, \frac{a_2b_1 - a_1b_2}{b_1c_2 - b_2c_1})$ is obvious from the fact that $g_1(u_*) = g_2(u_*) = 0$. We will next search for other solutions that are not this 'trivially' nontrivial.

6.4.2 Nontrivial constant solutions

Suppose now that $u_* = (u_1, \dots, u_m)$ is a nontrivial constant fixed point of T with $u_i \neq 0$ for all i. We will use the Leray–Schauder theorem to compute the local index of T at u_*. Since u_* is in the interior of $\mathbf{P}$, we do not need that T is positive as in the previous discussion so that we can take $K = 0$. The main result of this section, Theorem 6.4.8, yields a formula to compute the indices at nontrivial constant fixed points. In applications, the sum of these indices and those at semitrivial fixed points will provide the possibility for nontrivial and nonconstant fixed points to exist.

In the sequel, we will denote

$$d_A(u_*) = \operatorname{diag}[a_{11}(u_*), \dots, a_{mm}(u_*)]\,. \tag{6.4.6}$$

From the ellipticity assumption on A, we easily see that $a_{ii}(u_*) > 0$ for all i and thus $d_A(u_*)$ is invertible.

The following lemma describes the eigenspaces of $T'(u_*)$.

Lemma 6.4.7. *The solution space of $T'(u_*)\Phi = \mu\Phi$ is spanned by $\mathbf{c}_{i,j}\psi_i$ with $\mathbf{c}_{i,j}$ solving*

$$\hat{\lambda}_i[A(u_*) + (\mu - 1)d_A(u_*)]\mathbf{c} = \partial_u F(u_*)\mathbf{c}\,. \tag{6.4.7}$$

Proof. We have $D(T(u_*)) = Du_* = 0$ so that (6.3.29)–(6.3.31), noting that u_* is no longer a semitrivial solution, with $w = u_*$, shows that the i-th component W_i of $T'(u_*)\Phi$ solves

$$-\operatorname{div}(a_{ii}(u_*)DW_i + w_i \sum_{j\neq i} b_{ij}(u_*)D\phi_j) = \partial_{u_k} f_i(u_*)\phi_k .$$

Since $w_i b_{ij}(u_*) = u_{i,*}b_{ij}(u_*) = a_{ij}(u_*)$, the eigenvalue problem $T'(u_*)\Phi = \mu\Phi$ is, as $W = T'(u_*)\Phi$

$$-\operatorname{div}(a_{ii}(u_*)D(\mu\phi_i) + \sum_{j\neq i} a_{ij}(u_*)D\phi_j) = \partial_{u_k} f_i(u_*)\phi_k .$$

In matrix form, the above can be written as

$$-\operatorname{div}([A(u_*) + (\mu - 1)d_A(u_*)]D\Phi) = \partial_u F(u_*)\Phi . \tag{6.4.8}$$

Since u_* is a constant vector, by Lemma 6.4.1, we can write $\Phi = \sum \mathbf{c}_{i,j}\psi_i$ with $\mathbf{c}_{i,j}$ solving

$$\hat{\lambda}_i[A(u_*) + (\mu - 1)d_A(u_*)]\mathbf{c} = \partial_u F(u_*)\mathbf{c} \quad \forall i .$$

This is (6.4.7) and the lemma is proved. □

We now have the following explicit formula for $\operatorname{ind}(T, u_*)$.

Theorem 6.4.8. *Assume that*

$$\operatorname{Ker}(\hat{\lambda}_i A(u_*) - \partial_u F(u_*)) = \{0\} \quad \forall i . \tag{6.4.9}$$

For $i > 0$ let

$$\mathcal{A}_i = d_A(u_*)^{-1}[A(u_*) - \hat{\lambda}_i^{-1}\partial_u F(u_*)]$$

and assume that the (algebraic) multiplicity of any negative eigenvalue λ of $\mathcal{A}_i$ is equal to its geometric one $N_{i,\lambda} := \dim(\operatorname{Ker}(\mathcal{A}_i - \lambda I))$.

We then denote $N_i = \sum_{\lambda<0} N_{i,\lambda}$ and by M_i the (geometric) multiplicity of $\hat{\lambda}_i$, the i-th eigenvalue of the Laplacian operator with Neumann boundary condition.

Then there exists an integer L_0 such that

$$\operatorname{ind}(T, u_*) = (-1)^\gamma , \quad \gamma = \sum_{i\leq L_0} N_i M_i . \tag{6.4.10}$$

Proof. We will apply Leray–Schauder's theorem to compute $\operatorname{ind}(T, u_*)$. First of all, Lemma 6.4.7 and the assumption (6.4.9) show that $\Phi = 0$ is the only solution to $T'(u_*)\Phi = \Phi$ so that $\mu = 1$ is not an eigenvalue of $T'(u_*)$. By Leray–Schauder's theorem, we have that $\operatorname{ind}(T, u_*) = (-1)^\gamma$, where γ is the sum of (algebraic) multiplicities of eigenvalues μ of $T'(u_*)$ that are greater than 1. Remark 6.4.9 after the proof proves that γ_* is the sum of the dimensions of $Ker(T'(u_*) - \mu I)$ for $\mu > 1$. Lemma 6.4.7 then clearly shows that γ is the sum of the dimensions of solution spaces of (6.4.7) and therefore

$$\gamma := \sum_i \gamma_i^* M_i \,, \tag{6.4.11}$$

and $\gamma_i^* = \sum_{\mu>1} n_{i,\mu}$, where $n_{i,\mu}$ is the dimension of the solution space of

$$\hat{\lambda}_i[A(u_*) + (\mu - 1)d_A(u_*)]\mathbf{c} - \partial_u F(u_*)\mathbf{c} = 0 \,.$$

For $i = 0$ we have $\hat{\lambda}_0 = 0$ so that $n_{0,\mu} = \dim(\mathrm{Ker}(\partial_u F(u_*)))$, which is zero because of (6.4.9). For $i > 0$ and $\mu > 1$ let $\mathcal{A}_i$ be defined in the statement of this theorem, and $\lambda = 1 - \mu < 0$. The above equation is clearly equivalent to $\mathcal{A}_i\mathbf{c} = \lambda\mathbf{c}$ so that $\mathbf{c}$ is an eigenvector to a negative eigenvalue λ of the matrix $\mathcal{A}_i$. It is clear that the number γ_i^* in (6.4.11) is the sum of the dimensions of eigenspaces of $\mathcal{A}_i$ to negative eigenvalues. That is $\gamma_i^* = N_i = \sum_{\lambda<0} N_{i,\lambda}$.

From the ellipticity assumption on A, $d_A(u_*)^{-1}A(u_*)$ is positive definite. Therefore, it is not difficult to see that $\mathcal{A}_i$ is positive if $\hat{\lambda}_i$ is large. Thus, as $\lim_{i\to\infty} \hat{\lambda}_i = \infty$, there is an integer L_0 such that $\mathcal{A}_i$ has no negative eigenvalues if $i > L_0$. Hence, $\gamma_i^* = N_i = 0$ if $i > L_0$. The theorem then follows. □

Remark 6.4.9. We should note that γ is the sum of (algebraic) multiplicities of eigenvalues $\mu > 1$ but coincides with the sum of the dimensions of the eigenspaces in our case. To see this, we first recall [47, Proposition 8.18], which states that the algebraic and geometric multiplicities of a real eigenvalue μ of a compact linear operator L on Hilbert spaces are the same iff there are bases $\{x_i\}$, $\{x_i^*\}$ of $Ker(\mu I - L)$ and $Ker(\mu I - L^*)$ such that $\det((\langle x_i, x_j^* \rangle)) \neq 0$.

We consider $L := T'(u_*)$ and $\mu > 1$ as one of its eigenvalue. We see that $Ker(I - \mu L)$ and $Ker(I - \mu L^*)$ respectively have the bases $\{x_i\}$ and $\{x_i^*\}$ given by $x_i = \sum \mathbf{c}_{i,j}\psi_i$ and $x_i^* = \sum \mathbf{c}_{i,j}^*\psi_i$. The vectors $\mathbf{c}_{i,j}^*$ are the solutions of the adjoint of (6.4.7) or eigenvectors to a negative eigenvalue λ of $\mathcal{A}_i^*$. We then have $\langle x_i^*, x_j \rangle = \langle \mathbf{c}_{i,j)^*, \mathbf{c}_{i,j}} \langle \psi_i^*, \psi_j \rangle$. As we assume in the theorem that the algebraic and geometric multiplicity of λ are the same, we have $\det((\langle \mathbf{c}_i, \mathbf{c}_j^* \rangle)) \neq 0$. Hence, the determinant of the matrix $(\langle x_i^*, x_j \rangle)$ is nonzero so that the (algebraic) multiplicity of μ is equal to its geometric one, the dimension of $Ker(I - \mu L)$, again by [47, Proposition 8.18].

Remark 6.4.10. Since $\hat{\lambda}_0 = 0$, (6.4.9) implies that $\mathrm{Ker}(\partial_u F(u_*)) = \{0\}$ so that u_* is an isolated constant solution to $F(u_*) = 0$. Also, as $A(u_*)$ is positive definite and $\lim_{i\to\infty} \hat{\lambda}_i = \infty$, we see that (6.4.9) is true when i is large.

Combining Theorem 6.4.8 and Theorem 6.4.5, we obtain the following corollary.

Corollary 6.4.11. *Assume as in Theorem 6.4.5. Suppose further that there is only one nontrivial constant solution u_*. Let γ be as in* (6.4.10). *There is a nontrivial nonconstant solution $(u_1, u_2) > 0$ in the following situations:*

1) *γ is odd and a) or c) of the theorem hold.*

2) *γ is even and b) of the theorem holds.*

Proof. We have seen from the proof of Theorem 6.4.5 that the sum of the local indices at the trivial and semitrivial solutions is 0 in the cases a) and c) of the theorem. If γ is odd then $\mathrm{ind}(T, u_*) = -1$. Similarly, if b) holds then the sum of the indices at the trivial and semitrivial solutions is 2. If γ is even then $\mathrm{ind}(T, u_*) = 1$. Thus, the sum of the local indices at constant solutions is not 1 in both cases. Since $\mathrm{ind}(T, \mathbf{B} \cap \mathbf{X}) = 1$, a nonconstant and nontrivial solution must exist. □

In applications, the following result is useful to determine whether γ is odd by a simple mean. For simplicity, we consider systems of two equations (i.e., $m = 2$). Combining this result with Theorem 6.4.5, we obtain a more general conclusion and simpler proof than those in [32, Theorem 4.7].

Corollary 6.4.12. *Assume as in Theorem 6.4.8 and $m = 2$. Consider the numbers*

$$\Delta_i := \det \mathcal{A}_i\,, \quad \text{with } \mathcal{A}_i = d_A(u_*)^{-1}[A(u_*) - \hat{\lambda}_i^{-1}\partial_u F(u_*)]\,,$$

the 2×2 matrices defined in Theorem 6.4.8. Assume that $\Delta_i \neq 0$ for all i and define the set $I_\Delta := \{i \geq 1 : \Delta_i < 0\}$.

Then $\sum_{j \in I_\Delta} M_j \equiv \gamma (\mathrm{mod}\ 2)$. If $I_\Delta = \emptyset$ then γ is even.

Proof. First of all, the assumption $\Delta_i \neq 0$ implies 0 is not an eigenvalue of $\mathcal{A}_i$ so that (6.4.9) for all i.

We next recall the formula for γ in (6.4.10) of Theorem 6.4.8

$$\gamma = \sum_i N_i M_i, \quad N_i = \sum_{\lambda < 0} N_{i,\lambda}\,,$$

where $N_{i,\lambda} = \dim(Ker(\mathcal{A}_i - \lambda I))$. For any negative λ we will see that the oddness of $N_{i,\lambda}$ is determined by the sign of Δ_i.

Indeed, if $\Delta_i > 0$ and $\lambda < 0$ is an eigenvalue of $\mathcal{A}_i$ then either λ is a double eigenvalue or there is another negative eigenvalue $\lambda' \neq \lambda$. In the latter case, because $\mathcal{A}_i$ is 2×2, $N_{i,\lambda} = N_{i,\lambda'} = 1$. We see that $N_i = \sum_{\lambda<0} N_{i,\lambda}$ is even.

If $\Delta_i < 0$ and $\lambda < 0$ is an eigenvalue of $\mathcal{A}_i$ then the other λ' is positive so that $N_i = N_{i,\lambda} = 1$.

We write $\gamma = \sum_{j \in I_\Delta} N_j M_j + \sum_{j \notin I_\Delta} N_j M_j$. The above discussion and the definition of I_Δ show that the first sum is just $\sum_{j \in I_\Delta} M_j$ because $N_j = 1$, and the second one is even because N_i's are. Thus, $\sum_{j \in I_\Delta} M_j \equiv \gamma(\mathrm{mod}\ 2)$ and of course γ is even if $I_\Delta = \emptyset$. □

Remark 6.4.13. We consider a simpler situation. Suppose that i_0 is the first integer such that M_{i_0}, the multiplicity of $\hat{\lambda}_{i_0}$, is odd and that $\Delta_{i_0} < 0$ and $\Delta_i > 0$ for any $i > i_0$. Then γ is odd. Indeed, as $\gamma = \sum_i N_i M_i$, we now see that $N_i M_i$ is even when $i < i_0$ and $i > i_0$ (because either M_i or N_i is even). Thus, γ is odd because $N_{i_0} = 1$ and M_{i_0} is odd.

Remark 6.4.14. Of course, because A is positive definite so are the real parts of the eigenvalues of A and $\det(d_A(u_*) \geq 0$. We see that the sign of Δ_i is that of the determinant of $A(u_*) - \hat{\lambda}_i^{-1}\partial_u F(u_*)$.

Remark 6.4.15. We should remark that, when $m = 2$, we have the following formula, which holds for any 2×2 matrix:

$$\det(A + B) = \det A + \det B + \operatorname{trace} A^T B\,.$$

We will let $A = A(u_*)$ and $B = -\hat{\lambda}_i^{-1}\partial_u F(u_*)$. We then see that the sign of Δ_i is determined by that of the quadratic $\delta(x) := ax^2 + bx + c$ where $x = \hat{\lambda}_i$ and

$$a = \det A(u_*),\;\; b = -\operatorname{trace} A(u_*)^T F_u(u_*),\;\; c = \det F_u(u_*)\,.$$

Concerning the existence of the roots of $\delta(x)$, we note that $a > 0$, because A is positive definite, so they always exist if $c < 0$. If $c > 0$, we consider the case when the matrix $F_u(u_*)$ is positive definite. In this case, we note the inequality

$$4\det(AB) \leq \operatorname{trace}(AB)$$

as a consequence of [5, Prop. 8.4.14], which holds for any positive definite 2×2 matrix. This implies $b^2 - 4ac \geq 0$ so that $\delta(x)$ has real roots.

As $a > 0$, we see that $\delta(x) < 0$ if $x \in (x_{1,*}, x_{2,*})$, where $x_{1,*}, x_{2,*}$ are the two roots of $\delta(x)$. We then have

$$I_\Delta = \{l > 0 : \hat{\lambda}_i \in (x_{1,*}, x_{2,*})\}\,.$$

We recall a finding of Theorem 6.4.5 when $\hat{f}_i(u) = u_i g_i(u)$. As we see from the proof of the theorem, the stability of $u_{i,*}$ is determined by the sign of $g_i(u_{j,*})$, $j \neq i$. We know that $u_{i,*}$ is stable (respectively unstable) in its complement direction if $g_i(u_{j,*}) < 0$ (respectively $g_i(u_{j,*}) > 0$).

Concerning the (SKT) system, if the system is strongly competitive as in case b), that is $g_i(u_{i,*}) < 0$ for $i = 1, 2$, the constant semitrivial solutions $u_{i,*}$ are both stable in their complement directions, then we showed that the sum of the local fixed-point indices at the trivial and semitrivial solutions is 2. We see that a nontrivial solution always exists, but it could be the constant solution u_*. On the other hand, the total sum of the indices is 1. Thus, if the index at the constant solution u_* is 1 then there must be another *nonconstant* solution besides the constant solution u_*. This will be the case if we can show the number γ in Corollary 6.4.12 is even or I_Δ, the set of integers $i > 0$ such that $\Delta_i < 0$, is empty. We consider the latter case and find the conditions such that $\Delta_i > 0$ for all $i > 0$.

From Remark 6.4.15, we see that the sign of Δ_i is determined by the quadratic $\delta(x)$, which is increasing if $x \geq -b/(2a)$. Thus, if $\hat{\lambda}_1$ is the second eigenvalue of the Laplacian operator with Neumann boundary conditions and we have

$$\delta(\hat{\lambda}_1) > 0,\;\; \hat{\lambda}_1 \geq \frac{-b}{2a} = \frac{\operatorname{trace} A(u_*)^T F_u(u_*)}{\det A(u_*)}$$

then there is a nonconstant nontrivial solution. The above two conditions can, of course, be combined into one saying that $\hat{\lambda}_1$ is greater then $x_{2,*}$, the larger root of $\delta(x)$, because the sequence $\{\hat{\lambda}_i\}$ is increasing.

Of course, the above discussion can be generalized by assuming that the sum $\sum_{i\in I_\Delta} M_i$ is even.

6.4.3 Large diffusivity and some nonexistent results

We discuss some nonexistence results by showing that if the parameter λ_0 is sufficiently large then there are no nonconstant solutions. We consider the following system:

$$\begin{cases} -\operatorname{div}(A(u, Du) = f(u) + B(u, Du) & \text{in } \Omega\,, \\ \text{homogenenous Neumann boundary conditions} & \text{on } \partial\Omega\,, \end{cases} \tag{6.4.12}$$

We first have the following nonexistent result under a strong assumption on the uniform boundedness of solutions. This assumption will be relaxed later in Corollary 6.4.18.

Theorem 6.4.16. *Assume that A satisfies A) and $\hat{f}(u, Du) := f(u) + B(u, Du)$ for some $f \in C^1(\mathbb{R}^m, \mathbb{R}^m)$ and $B : \mathbb{R}^m \times \mathbb{R}^{mn} \to \mathbb{R}^m$ such that $|B(u, p)| \le b(u)|p|$ for some continuous nonnegative function b on $\mathbb{R}^m$. Suppose also that there is a constant C independent of λ_0 such that for any solutions of* (6.4.12)

$$\|u\|_{L^\infty(\Omega)} \le C\,.$$

If the constant λ_0 in A) is sufficiently large then there is no nonconstant solution to (6.4.12).

Proof. For any function g on Ω let us denote the average of g over Ω by g_Ω. That is,

$$g_\Omega = \frac{1}{|\Omega|}\int_\Omega g\,dx\,.$$

Integrating (6.4.12) and using Neumann boundary conditions, we have $f(u)_\Omega + B(u, Du)_\Omega = 0$. Thanks to this, we test the system with $u - u_\Omega$ to get

$$\int_\Omega \langle A(u, Du), Du\rangle\,dx = \int_\Omega [\langle f(u) - f(u)_\Omega, u - u_\Omega\rangle - \langle B(u, Du)_\Omega, u - u_\Omega\rangle]\,dx\,. \tag{6.4.13}$$

We estimate the terms on the right-hand side. First of all, by Hölder's inequality

$$\int_\Omega \langle f(u) - f(u)_\Omega, u - u_\Omega\rangle\,dx \le \left(\int_\Omega |f(u) - f(u)_\Omega|^2\,dx\right)^{\frac{1}{2}} \left(\int_\Omega |u - u_\Omega|^2\,dx\right)^{\frac{1}{2}}\,.$$

Applying Poincaré's inequality to the functions $f(u)$, u on the right-hand side of the above inequality, we can bound it by

$$\operatorname{diam}(\Omega)^2 \left(\int_\Omega |f_u(u)|^2|Du|^2\,dx\right)^{\frac{1}{2}} \left(\int_\Omega |Du|^2\,dx\right)^{\frac{1}{2}} \le F_* \operatorname{diam}(\Omega)^2 \|Du\|^2_{L^2(\Omega)}\,,$$

where $F_* := \sup_\Omega |f_u(u(x))|$. This number is finite because we are assuming that $\|u\|_{L^\infty(\Omega)}$ is bounded uniformly.

Similarly, we define $B_* := \sup_\Omega b(u(x))$. Using the fact that $|B(u, Du)| \le b(u)|Du| \le B_*|Du|$, we have $|B(u, Du)_\Omega| \le B_*|\Omega|^{-1}\|Du\|_{L^1(\Omega)}$. Furthermore, by Hölder's and Poincaré's inequalities,

$$\|Du\|_{L^1(\Omega)} \le |\Omega|^{\frac{1}{2}}\|Du\|_{L^2(\Omega)},\quad \|u - u_\Omega\|_{L^1(\Omega)} \le C|\Omega|^{\frac{1}{2}} \operatorname{diam}(\Omega)\|Du\|_{L^2(\Omega)}\,.$$

We then obtain

$$\left|\int_\Omega \langle B(u, Du)_\Omega, u - u_\Omega\rangle\,dx\right| \le B_* \operatorname{diam}(\Omega)\|Du\|^2_{L^2(\Omega)}\,.$$

Using the above estimates and the ellipticity condition A) in (6.4.13), we get

$$\lambda_0 \int_\Omega |Du|^2\,dx \le F_* \operatorname{diam}(\Omega)^2 \int_\Omega |Du|^2\,dx + B_* \operatorname{diam}(\Omega) \int_\Omega |Du|^2\,dx\,.$$

If λ_0 is sufficiently large then the above inequality clearly shows that $\|Du\|_{L^2(\Omega)} = 0$ and thus u must be a constant vector. □

Remark 6.4.17. If we assume Dirichlet boundary conditions and $f(0) \equiv 0$ then 0 is the only solution if λ_0 is sufficiently large. To see this we test the system with u and repeat the argument in the proof.

The assumption on the boundedness of the L^∞ norms of the solutions in Theorem 6.4.16 can be weakened if $\lambda(u)$ has a polynomial growth. We have the following result.

Corollary 6.4.18. *The conclusion of Theorem 6.4.16 continues to hold for the system if one has a uniform estimate for $\|u\|_{L^1(\Omega)}$ and $\lambda(u) \sim \lambda_0 + (1 + |u|)^k$ for some $k > 0$.*

Proof. We just need to show that the two assumptions in fact provide the uniform bound of L^∞ norms needed in the previous proof. From the growth assumption on λ, we see that the number Λ in (4.3.3) of Section 4.3 is now

$$\Lambda = \sup_{W\in\mathbb{R}^m} \frac{|\lambda_W(W)|}{\lambda(W)} \sim \sup_{W\in\mathbb{R}^m} \frac{(1 + |W|)^{k-1}}{\lambda_0 + (1 + |W|)^k}\,.$$

By considering the case where $(1 + |W|)^k$ is greater or less than λ_0, we can easily see that Λ can be arbitrarily small if λ_0 is sufficiently large. On the other hand, our assumptions yield that $\|u\|_{L^1(\Omega)}$ is uniformly bounded. Thus, we can fix a $R_0 > 0$ and use the fact that $\|u\|_{BMO(B_{R_0})} \leq C(R_0)\|u\|_{L^1(\Omega)}$ to see that the smallness condition (4.3.19) of Remark 4.3.19 holds if λ_0 is sufficiently large. We then have that the Hölder norms, and the L^∞ norms, of the solutions to (6.4.12) are uniformly bounded, independently of λ_0. This is the key assumption of the proof of Theorem 6.4.16 so that the proof can continue as before. □

6.4.4 Small diffusivity

In contrast to the previous example, we consider the case of porous media type cross-diffusion systems where the self-diffusivities are all very small. Both homogeneous Neumann and Dirichlet boundary conditions can be considered. We consider the system

$$-\Delta(\lambda_0 u_i + u_i\lambda_i(u)) = k_i u_i + u_i\bar{g}_i(u)\,, \tag{6.4.14}$$

where $i = 1, \ldots, m$ and $\lambda_0 > 0$. The vectorial form is

$$-\Delta([\lambda_0 Id + \lambda_d(u)]u) = [K_d + \bar{G}(u)]u\,, \tag{6.4.15}$$

where $\lambda_d(u) = \text{diag}[\lambda_i(u)]_{i=1}^m$, $K_d = \text{diag}[k_i]_{i=1}^m$, $\bar{G}(u) = \text{diag}[\bar{g}_i(u)]_{i=1}^m$. The existence of strong solutions to this system was established in Corollary 5.7.6 under the assumption that we can control the L^p norms of their solutions for some $p > N/2$. That is, if there are $r_0 > N/2$ and a constant c_0 such that any strong solution u to the systems (6.4.15) satisfies

$$\|u\|_{L^{r_0}(\Omega)} \leq c_0\,. \tag{6.4.16}$$

However, the obtained solution could be 0, an obvious solution to (6.4.15). We will first show that this is not the case if λ_0 is sufficiently small.

Theorem 6.4.19. *Assume that the uniform bound* (6.4.16) *holds for any strong solution u to the system* (6.4.15). *There is $\lambda_* > 0$ such that* (6.4.15) *has a nonzero strong solution u if $0 < \lambda_0 \leq \lambda_*$.*

Proof. We simply apply the theory in this chapter to the family of regular system (6.4.15) to find conditions such that (6.4.15) has nontrivial or semitrivial strong solutions, i.e., $u \neq 0$. We see that strong solutions to (6.4.15) are fixed points of the maps T associated to (6.3.1) for some appropriate matrix K. We can take K to be kId for some $k > 0$ and k is sufficiently large. As $A(u) = \lambda_0 Id + \partial_u(\lambda_d(u)u)$, we have seen before that a nontrivial or semitrivial solution exists if the trivial solution 0 is unstable. This is to say ((6.3.2)) that the eigenvalue problem of $T'(0)h = \lambda h$, which reads

$$-\text{div}(A(0)Dh) + Kh = \lambda^{-1}(\hat{f}_u(0) + K)h\,, \tag{6.4.17}$$

has a positive eigenfunction h to an eigenvalue $\lambda > 1$. In our present case $\hat{f}(u) = (K_d + \bar{G}(u))u$ and $\bar{G}(0) = 0$, so that $\hat{f}_u(0) = K_d$. It is clear that $A(0) = \lambda_0 Id$. We then consider the eigenvalue problem

$$-\operatorname{div}(\lambda_0 Dh) + Kh = \mu(K_d + K)h \tag{6.4.18}$$

and the assumption that it has a positive eigenfunction h for some $\mu < 1$. This holds for sufficiently small λ_0 and can be easily seen by using the variational characterization of the principal eigenvalue of (6.4.18). That is, the smallest μ satisfies

$$\begin{aligned}\mu &= \inf_{h\in W_0^{1,2}(\Omega)} \frac{\lambda_0\|Dh\|^2_{L^2(\Omega)} + \langle Kh, h\rangle_{L^2(\Omega)}}{\langle (K_d+K)h, h\rangle_{L^2(\Omega)}} \\ &= \lambda_0 \inf_{h\in W_0^{1,2}(\Omega)} \frac{\|Dh\|^2_{L^2(\Omega)}}{\langle (K_d+K)h, h\rangle_{L^2(\Omega)}} + \inf_{h\in W_0^{1,2}(\Omega)} \frac{\langle Kh, h\rangle_{L^2(\Omega)}}{\langle (K_d+K)h, h\rangle_{L^2(\Omega)}}\,.\end{aligned}$$

If Neumann boundary conditions are considered then the infimum is taken over $h \in W^{1,2}(\Omega)$. The first infimum on the far right is finite, as it describes the principal eigenvalue of the Laplacian, and the second infimum is strictly less than 1 because $k_i > 0$. We see that that is $\mu < 1$ if λ_0 is sufficiently small. The proof is complete. □

Finally, we would like to discuss the Hölder weak solution u obtained in the preceding chapter. In Chapter 5, we established the existence of such a solution as the limit of some sequence of strong solutions to the regularized systems with $\lambda_0 = \lambda_{0,n}$ tending to zero. This solution may be the trivial solution $u \equiv 0$ and the result is not interesting. We will show that this is not the case.

Theorem 6.4.20. *There exists a Hölder continuous nonzero weak solution to the system (6.4.14) when $\lambda_0 = 0$.*

Proof. Consider a sequence of strong solutions $u_n = (u_{i,n})$ obtained by Theorem 6.4.19, with $\lambda_0 = \lambda_{0,n}$ tending to zero. Assume that there is $\gamma > 0$ such that $\{u_n\}$ converges to 0 in $C^{0,\gamma}(\Omega)$.

We rewrite the system (6.4.14) as

$$-\Delta(W_{i,n}) = W_{i,n}h_{i,n}\,, \tag{6.4.19}$$

where $W_{i,n} := (\lambda_{0,n}\lambda_i^{-1}(u_n) + 1)u_{i,n}\lambda_i(u_{i,n})$ and $h_{i,n} := (k_i + \bar{g}_i(u_n))\lambda_i^{-1}(u_n)$ for $i = 1, \dots, m$. Testing these equation with $W_{i,n}$, we have

$$\int_\Omega |DW_{i,n}|^2\,dx = \int_\Omega h_{i,n}W_{i,n}^2\,dx\,.$$

Dividing this by $\|W_n\|_{W^{1,2}(\Omega)}$, $W_n = (W_{i,n})_{i=1}^m$, and setting $\bar{W}_n = (\bar{W}_{i,n})_{i=1}^m$ with $\bar{W}_{i,n} = W_{i,n}/\|W_n\|_{W^{1,2}(\Omega)}$, we obtain

$$\int_\Omega |D\bar{W}_n|^2\,dx = \int_\Omega \bar{W}_n^2 \sum_i h_{i,n}\,dx\,.$$

As $\|\bar{W}_n\|_{W^{1,2}(\Omega)} \leq 1$, the above yields

$$\left| \int_\Omega \bar{W}_n^2 \sum_i h_{i,n} \, dx \right| \leq C(|\Omega|) . \tag{6.4.20}$$

By compactness, $\bar{W}_n$ converges in $L^2(\Omega)$ to some $\bar{W} \neq 0$ because $\|\bar{W}_n\|_{W^{1,2}(\Omega)} = 1$. However, as $u \to 0$ in $L^\infty(\Omega)$, $\lambda_i(u) \to 0$ and we see that $h_i \to \infty$ on Ω, the left-hand side of the (6.4.20) is infinite as $n \to \infty$. We then obtain a contradiction. □

A The duality RBMO(μ)–$H^1(\mu)$

A.1 Some simple consequences from Tolsa's works

The RBMO(μ) space was introduced by Tolsa in [38, 39]. Tolsa considered *nondoubling* measure μ and defined

$$[f]_{*,\mu} := \sup_Q \fint_{\lambda Q} |f - f_Q|\, d\mu \tag{A.1.1}$$

for some constant $\lambda > 1$. This constant λ is not important as shown in [39]. The definition of RBMO(μ) spaces in [39] coincides with the $BMO(\mu)$, defined by (3.2.1), if μ is doubling. It was only assumed in [39] that

M.1) There are a constant C_μ and a fixed number $n \in (0, N]$ such that for any cube Q_r with side length $r > 0$

$$\mu(Q_r) \le C_\mu r^n\,. \tag{A.1.2}$$

The Hardy space $H^1(\mu)$ was introduced in [38] and the duality between $H^1(\mu)$ and RBMO(μ) was also established. For our purpose, we don't need such a full force generality and we just recall the following deep result in [39].

Lemma A.1.1 (The Main Lemma – [39, Lemma 4.1]). *Let $f \in$ RBMO(μ) with compact support and $\int_\Omega f\, d\mu = 0$. There exist functions $h_m \in L^\infty(\mu)$ and $\phi_{y;m}$, $m \ge 0$, such that*

$$f(x) = h_0(x) + \sum_{m=1}^{\infty} \int \phi_{y;m}(x) h_m(y) d\mu(y)\,; \tag{A.1.3}$$

with convergence in $L^1(\mu)$ and

$$\sum_{m=0}^{\infty} \|h_m\|_{L^\infty(\mu)} \le C[f]_{*,\mu}\,. \tag{A.1.4}$$

Importantly, the functions $\phi_{y;m}$ satisfy the properties in Lemma A.1.2 below.

It was shown in [39] that the functions $\phi_{y;m}$ satisfy the following properties.

Lemma A.1.2. *There is a constant C, depending also on C_μ, such that for any $y \in$ supp(μ) there is some cube $Q \subset \mathbb{R}^N$ centered at y:*

1) $\phi_{y;m} \in C^1_0(Q)$.

2) $0 \le \phi_{y;m}(x) \lesssim Cl(Q)^{-n}$ *for all $x \in Q$.*

3) $|D\phi_{y;m}(x)| \lesssim Cl(Q)^{-n-1}$ *or all $x \in Q$.*

Proof. In [39, Lemma 7.8], for suitable and fixed constants α, β and *some* cubes Q_1, Q_2 concentric with Q and $\alpha l(Q_1) \le l(Q) \le \beta l(Q_2)$, 1) comes from a) of [39, Lemma 7.8] as $\phi_{y;m} = 0$ outside Q_2. Similarly, 2) comes from [39, b) and c) of Lemma 7.8] if we note that $l(Q) \lesssim |y - x|$ for $x \in Q_2 \setminus Q_1$. Finally, 3) comes from [39, d) of Lemma 7.8]. □

https://doi.org/10.1515/9783110608762-007

Right after the statement of [39, Lemma 4.1], there is a short proof of the fact that the $H^1(\mu)$ norm of a function is bounded by $\|f\|_{L^1(\mu)} + \|M_\Phi f\|_{L^1(\mu)}$ ($M_\Phi f$ is defined in [39, Definition 1.1], which is generally larger than the one defined in (A.1.6) below). For our purpose in this paper, we need only estimate $\langle f, g\rangle$ with $g \in \mathrm{RBMO}(\mu)$. We then state the following lemma.

Lemma A.1.3. *Let* $f \in \mathrm{RBMO}(\mu)$ *with the representation* (A.1.3). *Let* $F \in L^1(\mu)$ *such that*

$$\int_\Omega F\,d\mu = 0\,, \quad M_{\hat{\Phi}}F \in L^1(\mu)\,, \tag{A.1.5}$$

where

$$M_{\hat{\Phi}}F(y) = \sup_{m\geq 1} \int_\Omega \phi_{y;m}(x)F(x)d\mu(x)\,. \tag{A.1.6}$$

Then

$$|\langle F, f\rangle| \leq C(\|F\|_{L^1(\mu)} + \|M_{\hat{\Phi}}F\|_{L^1(\mu)})[f]_{*,\mu}\,. \tag{A.1.7}$$

Proof. We repeat the argument right after the statement of [39, Lemma 4.1]. From (A.1.3), we have

$$|\langle F, f\rangle| \leq \left|\int_\Omega Fh_0\,d\mu\right| + \left|\sum_{m=1}^{\infty} \int_\Omega \int_\Omega F\phi_{y;m}(x)h_m(y)d\mu(y)d\mu(x)\right|\,.$$

Since

$$\left|\int_\Omega F\phi_{y;m}(x)h_m(y)d\mu(y)\right| \leq M_{\hat{\Phi}}F(x)\|h_m\|_{L^\infty(\mu)}\,, \tag{A.1.8}$$

by the definition (A.1.6) of $M_{\hat{\Phi}}F$, we have

$$|\langle F, f\rangle| \leq \|F\|_{L^1(\mu)}\|h_0\|_{L^\infty(\mu)} + \|M_{\hat{\Phi}}F\|_{L^1(\mu)} \sum_{m=1}^{\infty} \|h_m\|_{L^\infty(\mu)}\,.$$

By (A.1.4), the above gives the lemma. □

Inspired by Lemma A.1.2, we introduce the following definition.

Definition A.1.4. A function $\phi \in C^1(\mathbb{R}^N)$ is said to be in $\check{\Phi}$ if for any $y \in \mathbb{R}^N$ and *some* cube $Q \subset \mathbb{R}^N$ centered at y and the constant C as in Lemma A.1.2:
f.1) $0 \leq \phi(x) \lesssim Cl(Q)^{-n}$ for all $x \in Q$.
f.2) $\phi \in C_0^1(Q)$ and $|D\phi(x)| \lesssim Cl(Q)^{-n-1}$ or all $x \in Q$.
For any $F \in L^1(\mu)$ we define

$$M_{\check{\Phi}}F(y) = \sup_{\phi\in\check{\Phi}} \int_\Omega \phi(x)f(x)d\mu(x) \in L^1(\mu)\,, \tag{A.1.9}$$

$$\|F\|_{\check{\Phi}} = \|F\|_{L^1(\mu)} + \|M_{\check{\Phi}}F\|_{L^1(\mu)}\,. \tag{A.1.10}$$

By Lemma A.1.2, the functions $\phi_{y;m}$ belong to $\check{\Phi}$ so that $M_{\hat{\Phi}}F(y) \leq M_{\check{\Phi}}F(y)$. We now have from Lemma A.1.3 the following result.

Lemma A.1.5. *Let $f \in BMO(\mu)$ and $F \in L^1(\mu)$ such that*

$$\int_\Omega F\,d\mu = 0\,. \tag{A.1.11}$$

Then

$$|\langle F, f\rangle| \leq C\|F\|_{\check{\Phi}}[f]_{*,\mu}\,. \tag{A.1.12}$$

B Some algebraic inequalities

B.1 On the spectral gap conditions

Let m, l be any integers. For $X = [X_i]_{i=1}^m$, $X_i \in \mathbb{R}^l$ and for any C^1 function $k\colon \mathbb{R}^+ \to \mathbb{R}$ we consider the maps

$$K(X) = k(|X|)|X|^{-1}X, \quad \zeta = |X|^{-1}X = [\zeta_i]_{i=1}^m . \tag{B.1.1}$$

We see that $D_X(|X|) = \zeta^T$ and $D_X\zeta = |X|^{-1}(I - \zeta\zeta^T)$, where $\zeta\zeta^T = [\langle\zeta_i, \zeta_j\rangle]$. Hence,

$$D_XK(X) = k(|X|)D_X\zeta + D_Xk(|X|)\zeta^T = k(|X|)|X|^{-1}(I - \zeta\zeta^T) + k'(|X|)\zeta\zeta^T .$$

We then introduce the notations

$$b = k'(|X|)|X|/k(|X|), \quad \mathcal{K}(\zeta) = I + (b-1)\zeta\zeta^T . \tag{B.1.2}$$

Therefore, the calculation for $D_XK(X)$ yields

$$D_XK(X) = k(|X|)|X|^{-1}(I + (b-1)\zeta\zeta^T) = k(|X|)|X|^{-1}\mathcal{K}(\zeta) . \tag{B.1.3}$$

If $k(|X|) \neq 0$ and $k'(|X|) \neq 0$ then $\mathcal{K}(\zeta)$ is invertible. We can use the Sherman–Morrison formula $(I + wv^T)^{-1} = I - (1 + v^Tw)^{-1}wv^T$, setting $w = (b-1)\zeta$ and $v = \zeta$, to see that

$$(D_XK(X))^{-1} = \frac{|X|}{k(|X|)}(I + (b^{-1} - 1)\zeta\zeta^T) . \tag{B.1.4}$$

Otherwise, if $k(|X|) = 0$ (respectively $k'(|X|) = 0$) then $D_XK(X) = k'(|X|)\zeta\zeta^T$ (respectively $D_XK(X) = k(|X|)|X|^{-1}(I - \zeta\zeta^T)$) and $D_XK(X)$ is not invertible.

The following lemmas on K can provide different ways to obtain energy estimates for $X = Du$, a key ingredient of our proof.

Lemma B.1.1. *Let $X_i\colon \Omega \to \mathbb{R}^l$ be C^1 maps on a domain $\Omega \subset \mathbb{R}^n$ for $i = 1, \dots, m$. If $b = k'(|X|)|X|/k(|X|) > 0$ then*

$$\langle DX, D(K(X))\rangle \geq 0 . \tag{B.1.5}$$

Moreover, for $\alpha = 1 - (\frac{b-1}{b+1})^2$

$$\langle DX, D(K(X))\rangle \geq \alpha^{\frac{1}{2}}|DX||D(K(X))| . \tag{B.1.6}$$

Proof. By (B.1.3) $D(K(X)) = \mathbf{k}(|X|)\mathcal{K}(\zeta)DX$, where $\mathbf{k}(t) := k(t)t^{-1}$, and so

$$\langle DX, D(K(X))\rangle = \mathbf{k}(|X|)(|DX|^2 + (b-1)\langle DX, \zeta\zeta^T DX\rangle) .$$

Note that $|\zeta\zeta^T| \leq 1$ so that $\langle DX, D(K(X))\rangle \geq 0$ if $s = b - 1 > -1$. This gives (B.1.5).

https://doi.org/10.1515/9783110608762-008

Since $\zeta\zeta^T$ is a projection, i.e., $(\zeta\zeta^T)^2 = \zeta\zeta^T$, we have, setting $J = \langle \zeta\zeta^T DX, DX\rangle$,

$$\begin{aligned}|D(K(X))|^2 &= |K_X(X)DX|^2 = \mathbf{k}^2(|X|)\langle\mathcal{K}\rangle(\zeta)\mathcal{D}\mathcal{X}, \mathcal{K}(\zeta)\mathcal{D}\mathcal{X} \\ &= \mathbf{k}^2(|X|)(|DX|^2 + (2s + s^2)J)\,.\end{aligned}$$

Hence, we can write $\langle DX, D(K(X))\rangle^2 - \alpha|DX|^2|D(K(X))|^2$ as

$$\mathbf{k}^2(|X|)[(1-\alpha)|DX|^4 + (2s - \alpha(2s + s^2))|DX|^2 J + s^2 J^2]\,.$$

If we choose $\alpha = 1 - (\frac{s}{s+2})^2$ then the above is $\mathbf{k}^2(|X|)\left(\frac{s}{s+2}|DX|^2 - sJ\right)^2 \geq 0$. Therefore, $\langle DX, D(K(X))\rangle^2 \geq \alpha|DX|^2|D(K(X))|^2$. This and (B.1.5) yield (B.1.6). □

We now consider a matrix A satisfying for some positive λ, Λ and any vector χ

$$\langle A\chi, \chi\rangle \geq \lambda|\chi|^2\,, \quad |A\chi| \leq \Lambda|\chi|\,. \tag{B.1.7}$$

Lemma B.1.2. *Assume* (B.1.7) *and that* $b > 0$ *((B.1.2)). For* $\kappa = \lambda/\Lambda^2$ *and* $\nu = \lambda/\Lambda$ *we have*

$$\langle \kappa ADX, D(K(X))\rangle \geq (\alpha^{\frac{1}{2}} - (1-\nu^2)^{\frac{1}{2}})|DX||D(K(X))|\,.$$

Proof. From (B.1.7) with $\chi = DX$, we note that

$$\begin{aligned}|\kappa ADX - DX|^2 &= \kappa^2|ADX|^2 - 2\kappa\langle ADX, DX\rangle + |DX|^2 \\ &\leq (\kappa^2\Lambda^2 - 2\kappa\lambda + 1)|DX|^2 = (1-\nu^2)|DX|^2\,.\end{aligned}$$

Therefore, using (B.1.6),

$$\begin{aligned}\langle \kappa ADX, D(K(X))\rangle &= \langle \kappa ADX - DX, D(K(X))\rangle + \langle DX, D(K(X))\rangle \\ &\geq -|\kappa ADX - DX||DK(X)| + \alpha^{\frac{1}{2}}|DX||D(K(X))|\,.\end{aligned}$$

As $|\kappa ADX - DX|^2 \leq (1-\nu^2)|DX|^2$, we obtain the lemma. □

Let $k(t) = |t|^{s+1}$, then $K(X) = |X|^s X$ and $b = s + 1$. The above lemma then gives the following result, which was used in deriving our energy estimates for $X = Du$.

Lemma B.1.3. *Assume* (B.1.7). *If* $s > -1$ *and* $\nu = \frac{\lambda}{\Lambda} > \frac{s}{s+2}$, *then*

$$\langle ADX, D(|X|^s X)\rangle \geq c_0\frac{\Lambda^2}{\lambda}|X|^s|DX|^2\,, \tag{B.1.8}$$

where $c_0 = (1 - (\frac{s}{s+2})^2)^{\frac{1}{2}} - (1-\nu^2)^{\frac{1}{2}} > 0$.

In particular, if $p > 1/2$ *and* $1 - 1/p < \nu$ *then there is* $c_1 \sim c_0$ *such that*

$$\langle ADX, D(|X|^{2p-2}X)\rangle \geq c_1\nu^{-2}\lambda|X|^{2p-2}|DX|^2\,. \tag{B.1.9}$$

Because $|D(K(X))| \sim |X|^s|DX|$, the first assertion comes from Lemma B.1.2. The second assertion follows by taking $s = 2p - 2$ and noting that $s/(s+2) = 1 - 1/p$ and $s > -1$ if $p > 1/2$.

Next, the following lemma was used in the checking of the condition K) for the map $K_{\varepsilon_0}(u)$ in the proof of Remark 4.4.5.

Lemma B.1.4. *For any* $\varepsilon_0, \lambda_0 > 0$ *let* $k(t) = |\log(t)| + \varepsilon_0$ *and* $X(u) = [\lambda_0 + |u_i|]_{i=1}^m$ *in* (B.1.1). *There exists a constant* $C(\varepsilon_0)$ *such that the map* $\mathbb{K}(u) = (K_u(X(u))^{-1})^T$ *satisfies*

$$|\mathbb{K}(u)| \le C(\varepsilon_0)|X|, \quad \|\mathbb{K}_u(u)\|_{L^\infty(\mathbb{R}^m)} \le C(\varepsilon_0)\,. \tag{B.1.10}$$

Proof. As $k'(t) = \mathrm{sign}(\log t)t^{-1}$, we have

$$b^{-1} = k(|X|)(k'(|X|)|X|)^{-1} = \mathrm{sign}(\log(|X|))(|\log(|X|)| + \varepsilon_0)\,.$$

Define $X_u = \mathrm{diag}[\mathrm{sign}\, u_i]$. We have from (B.1.4) and the definition $X = [\lambda_0 + |u_i|]_{i=1}^m$ that

$$\mathbb{K}(u) = (X_u)^{-1}\frac{|X|}{|\log(|X|)| + \varepsilon_0}(I + (\mathrm{sign}(\log(|X|))(|\log(|X|)| + \varepsilon_0) - 1)\zeta\zeta^T)\,.$$

As $\varepsilon_0 > 0$, we easily see that $|\mathbb{K}(u)| \le C(\varepsilon_0)|X|$ for some constant $C(\varepsilon_0)$. A straightforward calculation also shows that $\|\mathbb{K}_u(u)\|_{L^\infty(\mathbb{R}^m)} \le C(\varepsilon_0)$. □

Remark B.1.5. If $\lambda(u) \sim (\lambda_0 + |u|)^k \sim |X(u)|^k$, with $k \ne 0$, $\Lambda(u) = \lambda^{\frac{1}{2}}(u)$ and $\Phi(u) = |\Lambda_u(u)|$. We then have $\Lambda(u)\Phi^{-1}(u) \sim |X(u)|$ and $\Phi(u)|\Phi_u(u)|^{-1} \sim |X(u)|$. We obtain from (B.1.10) that $|\mathbb{K}(u)| \lesssim \Lambda(u)\Phi^{-1}(u)$ and $|\mathbb{K}(u)||\Phi_u(u)| \lesssim \Phi(u)$. Thus, the assumptions on the map K for the local Gagliardo–Nirenberg inequality are verified here.

We then need that $K(U)$ is BMO and $\mathbf{W}^\alpha$ is a weight. By (B.1.2),

$$K_U(U) = \frac{|\log(|X|)| + \varepsilon_0}{|X|}\left(I + \left(\frac{X_s}{|\log(|X|)| + \varepsilon_0} - 1\right)\zeta\zeta^T\right)X_U\,, \quad X = [\lambda_0 + |U_i|]_{i=1}^m\,.$$

Recall that

$$\mathbf{W} = \Lambda^{p+1}(U)\Phi^{-p}(U) \sim |U|^p\lambda^{\frac{1}{2}}(U), \ \alpha > 2/(p+2), \ \beta < p/(p+2)\,.$$

Finally, we present an elementary iteration result, Lemma B.1.7 below, which was used to improve the energy estimates for Du.

We recall the following well-known iteration lemma (e.g., see [16, Lemma 6.1, p.192]).

Lemma B.1.6. *Let* f, g, h *be bounded nonnegative functions in the interval* $[\rho, R]$ *with* g, h *increasing. Assume that for* $\rho \le s < t \le R$ *we have*

$$f(s) \le \varepsilon_0 f(t) + [(t-s)^{-\alpha}g(t) + h(t)]$$

with $\alpha > 0$ *and* $0 \le \varepsilon_0 < 1$. *Then*

$$f(\rho) \le c(\alpha, \varepsilon_0)[(R-\rho)^{-\alpha}g(R) + h(R)]\,.$$

The constant $c(\alpha, \varepsilon_0)$ *can be taken to be* $(1-\nu)^{-\alpha}(1-\nu^{-\alpha}\nu_0)^{-1}$ *for any* ν *satisfying* $\nu^{-\alpha}\nu_0 < 1$.

We then have the following lemma.

Lemma B.1.7. *Let F, G, g, h be bounded nonnegative functions in the interval $[\rho, R]$ with g, h increasing. Assume that for $\rho \le s < t \le R$ we have*

$$F(s) \le \varepsilon[F(t) + G(t)] + [(t-s)^{-\alpha} g(t) + h(t)]\,, \tag{B.1.11}$$

$$G(s) \le C[F(t) + (t-s)^{-\alpha} g(t) + h(t)] \tag{B.1.12}$$

with $C \ge 0$, $\alpha, \varepsilon > 0$.

If $2C\varepsilon < 1$. then there is constant $c(C, \alpha, \varepsilon)$ such that

$$F(s) + G(s) \le c(C, \alpha, \varepsilon)[(t-s)^{-\alpha} g(t) + h(t)] \quad \rho \le s < t \le R\,. \tag{B.1.13}$$

Proof. Letting $\varepsilon_0 = 2C\varepsilon$, we obtain from (B.1.11)

$$CF(s) \le \frac{\varepsilon_0}{2}[F(t) + G(t)] + C[(t-s)^{-\alpha} g(t) + h(t)]\,. \tag{B.1.14}$$

Let $t_1 = (s+t)/2$ and use (B.1.14) with s being t_1 and (B.1.12) with t being t_1 to obtain a constant C_1 such that

$$G(s) \le \frac{\varepsilon_0}{2}[F(t) + G(t)] + C_1[(t-s)^{-\alpha} g(t) + h(t)]\,. \tag{B.1.15}$$

Of course, we can assume that $C \ge 1$ so that (B.1.14) and (B.1.15) give

$$F(s) + G(s) \le \varepsilon_0[F(t) + G(t)] + C_1[(t-s)^{-\alpha} g(t) + h(t)]\,. \tag{B.1.16}$$

Thus, if $\varepsilon_0 < 1$ or $2C\varepsilon < 1$ then the previous lemma applies with $f(t) = F(t) + G(t)$ to give

$$F(\rho) + G(\rho) \le c(\alpha, \varepsilon)[(R-\rho)^{-\alpha} g(R) + h(R)]\,.$$

Obviously, the above argument holds if we replace the interval $[\rho, R]$ by any subinterval $[s, t]$. The above inequality then gives (B.1.13). □

C Partial regularity

We sketch the main ideas of the proof of the partial regularity theory of bounded weak solutions to the parabolic system

$$u_t = \operatorname{div}(A(u)Du) + \hat{f}(u, Du) \quad \text{in } Q\,. \tag{C.0.1}$$

Roughly speaking, the conclusion is that there is $\varepsilon_0 > 0$ such that a bounded weak solution u of (C.0.1) is Hölder continuous on the regular open set Q_0 defined by

$$Q_0 := \left\{ z \in Q : \liminf_{R\to 0} ⨌_{Q_R(z)} |u - u_{z,R}|^2 \, dz \le \varepsilon_0 \right\}. \tag{C.0.2}$$

Moreover, if $A(u)$ is Hölder continuous in u then Du is also Hölder continuous in Q_0.

We rely on the paper by Giaquinta and Struwe [14]. Many other works on this topic, in various settings, are essentially following the same main ideas of this paper: using the higher integrability of Du and a perturbation argument from 'frozen coefficients' systems.

Firstly, we have the higher integrability of Du: there is $p > 2$ such that for any $Q_R \subset Q$

$$\left(⨌_{Q_R} |Du|^p \, dz \right)^{\frac{1}{p}} \le C \left(⨌_{Q_{2R}} |Du|^2 \, dz \right)^{\frac{1}{2}}. \tag{C.0.3}$$

The proof of this fact is now quite standard. One relies on the two basic inequalities: the Poincaré and a Caccioppoli type inequalities

$$\sup_{t\in(t_0-R^2,t_0)} \int_{B_R} |u - u_{Q_R}|^2 \, dx \le C \iint_{Q_{2R}} |Du|^2 \, dz\,,$$

and

$$\iint_{Q_R} |Du|^2 \, dz \le C \frac{1}{R^2} \iint_{Q_{2R}} |u - u_{2R}|^2 \, dz\,.$$

These two inequalities are easily obtained by testing the system (C.0.1) with $(u - u_R)\psi^2\eta$, where ψ and η are appropriate cutoff functions in x and t.

The higher integrability estimate (C.0.3) then follows from the famous Gehring lemma [16].

Now comes the perturbation argument. We 'freeze' the system (C.0.1) by considering the following linear parabolic system with constant coefficients:

$$\begin{cases} v_t = \operatorname{div}(A(u_R)Dv) & \text{in } Q_R\,, \\ v = u & \text{on } S_R\,, \end{cases} \tag{C.0.4}$$

https://doi.org/10.1515/9783110608762-009

where $S_R = B_R \cup (\partial B_R \times (t_0 - R^2, t_0)$ is the parabolic boundary of Q_R. For any $\rho < R$ we have the following Campanato estimates for this solution:

$$\iint_{Q_\rho} |Dv|^2 \, dz \le C\left(\frac{\rho}{R}\right)^{N+2} \iint_{Q_R} |Dv|^2 \, dz \,, \tag{C.0.5}$$

$$\iint_{Q_\rho} |Dv - (Dv)_\rho|^2 \, dz \le C\left(\frac{\rho}{R}\right)^{N+4} \iint_{Q_R} |Dv - (Dv)_R|^2 \, dz \,. \tag{C.0.6}$$

Using (C.0.5) and letting $w = u - v$, we easily see that

$$\iint_{Q_\rho} |Du|^2 \, dz \le C\left(\frac{\rho}{R}\right)^{N+2} \iint_{Q_R} |Du|^2 \, dz + C\iint_{Q_R} |Dw|^2 \, dz \,. \tag{C.0.7}$$

Subtracting the equations (C.0.1) and (C.0.4) and testing the result with w, we get

$$\iint_{Q_R} |Dw|^2 \, dz \le C\iint_{Q_R} |A(u) - A(u_R)|^2 |Du|^2 \, dz + \iint_{Q_R} |w||\hat{f}(u, Du)| \, dz \,.$$

That is,

$$\iint_{Q_R} |Dw|^2 \, dz \le C\iint_{Q_R} |A(u) - A(u_R)|^2 |Du|^2 \, dz + \iint_{Q_R} |w||Du|^2 \, dz \,. \tag{C.0.8}$$

As A is continuous, we can find a continuous concave function ω such that $|A(u) - A(u_R)| \le \omega(|u - u_R|^2)$. Hence, Hölder's inequality, the high integrability estimate (C.0.3) and the concavity of ω give

$$\begin{aligned}
&\iint_{Q_R} |A(u) - A(u_R)|^2 |Du|^2 \, dz \\
&\le C\left(\iint_{Q_R} |Du|^p \, dz\right)^{\frac{2}{p}} \left(\iint_{Q_R} \omega \, dz\right)^{1-\frac{2}{p}} \\
&\le C\iint_{Q_{2R}} |Du|^2 \, dz \left(⨍\!\!⨍_{Q_R} \omega \, dz\right)^{1-\frac{2}{p}} \\
&\le C\iint_{Q_{2R}} |Du|^2 \, dz \left(\omega\left(⨍\!\!⨍_{Q_R} |u - u_R|^2 \, dz\right)\right)^{1-\frac{2}{p}} \,.
\end{aligned}$$

In the same way, we estimate the second integral on the right-hand side of (C.0.8).

Combining this with (C.0.7) and the smallness condition (C.0.2), we derive

$$\iint_{Q_\rho} |Du|^2 \, dz \le C[\left(\frac{\rho}{R}\right)^{N+2} + \chi(\varepsilon_0, R)] \iint_{Q_\rho} |Du|^2 \, dz \,,$$

where $\chi(\varepsilon_0, R)$ can be very small as $R \to 0$ and ε_0 are small. Setting

$$\phi(R) = ⨍\!\!⨍_{Q_R} |u - u_R|^2 \, dz \quad \text{and} \quad \tau = \frac{\rho}{R}$$

and using the Poincaré inequality, we easily derive

$$\phi(\tau R) \le C\tau^2 [1 + \tau^{-n-4} \chi(\varepsilon_0, R)] \phi(R) \,.$$

The fact that $\chi(\varepsilon_0, R)$ can be very small as $R \to 0$ and the above then give for any given $\alpha \in (0,1)$ that $\phi(\tau R) \le \tau^{2\alpha}\phi(R)$. That is,

$$\phi(\rho) \le \left(\frac{\rho}{R}\right)^{2\alpha} \phi(R)\,.$$

The definition of ϕ and the Campanato–Morrey embedding theorem then show that, on Q_0, u is Hölder continuous for any exponent $\alpha \in (0,1)$.

In the same way, we assume that A is Hölder continuous with some exponent $\sigma \in (0,1)$ and use (C.0.6) instead of (C.0.5) to derive that

$$\iint_{Q_\rho} |Du - (Du)_\rho|^2\, dz \le C\left(\frac{\rho}{R}\right)^{N+4} \iint_{Q_R} |Du - (Du)_R|^2\, dz + CR^{N+2\alpha+2\alpha\sigma(1-\frac{2}{p})}\,.$$

Again, the above provides some $\gamma > 0$ such that

$$\iint_{Q_\rho} |Du - (Du)_\rho|^2\, dz \le C\left(\frac{\rho}{R}\right)^{N+2+2\gamma} \iint_{Q_R} |Du - (Du)_R|^2\, dz\,.$$

This shows that Du is Hölder continuous in Q_0.

Bibliography

[1] H. Amann. Dynamic theory of quasilinear parabolic equations II. Reaction-diffusion systems. *Differential Integral Equations*, 3(1), 1990, 13–75.

[2] H. Amann. Dynamic theory of quasilinear parabolic systems III. Global existence. *Math Z.*, 202, 1989, 219–250.

[3] H. Amann. Fixed-point problems and nonlinear eigenvalue problems in ordered Banach spaces. *SIAM Rev.*, 18, 1976, 620–709.

[4] S. Ahmad and D. Le. Global and Blow Up Solutions to Cross Diffusion Systems. *Advances in Nonlinear Analysis*, 4(3), 2015, 209–219.

[5] D. Bernstein. *Matrix Mathematics*. Princeton University Press, 2005.

[6] O. H. Brezis and M. Crandall. Uniqueness of solutions of the initial value problem for $ut - \Delta\phi(u) = 0$. *J. Math. Pures Appl.*, 58, 1979, 153–163.

[7] J. Dugunji. *Topology*. Allyn and Bacon Boston, 1966.

[8] C. Fefferman and E. M. Stein. Hp spaces of several variables. *Acta Math.*, 129, 1972, 137–193.

[9] D. G. de Figueiredo and E. Mitidieri. Maximum principles for linear elliptic systems. *Rend. Istit. Mat. Univ. Trieste*, 22(1–2), 1992, 36–66.

[10] J. Fleckinger-Pelle and P. Takac. Maximum and anti-maximum principles for some elliptic problems, in: *Advances in Diff. Eqns. and Math. Phys.*, Atlanta, GA, 1997, in: *Contemp. Math.*, vol. 217. Amer. Math. Soc. Providence, RI, 1998, 19–32.

[11] B. Franchi, C. Perez and R. L. Wheeden. Self-Improving Properties of John Nirenberg and Poincaré Inequalities on Spaces of Homogeneous Type. *J. Functional Analysis*, 153, 108–146, 1998.

[12] A. Friedman and S. Kamin. The asymptotic behavior of gas in an N-dimensional porous medium. *Trans. Amer. Math. Soc.*, 262, 1980, 551–563.

[13] A. Friedman. *Partial Differential Equations*. New York, 1969.

[14] M. Giaquinta and M. Struwe. On the partial regularity of weak solutions of nonlinear parabolic systems. *Math. Z.*, 179, 1982, 437–451.

[15] D. Gilbarg and N. S. Trudinger. *Elliptic Partial Differential Equations of Second Order*, Springer Berlin Heidelberg, 2001.

[16] E. Giusti. *Direct Methods in the Calculus of Variations*. World Scientific, Singapore, 2003.

[17] L. Grafakos. *Modern Fourier Analysis*, 2nd edn. Springer-Verlag New York, 2010.

[18] G. Degla. An overview of semi-continuity results on the spectral radius and positivity. *J. Math. Anal. Appl.*, 338, 2008, 101–110.

[19] P. Hajlasz and P. Koskela. Sobolev met Poincaré. *Memoirs of AMS*, 145(688), 2000.

[20] R. L. Johnson and C. J. Neugebauer. Properties of BMO functions whose reciprocals are also BMO. *Z. Anal. Anwendungen*, 12(1), 1993, 3–11.

[21] O. John and J. Stara. On the regularity of weak solutions to parabolic systems in two spatial dimensions. *Comm. P.D.E.*, 27, 1998, 1159–1170.

[22] J. Jost. *Riemannian Geometry and Geometric Analysis*. Springer, 2005.

[23] K. H. W. Küfner. Invariant regions for quasilinear reaction-diffusion systems and applications to a two population model. *NoDEA*, 3, 1996, 421–444.

[24] L. Korobenko, D. Maldonado and C. Rios. From Sobolev inequality to doubling. *Proc. AMS*, 143(9), 2015, 4017–4028.

[25] O. A. Ladyzhenskaya, V. A. Solonnikov and N. N. Uraltseva]. *Linear and Quasi-linear Equations of Parabolic Type*. Translations of Mathematical Monographs, AMS, 1968.

[26] D. Le. Global Existence and Regularity Results for Strongly Coupled Nonregular Parabolic Systems via Iterative Methods. *to appear on DCDS*.

https://doi.org/10.1515/9783110608762-010

[27] D. Le. Global existence results for near triangular nonlinear parabolic systems. *Adv. Nonlinear Studies*, 13(4), 2013, 933–944.

[28] D. Le. Regularity of BMO weak solutions to nonlinear parabolic systems via homotopy. *Trans. Amer. Math. Soc.*, 365 (5), 2013, 2723–2753.

[29] D. Le. Weighted Gagliardo–Nirenberg Inequalities Involving BMO Norms and Solvability of Strongly Coupled Parabolic Systems. *Adv. Nonlinear Stud.*, 16(1), 2016, 125–146.

[30] D. Le. Existence of Strong and Nontrivial Solutions to Strongly Coupled Elliptic Systems. *J. Funct. Anal.*, 272(11), 2017, 4407–4459.

[31] D. Le and V. Nguyen. Global solutions to cross diffusion parabolic systems on 2D domains. *Proc. Amer. Math. Soc.*, 143(7), 2015, 2999–3010.

[32] D. Le, L. Nguyen and T. Nguyen. Coexistence in Cross Diffusion systems. *Indiana Univ. J. Math.*, 56(4), 2007, No. 4, 1749–1791.

[33] J. Lehrbäck. Weighted Hardy inequalities and the size of the boundary. *Manuscripta Math.*, 127(2), 2008, 249–273.

[34] G. M. Lieberman. *Second Order Parabolic Differential Equations*. World Scientific, 1998.

[35] Y. Lou and W. Ni. Diffusion, Self-Diffusion and Cross-Diffusion. *JDE*, 131(1), 1996, 79–131.

[36] B. Muckenhoupt. Weighted norm inequalities for the Hardy maximal function. *Trans. Amer. Math. Soc.*, 165, 1972, 207–226.

[37] J. Necas and V. Sverak. On regularity of solutions of nonlinear parabolic systems. *Ann. Scuola Norm. Sup. Pisa Cl. Sci.* Série 4, 18(1), 1991, 1–11.

[38] X. Tolsa. *BMO*, H^1, and Calderon–Zygmund operators for non doubling measures. *Math. Ann.*, 319, 2001, 89–149.

[39] X. Tolsa. The atomic space H^1 for non doubling measures in terms of a maximal operator. *Trans. Amer. Math. Soc.*, 355, 2003, 315–348.

[40] J. Orobitg and C. Pérez. A_p weights for nondoubling measures in $\mathbb{R}^n$ and applications. *Trans. Amer. Math. Soc.*, 354, 2002, 2013–2033.

[41] R. Redlinger. Existence of the global attractor for a strongly coupled parabolic system arising in population dynamics. *J. Diff. Eqns.*, 118, 1995, 219–252.

[42] T. Riviere and P. Strzelecki. A sharp nonlinear Gagliardo Nirenberg type estimate and applications the regularity of elliptic systems. *Comm. P.D.E.*, 30, 2005, 589–604.

[43] P. Strzelecki. Gagliardo Nirenberg inequalities with a BMO term. *Bull. London Math. Soc.*, 38, 2006, 294–300.

[44] N. Shigesada, K. Kawasaki and E. Teramoto. Spatial segregation of interacting species. *J. Theor. Biol.*, 79, 1979, 83–99.

[45] E. M. Stein. *Harmonic Analysis, Real Variable Methods, Orthogonality and Oscillatory Integrals*. Princeton Univ. Press Princeton, NJ, 1993.

[46] A. Yagi. Global solution to some quasilinear parabolic systems in population dynamics. *Nonlin. Anal.*, 21, 1993, 603–630.

[47] E. Zeidler. *Nonlinear Functional Analysis and its Applications, Vol. I, Fixed-Points Theorems*. Springer-Verlag New York, 1986.

Index

https://doi.org/10.1515/9783110608762-011

Also of Interest

Volume 27
Blow-Up in Nonlinear Equations of Mathematical Physics. Theory and Methods
Maxim Olegovich Korpusov, Alexey Vital'evich Ovchinnikov, Alexey Georgievich Sveshnikov, Egor Vladislavovich Yushkov, 2018
ISBN 978-3-11-060108-4, e-ISBN (PDF) 978-3-11-060207-4,
e-ISBN (EPUB) 978-3-11-059900-8

Volume 26
Implicit Fractional Differential and Integral Equations. Existence and Stability
Saïd Abbas, Mouffak Benchohra, John R. Graef, Johnny Henderson, 2018
ISBN 978-3-11-055313-0, e-ISBN (PDF) 978-3-11-055381-9,
e-ISBN (EPUB) 978-3-11-055318-5

Volume 25
Function Spaces. Volume 2
Luboš Pick, Alois Kufner, Oldřich John, Svatopluk Fucík, 2018
ISBN 978-3-11-027373-1, e-ISBN (PDF) 978-3-11-032747-2,
e-ISBN (EPUB) 978-3-11-038221-1

Volume 24
Singular Solutions of Nonlinear Elliptic and Parabolic Equations
Alexander A. Kovalevsky, Igor I. Skrypnik, Andrey E. Shishkov, 2016
ISBN 978-3-11-031548-6, e-ISBN (PDF) 978-3-11-033224-7,
e-ISBN (EPUB) 978-3-11-039008-7

Volume 23/2
Nonlinear Equations with Small Parameter. Volume 2: Partial Differential Equations
Sergey G. Glebov, Oleg M. Kiselev, Nikolai N. Tarkhanov, 2018
ISBN 978-3-11-053383-5, e-ISBN (PDF) 978-3-11-053497-9,
e-ISBN (EPUB) 978-3-11-053390-3

Volume 23/1
Nonlinear Equations with Small Parameter. Volume 1: Ordinary Differential Equations
Sergey G. Glebov, Oleg M. Kiselev, Nikolai N. Tarkhanov, 2017
ISBN 978-3-11-033554-5, e-ISBN (PDF) 978-3-11-033568-2,
e-ISBN (EPUB) 978-3-11-038272-3

Bei Fragen zur Produktsicherheit wenden Sie sich bitte an:
If you have any questions regarding product safety,
please contact:

Walter de Gruyter GmbH
Genthiner Straße 13
10785 Berlin
productsafety@degruyterbrill.com